Springer INdAM Series

Volume 32

Springer INdAM Series

This series will publish textbooks, multi-authors books, thesis and monographs in English language resulting from workshops, conferences, courses, schools, seminars, doctoral thesis, and research activities carried out at INDAM - Istituto Nazionale di Alta Matematica, http://www.altamatematica.it/en. The books in the series will discuss recent results and analyze new trends in mathematics and its applications.
THE SERIES IS INDEXED IN SCOPUS

More information about this series at http://www.springer.com/series/10283

Fatiha Alabau-Boussouira ·
Fabio Ancona · Alessio Porretta ·
Carlo Sinestrari
Editors

Trends in Control Theory and Partial Differential Equations

Springer

Editors
Fatiha Alabau-Boussouira
Laboratoire Jacques-Louis Lions UMR 7598
Sorbonne Université
Paris, France

Département de Mathématiques
Université de Lorraine
Metz, France

Alessio Porretta
Dipartimento di Matematica
Università di Roma "Tor Vergata"
Rome, Italy

Fabio Ancona
Dipartimento di Matematica
"Tullio Levi-Civita"
Università degli Studi di Padova
Padua, Italy

Carlo Sinestrari
Dipartimento di Ingegneria Civile e
Ingegneria Informatica
Università di Roma "Tor Vergata"
Rome, Italy

ISSN 2281-518X ISSN 2281-5198 (electronic)
Springer INdAM Series
ISBN 978-3-030-17948-9 ISBN 978-3-030-17949-6 (eBook)
https://doi.org/10.1007/978-3-030-17949-6

This Springer imprint is published by the registered company Springer Nature Switzerland AG
The registered company address is: Gewerbestrasse 11, 6330 Cham, Switzerland

Preface

This volume collects contributions from participants in the INdAM Workshop "New trends in control theory and partial differential equations", which was held in Rome on July 3–7, 2017. The aim of the workshop was to present recent developments in the field of partial differential equations (PDEs) and control theory, and to underline connections and interactions between the different areas of this vast subject.

Control theory is a branch of mathematics which has been widely studied since the second half of the past century, and finds application in many different fields, such as mechanics, engineering, finance, medicine, and climatology. On the other hand, it often poses challenging problems on the theoretical side, and has stimulated the development of new theories which have become useful in other areas of mathematics as well.

Control theory has greatly influenced the study of PDEs over the decades. As a classical example, the approach of dynamic programming has established a connection between optimization problems for finite- dimensional control systems and the theory of first-order Hamilton–Jacobi equations, which typically do not have smooth solutions. Similarly, the analysis of stochastic control problems has led to the study of second-order, possibly degenerate, nonlinear PDEs. A rigorous treatment of these equations has motivated the introduction of suitable kinds of generalized solutions, such as the class of viscosity solutions, which has found applications in many other branches of the PDE theory. In this context, it is important to study the regularity properties of the value function, such as Lipschitz continuity or semi-concavity, in order to build feedback controls in terms of the generalized gradients.

In a different direction, one can consider control systems in infinite dimension, which are governed by PDEs. Common examples are evolutionary PDEs, where the control either acts as a source term inside the domain or appears in the boundary conditions. A typical interesting issue is the controllability of the system, i.e., the ability to steer the system from one given state to another. Another related topic is observability, where one wishes to analyze whether the solution is determined by its values on a given subset. This kind of issue has been widely studied for the most common linear evolution equations, but there are many other interesting cases where less is known, e.g., in the presence of degeneracy of the operators or singularities of the data.

The contributions in this volume treat different aspects of the topics mentioned above and of related ones. Concerning the study of finite-dimensional systems, P. Albano studies a first-order equation of eikonal type, whose solution is the value function of a minimum time-optimal control problem, and gives sufficient conditions for the Lipschitz continuity. V. Basco and H. Frankowska consider an optimal control problem with infinite horizon and state constraints, and give sufficient conditions for the Lipschitz continuity of the value function. P. Cannarsa, W. Cheng, K. Wang, and J. Yan analyze the classical Herglotz' variational principle in the calculus of variations and give a Lipschitz estimate of the minimizers. M. Mazzola and K. T. Nguyen give a new proof of Lyapunov's theorem in convex analysis based on a Baire category approach. Focusing on a PDE perspective, I. Capuzzo Dolcetta gives an overview of results on the maximum principle for weakly elliptic equations, a class which includes, in particular, the Bellman–Isaacs equations associated to the optimal control of degenerate diffusion processes as well as differential games.

Other contributions in this volume deal with the controllability of PDEs. E. Fernández-Cara and D. Souza study the null controllability of a family of equations which include the Camassa–Holm and α-Burgers equations. In a more applied direction, G. Leugering, T. Li, and Y. Wang consider a system modeling the motion of nonlinear elastic strings and study the boundary controllability and stabilizability. D. Pighin and E. Zuazua study various cases of controllability for wave equations in general dimension, under non-negativity constraints on the controls. J. Vancostenoble proves the approximate controllability for a class of parabolic equations in the presence of a singular potential.

Further topics related to control of PDEs are treated in the volume. P. Cannarsa, G. Floridia, and M. Yamamoto prove a Carleman estimate for a transport equation and use it to deduce an observability inequality in terms of the boundary data. P. Loreti and D. Sforza consider a semilinear wave equation with an integral memory term, proving the existence of solutions and an estimate on the boundary values called hidden regularity, a property which is of interest in the study of controllability.

Finally, two contributions deal with differential systems which model the behavior of multiagent phenomena, a topic which is strictly connected to control theory and has attracted much attention in recent years. P. Cardaliaguet considers a system of N weakly coupled Hamilton–Jacobi equations with increasingly singular coupling and proves convergence as $N \to +\infty$ to a mean field games system. C. Pignotti and I. Reche Vallejo study the asymptotic behavior on a Cucker–Smale system, proving convergence to consensus under suitable assumptions.

The workshop provided an occasion to celebrate the 60th birthday of Piermarco Cannarsa, who has made fundamental contributions to these fields and has played an important part in the mathematical life of the people involved in the workshop and in this volume, as a teacher and/or as a collaborator and friend.

We thank the Istituto Nazionale di Alta Matematica "Francesco Severi" for hosting and financing the workshop, and for the efficient support provided by the staff of the institute. We also thank the European research group GDRE CONEDP and the Department of Mathematics "Tullio Levi-Civita" of the University of Padova, who contributed to financing the event.

Paris, France Fatiha Alabau-Boussouira
Padua, Italy Fabio Ancona
Rome, Italy Alessio Porretta
Rome, Italy Carlo Sinestrari

Contents

About the Editors

Fatiha Alabau-Boussouira gained her Ph.D. at the Pierre and Marie Curie University, Paris, in 1987. She was an Assistant Professor at the University of Bordeaux from 1988 until 1997, when she became a Full Professor at the Louis Pasteur University in Strasbourg; she subsequently moved to the University of Metz in 1999 (which became the University of Lorraine in 2012). She carries out her research at the Jacques-Louis Lions Laboratory at Sorbonne University. She is the author of more than 60 papers in mathematics analysis and applications. Her scientific interests mainly focus on the theory of control and stabilization of partial differential equations. From 2014 to 2017, she was President of the French Society of Applied and Industrial Mathematics (SMAI). She is now the head for France of the French-German-Italian LIA COPDESC on Applied Analysis.

Fabio Ancona obtained his Ph.D. in Mathematics in 1993 at University of Colorado at Boulder. He was a Research Associate in Mathematical Analysis at University of Bologna from 1995 until 2001, when he became Associate Professor. In 2008 he moved to the University of Padua, where he became a Full Professor in 2017. He is the author of more than 40 papers in mathematics analysis and applications. His scientific interests mainly focus on the theory of control and of hyperbolic partial differential equations.

Alessio Porretta is a Full Professor of Mathematical Analysis at the University of Rome Tor Vergata. His research activity focuses mainly on convection-diffusion equations, Hamilton–Jacobi, control theory, and mean field games. He has given seminars in more than 20 universities in Italy and abroad and has recently been invited to give courses on mean field games in Paris, Chicago, and ETH Zurich. He has authored over 70 research papers, with nearly 1000 citations in math journals.

Carlo Sinestrari received his Ph.D. from the University of Rome Tor Vergata, where he later became a Full Professor in Mathematical Analysis. His research interests include the analysis of nonlinear first-order partial differential equations and the formation of singularities for geometric evolution equations. He is the author of more than 40 papers in research journals. Together with P. Cannarsa, he has written a monograph on semi-concave functions and their applications to optimal control theory.

Some Remarks on the Dirichlet Problem for the Degenerate Eikonal Equation

Paolo Albano

Abstract In a bounded domain, we consider the viscosity solution of the homogeneous Dirichlet problem for the degenerate eikonal equation. We provide some sufficient conditions for the (local) Lipschitz regularity of such a function.

Keywords Eikonal equation · Degenerate equations · Viscosity solutions · Symplectic geometry

2010 Mathematics Subject Classification 35F30 · 35F21 · 35D40

1 Introduction and Statement of the Results

Let $\Omega \subset \mathbb{R}^n$ be an open bounded set with boundary, Γ, and let T be the viscosity solution of the Dirichlet problem

$$\begin{cases} \langle A(x)DT(x), DT(x)\rangle = 1 & \text{in} \quad \Omega, \\ T = 0 \quad \text{on} \quad \Gamma. \end{cases} \tag{1.1}$$

We recall that a continuous function $T : \overline{\Omega} \longrightarrow \mathbb{R}$ is a *viscosity solution* of (1.1) if

(1) T is a *viscosity subsolution*, i.e. $\langle A(x)D\varphi(x), D\varphi(x)\rangle \leq 1$, for every φ of class C^1 such that $T - \varphi$ has a local maximum at x;
(2) T is a *viscosity supersolution*, i.e. $\langle A(x)D\varphi(x), D\varphi(x)\rangle \geq 1$, for every φ of class C^1 such that $T - \varphi$ has a local minimum at x;
(3) $T(x) = 0$ for every $x \in \Gamma$.

We are interested in the Lipschitz regularity of T assuming that the map $x \mapsto A(x)$ takes values in the set of the $n \times n$ positive semidefinite symmetric matrices.

P. Albano
Dipartimento di Matematica, Università di Bologna, Piazza di Porta San Donato 5, 40127 Bologna, Italy
e-mail: paolo.albano@unibo.it

F. Alabau-Boussouira et al. (eds.), *Trends in Control Theory and Partial Differential Equations*, Springer INdAM Series 32,
https://doi.org/10.1007/978-3-030-17949-6_1

We recall that, if the equation is non degenerate (i.e. if $A(\cdot)$ is positive definite), then T is locally Lipschitz continuous on Ω. In fact a subsolution of a non degenerate eikonal equation is locally Lipschitz continuous.

We observe that, without more restrictive assumptions on the data, Eq. (1.1) may have no solution. Indeed, it is well known that, for a continuous function, the set

$$\{x \in \Omega \ : \ T - \varphi \text{ has a local minimum at } x \text{ for a suitable } \varphi \in C^1\}$$

is dense in Ω. Hence, if $A(\cdot)$ vanishes on an open subset of Ω, then Eq. (1.1) admits no viscosity supersolutions and, in particular, (1.1) has no viscosity solutions.

We point out that, in order to have the existence of a continuous solution of (1.1), it is not enough to assume that $A(\cdot)$ is not totally degenerate on $\overline{\Omega}$ (i.e. that $\operatorname{rank} A(x) \geq 1$ for every $x \in \overline{\Omega}$). For instance, in the plane with coordinates (x, y), the Dirichlet problem

$$\begin{cases} (\partial_x T(x, y))^2 = 1 & \text{in} \quad \Omega =]0, 1[\times]0, 1[, \\ T = 0 \quad \text{on} \quad \partial\Omega, \end{cases}$$

admits no continuous viscosity solutions.

We assume

(H1) $\Omega \subset \mathbb{R}^n$ is an open bounded set with boundary of class C^1, Γ.

(H2) Let $k \leq n$ be a positive integer and let

$$\overline{\Omega} \ni x \mapsto B(x)$$

be a Lipschitz continuous map on $\overline{\Omega}$ taking values in the set of the $k \times n$ matrices. We denote by L_B the Lipschitz constant of $B(\cdot)$ in $\overline{\Omega}$.

Set

$$A(x) = {}^t B \, B(x) \qquad (x \in \overline{\Omega}).$$

(Here ${}^t B$ is the transpose of the matrix B.) For $x \in \Gamma$, let $\nu(x)$ be the outward unit normal to Γ at x; we say that $x \in \Gamma$ is a *characteristic boundary point* if

$$\langle A(x)\nu(x), \nu(x)\rangle = 0.$$

Then, the set of all the characteristic boundary points is defined as

$$E = \{x \in \Gamma \ : \ \langle A(x)\nu(x), \nu(x)\rangle = 0\}.$$

We have the following global regularity result.

Theorem 1.1 *Under assumptions* $(H1)$ *and* $(H2)$, *let* $T : \overline{\Omega} \longrightarrow \mathbb{R}$ *be the continuous viscosity solution of the homogeneous Dirichlet problem (1.1) and suppose that* $E = \emptyset$. *Then,* T *is Lipschitz continuous on* $\overline{\Omega}$.

Remark 1.1 We point out that, in the statement above, we are assuming that there exists a continuous viscosity solution of Eq. (1.1): as already observed our assumptions are not strong enough to guarantee the existence of a viscosity solution of Eq. (1.1). On the other hand, Theorem 1.1 is in essence a result on the propagation of the regularity from Γ towards Ω.

Proof of Theorem 1.1 We observe that

$$T \geq 0 \quad \text{on} \quad \overline{\Omega}. \tag{1.2}$$

Let us consider the Kružkov transformation of u:

$$v(x) = 1 - e^{-\lambda T(x)} \qquad (x \in \overline{\Omega}), \tag{1.3}$$

with $\lambda = 1 + L_B$. By (1.2) and (1.3), we have

$$0 \leq v(x) < 1 \qquad x \in \overline{\Omega}. \tag{1.4}$$

Furthermore, because of T is a viscosity solution of (1.1), v is a viscosity solution of the equation

$$\begin{cases} v(x) + |\lambda^{-1} B(x) D v(x)| = 1 & \text{in} \quad \Omega, \\ v = 0 \quad \text{on} \quad \Gamma. \end{cases} \tag{1.5}$$

We remark that

$$|v(x) - v(y)| \leq L|x - y| \qquad x, y \in \overline{\Omega} \tag{1.6}$$

implies

$$|T(x) - T(y)| \leq \frac{L}{\lambda} e^{\lambda \|T\|_{L^\infty(\Omega)}} |x - y| \qquad x, y \in \overline{\Omega}.$$

Hence the proof reduces to show that Estimate (1.6) holds true.

Since Γ is of class C^1 there exists a positive number, ρ, such that, denoting by $d(x)$ the Euclidean distance function of x from Γ, we have that d is differentiable on the set

$$\Omega_\rho = \{x \in \Omega \mid d(x) < \rho\}.$$

Furthermore, because of Γ is noncharacteristic, we have that there exists a positive constant c such that

$$d(x) + |\lambda^{-1} B(x) D d(x)| \geq \frac{c}{\lambda} \qquad x \in \Omega_\rho.$$

We claim that Estimate (1.6) holds with

$$L = \frac{\lambda}{c} + \frac{1}{\rho}.$$

For this purpose, we will use a standard method of the theory of viscosity solutions: let us introduce the auxiliary function

$$\Phi(x, y) = v(x) - v(y) - L|x - y| \qquad x, y \in \overline{\Omega}.$$

Clearly, in order to verify (1.6), it suffices to show that for every $x, y \in \overline{\Omega}$ we have that $\Phi(x, y) \leq 0$. Let us define

$$M := \max_{x,y\in\overline{\Omega}} \Phi(x, y) = \Phi(\hat{x}, \hat{y}) \geq \Phi(\hat{x}, \hat{x}) = 0, \tag{1.7}$$

for a suitable $\hat{x}, \hat{y} \in \overline{\Omega}$.

We want to show that $M \leq 0$.

If $\hat{x} \in \Gamma$ and $\hat{y} \in \overline{\Omega}$ then, by the homogeneous Dirichlet condition and (1.4), we deduce that $M \leq 0$.

Hence, let us suppose that $\hat{x} \in \Omega$ and $\hat{y} \in \overline{\Omega}$. There are two cases either $\hat{y} \in \Gamma$ or $\hat{y} \in \Omega$. First, let us consider the case of $\hat{y} \in \Gamma$. Then $v(\hat{y}) = 0$ and, by (1.4) and (1.7), we deduce that $L|\hat{x} - \hat{y}| \leq 1$, i.e.

$$d(\hat{x}) \leq |\hat{x} - \hat{y}| \leq 1/L < \rho.$$

By construction $L > \lambda/c$, the function $Ld(\cdot)$ is a supersolution of (1.5) in Ω_ρ and $v \leq Ld$ on the boundary of Ω_ρ. Hence, by the comparison principle, we deduce that

$$v(x) - Ld(x) \leq 0 \qquad x \in \Omega_\rho,$$

and we find

$$M \leq v(\hat{x}) - Ld(\hat{x}) \leq 0.$$

It remains to consider the last case: $\hat{x}, \hat{y} \in \Omega$ and $\hat{x} \neq \hat{y}$. Since v is a viscosity solution of (1.5), we have

$$\begin{cases} v(\hat{x}) + |\lambda^{-1}B(\hat{x})L(\hat{x} - \hat{y})/|\hat{x} - \hat{y}|| \leq 1, \\ v(\hat{y}) + |\lambda^{-1}B(\hat{y})L(\hat{x} - \hat{y})/|\hat{x} - \hat{y}|| \geq 1. \end{cases}$$

Then, we find

$$v(\hat{x}) - v(\hat{y}) \leq |\lambda^{-1}B(\hat{y})L(\hat{x} - \hat{y})/|\hat{x} - \hat{y}|| - |\lambda^{-1}B(\hat{x})L(\hat{x} - \hat{y})/|\hat{x} - \hat{y}||$$

and, recalling that $\lambda = 1 + L_B$, we deduce

$$v(\hat{x}) - v(\hat{y}) \leq \lambda^{-1}L_B L|\hat{x} - \hat{y}| \leq L|\hat{x} - \hat{y}|,$$

hence $M \leq 0$. This completes our proof. □

A particular class of degenerate eikonal equations in which the Dirichlet problem (1.1) admits a continuous viscosity solution is the one associated with a system of Hörmander vector fields. Hereafter we assume

(H) (i) Ω is an open bounded set and Γ is a smooth manifold (without boundary) of dimension $n-1$.

(ii) $\{X_1, \ldots, X_N\}$ is a system of smooth real vector fields on Ω', an open neighborhood of Ω.

(iii) The system $\{X_1, \ldots, X_N\}$ satisfies Hörmander's bracket generating condition on Ω', i.e. the Lie algebra generated by the vector fields as well as by their commutators has, in every point, dimension n.

We consider the viscosity solution of the problem

$$\begin{cases} \sum_{j=1}^{N} (X_j T)^2(x) = 1 \quad \text{in} \quad \Omega, \\ T = 0 \quad \text{on} \quad \Gamma, \end{cases} \tag{1.8}$$

and, once more, we are interested to the Lipschitz regularity of the solution of (1.8).

It is well-known that, for instance in the case of $X_1, \ldots, X_N$ linearly independent vector fields, the solution of Eq. (1.8) has a geometrical interpretation: it is the sub-Riemannian distance function from the boundary of the set Ω.

A first Lipschitz regularity result for T was given in [1] for a very special class of Dirichlet problems on an unbounded domain. More precisely, let $M > 0$, let k be a positive integer define

$$\Omega = \{(x, y) \in \mathbb{R}^n \times \mathbb{R} \,:\, y > M|x|^{k+1}\}, \tag{1.9}$$

and consider the homogeneous Dirichlet problem

$$\begin{cases} |\nabla_x T(x, y)|^2 + |x|^{2k}(\partial_y T(x, y))^2 = 1 \quad \text{in } \Omega \\ T = 0, \qquad \text{on } \Gamma = \partial\Omega. \end{cases} \tag{1.10}$$

In [1] it is proved the following

Theorem 1.2 *The nonnegative viscosity solution of the Dirichlet problem (1.10) is locally Lipschitz continuous in* Ω. *Furthermore,* T *is Hölder continuous of exponent* $1/(k+1)$ *at* $(0, 0)$.

Remark 1.2 The exponent $1/(k+1)$ in the above result is optimal.

The obstruction to the Lipschitz regularity of the solution of (1.8) is due to the presence of characteristic boundary points, in the present case

$$x \in E \iff X_1(x), \ldots, X_N(x) \quad \text{are tangent to } \Gamma \text{ at } x.$$

Example 1 In the case of Eq. (1.10), we have $2n$ vector fields

$$X_1 = \frac{\partial}{\partial x_1}, \ldots, X_n = \frac{\partial}{\partial x_n}, X_{n+1} = x_1 \frac{\partial}{\partial y}, \ldots, X_{2n} = x_n \frac{\partial}{\partial y}$$

and the origin $(0, 0)$ is the only characteristic boundary point on the domain (1.9).

In general, near a noncharacteristic boundary point, T is Lipschitz continuous whilst T fails to be Lipschitz continuous at a characteristic boundary point (see e.g. [3, Theorem 4.2 (2)]). Due to the smoothness of the data a stronger regularity result can be proved: T is smooth near a noncharacteristic boundary point (see e.g. [3, Theorem 4.2 (1)]).

Furthermore, as a consequence of Theorem 1.1, if the whole boundary Γ is noncharacteristic, then T is Lipschitz continuous on $\overline{\Omega}$.

It is well known that T can be interpreted as the value function of a suitable time optimal control problem; in particular certain time–optimal trajectories–the so called *singular time–optimal trajectories*–play a key role in the study of the regularity of T (see Definition 2.2 below). More precisely, in [3], it is shown that T is locally Lipschitz continuous if and only if the minimum time problem admits no "singular" time optimal trajectories. We recall that the singular trajectories can be characterized as the time–optimal trajectories reaching a characteristic boundary point. One can show that E is a closed subset of Γ of $(n-1)$-dimensional Hausdorff measure zero (see e.g. [6]).

Hence, one should expect that the solution of (1.8) may fail to be Lipschitz continuous on a "small" set. A more precise result can be proved. Indeed, in [4], a partial regularity result is given: T is of class C^∞ on the complement of a closed set of measure zero. In other words, for the viscosity solution of (1.8)

- there can be a lack of regularity only on a set of measure zero;
- in some cases there is a sort of propagation of Lipschitz singularities, along the singular time optimal trajectories, from E towards the interior of Ω.

It is a difficult task to verify whether a time optimal trajectory is singular: the main tool available being the Pontryagin Maximum Principle (a set of necessary optimality conditions). We will provide a sufficient (geometrical) condition ensuring the absence of singular time optimal trajectories. For this purpose, let us begin by recalling some basic notions, see [7, Chap. XXI].

1.1 Basic Objects

We associate to a vector field $X_j, j = 1, \ldots, N$, its principal symbol, $X_j(x, \xi)$, which is a function invariantly defined in the cotangent bundle, $T^*\Omega'$. We recall that we can identify $T^*\Omega'$ with $\Omega' \times \mathbb{R}^n$ by means of the canonical coordinates induced by the coordinates on Ω'. (Hereafter we will use such an identification.) More precisely, given the vector field

$$X_j(x) = \sum_{i=1}^{n} a_{ji}(x) \frac{\partial}{\partial x_i},$$

we have that

$$X_j(x, \xi) = \sum_{i=1}^{n} a_{ji}(x)\xi_i \tag{1.11}$$

is the *principal symbol*[1] of X_j. Here $a_{ji} \in C^\infty(\Omega')$, $i = 1, \dots, n$ and $j = 1, \dots, N$.

Let us denote denote the *Hamiltonian* by

$$h(x, \xi) = \sum_{j=1}^{N} X_j(x, \xi)^2, \quad (x, \xi) \in \Omega' \times \mathbb{R}^n,$$

and we define the *characteristic set* as

$$\Sigma = \{(x, \xi) \in \Omega' \times (\mathbb{R}^n \setminus \{0\}) \; : \; h(x, \xi) = 0\}. \tag{1.12}$$

We recall that in $\Omega' \times \mathbb{R}^n$ is naturally defined the *symplectic form*, i.e. a closed non–degenerate smooth 2-form, which (in local canonical coordinates) can be written as

$$\sigma = \sum_{j=1}^{n} d\xi_j \wedge dx_j.$$

Given a smooth submanifold $V \subset \Omega' \times \mathbb{R}^n$ and a point $\rho \in V$, the symbol $(T_\rho V)^\sigma$ stands for the orthogonal with respect to the symplectic form of the tangent space to V at ρ. A submanifold $V \subset \Omega' \times \mathbb{R}^n$ is *symplectic* if

$$\operatorname{rank} \sigma|_V = \dim V \tag{1.13}$$

or equivalently $T_\rho V \cap (T_\rho V)^\sigma = \{0\}$, for every $\rho \in V$.

Remark 1.3 The cotangent bundle of a smooth manifold is naturally a symplectic manifold. We recall that, by the Darboux Theorem, all the symplectic manifolds have locally the same geometry (in contrast with the case of the Riemannian manifolds).

Example 2 Consider the system of vector fields given in Example 1. Then, we have that

$$\Sigma = \{(0, y, 0, \eta) \in \mathbb{R}^n \times \mathbb{R} \times \mathbb{R}^n \times \mathbb{R} \; : \; y \in \mathbb{R}, \quad \eta \neq 0\}$$

is a symplectic manifold of dimension 2.

The phenomenon observed in Theorem 1.2 is clarified by the following

[1] We observe that our terminology slightly differs from the standard one used by people working on pdes: usually (1.11) stands the principal symbol of the operator X_j/i.

Theorem 1.3 *Assume (H) and let Σ be a symplectic manifold. Then T is locally Lipschitz continuous on Ω.*

(See [3, Theorem 4.4].) In other words, if Σ is a symplectic manifold, no propagation of Lipschitz singularities occur from E towards the interior of Ω.

Remark 1.4 We observe that, for a large class of equations, the assumption "Σ is a manifold" is automatically satisfied: if $X_1, \ldots, X_N$ are linearly independent on Ω' (i.e. we deal with a vectorial distribution), then Σ is a manifold of codimension N. (Here we are implicitly assuming that $N \leq n$.) In such a class, the only assumption of Theorem 1.3 is that Σ is symplectic. We recall that a distribution X on a manifold Ω' is called *strongly bracket generating* if for any nonzero section S of the distribution, the tangent bundle to Ω', $T\Omega'$, is generated by S and the commutators $[S, X]$ (see [10, 11], see also [5]). It is easy to see that, given a distribution, the following assertions are equivalent

(1) the distribution is strongly bracket generating;
(2) Σ is a symplectic manifold.

(See e.g. [9, Sect. 5.6] see also [2, Theorem 3.2].)

Then we have the following scheme:

- if $E = \emptyset$, then T is locally Lipschitz continuous on $\overline{\Omega}$;
- if $E \neq \emptyset$ and Σ is a symplectic manifold, then T is locally Lipschitz continuous on Ω (and T is not Lipschitz continuous at any point of E).

A next natural question is to analyze the case of $E \neq \emptyset$ and Σ is a manifold which is not symplectic.

For this purpose, we need one more geometrical notion: the *conormal bundle* to Γ, in coordinates, can be written as

$$N^*\Gamma = \{(x, \xi) \ : \ x \in \Gamma, \ \xi = \lambda \nu(x), \ \lambda \neq 0\},$$

where $\nu(x)$ is the outward inner normal to Γ at x. (We point out that the set above should be understood as a subset of $T^*\Omega'$.) Clearly, the following characterization holds

$$E \neq \emptyset \iff N^*\Gamma \cap \Sigma \neq \emptyset.$$

In order to state our "propagation" result we need the notion of *Hamiltonian leaf*.

Definition 1.1 Let $V \subset T^*\Omega'$ be a smooth submanifold and let $\rho_0 \in V$. Consider the (vectorial) distribution

$$V \ni \rho \mapsto T_\rho V \cap (T_\rho V)^\sigma, \tag{1.14}$$

and suppose that (1.14) has constant (positive) rank $< \dim V$, i.e.

$$V \ni \rho \mapsto \dim\{T_\rho V \cap (T_\rho V)^\sigma\}$$

is a constant function. The Hamiltonian leaf going through the point ρ_0, $F_{\rho_0} \subset V$, is the integral manifold of the distribution (1.14), with $\rho_0 \in F_{\rho_0}$.

We point out that the existence of an Hamiltonian leaf is granted by the Frobenius Theorem (for the reader convenience we provide some more details in Appendix 2.2). In the sequel, we will apply Definition 1.1 mainly to the special case of Σ is a manifold (and we take $V = \Sigma$).

Remark 1.5 If, in Definition 1.1, we drop the assumption that the rank of σ is less than dim V, then V is a symplectic submanifold of $T^*\Omega'$, $T_\rho V \cap (T_\rho V)^\sigma \equiv \{0\}$, for ρ near ρ_0 and F_ρ reduces to the singleton $\{\rho_0\}$.

In order to clarify the notion of Hamiltonian leaf, let us discuss some examples.

Example 3 (Baouendi–Goulaouic). In $\mathbb{R}^3$ consider the vector fields

$$X_1 = \partial_x, \quad X_2 = \partial_y, \quad X_3 = x^k \partial_t.$$

(Here k is a positive integer.) Then, we have that

$$\Sigma = \{(0, y, t, 0, 0, \tau) \; : \; y, t \in \mathbb{R}, \; \tau \neq 0\}$$

is a manifold of dimension 3 and the rank of the restriction of the symplectic form to Σ is 2. Let $\rho_0 = (0, y_0, t_0, 0, 0, \tau_0) \in \Sigma$, then

$$F_{\rho_0} = \{(0, y, t_0, 0, 0, \tau_0) \; : \; y \in \mathbb{R}\},$$

i.e. F_{ρ_0} is a manifold of dimension 1.

Example 4 (Métivier). In $\mathbb{R}^2$ we consider the vector fields

$$X_1 = \partial_x, \quad X_2 = x^k \partial_y, \quad X_3 = y^\ell \partial_y.$$

(Here k, ℓ are positive integers.) Then, we have

$$\Sigma = \{(0, 0, 0, \eta) \; : \; \eta \neq 0\}$$

is a one dimensional manifold and the rank of the restriction of the symplectic form to Σ is zero. If $\rho_0 = (0, 0, 0, \eta_0) \in \Sigma$, then we have

$$F_{\rho_0} = \{(0, 0, 0, \eta) \; : \; \eta \neq 0\},$$

i.e. the F_{ρ_0} is a one dimensional manifold along the fiber.

Example 5 (Liu–Sussmann). In $\mathbb{R}^3$ consider the vector fields

$$X_1 = \partial_x, \quad X_2 = (1 - x)\partial_y + x^2 \partial_t,$$

then the characteristic set is

$$\Sigma = \{(x, y, t, 0, \eta, \tau) \ : \ x, y, t \in \mathbb{R}, \quad (1-x)\eta + x^2\tau = 0, \quad (\eta, \tau) \neq 0\}.$$

Then, we have the decomposition

$$\Sigma = (\Sigma \setminus (\Sigma_1 \cup \Sigma_2)) \cup \Sigma_1 \cup \Sigma_2$$

with

$$\Sigma_1 = \{(0, y, t, 0, 0, \tau) \ : \ y, t \in \mathbb{R}, \quad \tau \neq 0\}$$

and

$$\Sigma_2 = \{(2, y, t, 0, 0, \tau) \ : \ y, t \in \mathbb{R}, \quad \tau \neq 0\}.$$

Then, rank $\sigma|_{\Sigma\setminus(\Sigma_1\cup\Sigma_2)} = 4 = \dim(\Sigma \setminus (\Sigma_1 \cup \Sigma_2))$ (i.e. $\Sigma \setminus (\Sigma_1 \cup \Sigma_2)$ is a symplectic manifold). Let $\rho_0 = (0, y_0, t_0, 0, 0, \tau_0) \in \Sigma_1$, then we have

$$F_{\rho_0} = \{(0, y, t_0, 0, 0, \tau_0) \ : \ y \in \mathbb{R}\}.$$

Example 6 In $\mathbb{R}^3$ consider the vector fields

$$X_1 = \partial_x - y^k\partial_t, \quad X_2 = \partial_y + x^k\partial_t.$$

(Here k is an odd number.) Then

$$\Sigma = \{(x, y, t, y^k\tau, -x^k\tau, \tau) \ : \ x, y, t \in \mathbb{R}, \quad \tau \neq 0\}$$

is a manifold of dimension 4 and

$$\Sigma = (\Sigma \setminus \Sigma_1) \cup \Sigma_1$$

with

$$\Sigma \setminus \Sigma_1 = \{(x, y, t, y^k\tau, -x^k\tau, \tau) \ : \ t \in \mathbb{R}, \quad x^{k-1} + y^{k-1} \neq 0, \quad \tau \neq 0\}$$

and

$$\Sigma_1 = \{(0, 0, t, 0, 0, \tau) \ : \ t \in R, \quad \tau \neq 0\}.$$

Then, both $\Sigma \setminus \Sigma_1$ and Σ_1 are symplectic manifolds and $F_\rho = \{\rho\}$, for every $\rho \in \Sigma$.

For some model problems in [3], it was observed that the regularity of the solution of (1.8) is influenced by the interaction of the characteristic set Σ with the set of the characteristic boundary points E. The main result of the present paper is a geometrization (and a localization) of this fact.

We denote by

$$\pi : \Omega' \times \mathbb{R}^n \longrightarrow \Omega', \quad \pi(x, \xi) = x$$

the projection on the base. For $\delta > 0$, let $B_\delta(x_0) \subset \mathbb{R}^n$ be the (open) Euclidean ball with center at x_0 and radius δ.

Theorem 1.4 *Assume (H) and let $x_0 \in \Gamma$, $\delta > 0$ with $B_\delta(x_0) \subset \Omega'$. Suppose that for every $\rho \in N^*\Gamma \cap \Sigma$, with $\pi(\rho) \in B_\delta(x_0)$, there exists the Hamiltonian leaf going through ρ, F_ρ, and that*

$$\pi(F_\rho) \cap \Omega = \emptyset. \tag{1.15}$$

Then, possibly reducing δ, the solution T of (1.8) is locally Lipschitz continuous on $\Omega \cap B_\delta(x_0)$.

In other words, if the projection of the Hamiltonian leaves through the characteristic (conormal) points does not intersect Ω, then T is locally Lipschitz continuous.

Remark 1.6 (i) The system of vector fields given in Example 4, for every smooth domain Ω, trivially satisfies the assumptions of Theorem 1.4. Then, the solution of (1.8) is locally Lipschitz continuous on Ω.

(ii) We recall that, once it is established that T is locally Lipschitz continuous, automatically, we get a stronger regularity. Indeed, as shown in Theorem 4.1 of [3], for a solution of (1.8) the local Lipschitz continuity is equivalent to the local semiconcavity (i.e. T can be locally written as the sum of a smooth with a concave function).

(iii) We observe that if Σ is a symplectic manifold or $E = \emptyset$ or in the case of Example 6 no propagation occur and, by [3, Theorem 4.4], T is locally Lipschitz continuous on Ω.

2 Proof of Theorem 1.4

2.1 Preliminaries

It is well known that (1.8) admits a unique continuous viscosity solution $T : \overline{\Omega} \to \mathbb{R}$. Indeed, taking Γ as the target set, the minimum time function associated with $\{X_1, \ldots, X_N\}$ is a solution of the above equation. Such a function is defined as follows. Given $x \in \overline{\Omega}$ and a measurable control

$$u = (u_1, \ldots, u_N) : [0, +\infty[\to \mathbb{R}^N,$$

taking values in the closed unit ball of $\mathbb{R}^N$, $\overline{B}_1(0)$, let us denote by $y^{x,u}(\cdot)$ the unique solution of the Cauchy problem

$$\begin{cases} y'(t) = \sum_{j=1}^{N} u_j(t) X_j(y(t)) & (t \geq 0) \\ y(0) = x. \end{cases} \tag{2.16}$$

Define the *transfer time* to Γ as

$$\tau_\Gamma(x, u) = \inf \left\{ t \geq 0 \ : \ y^{x,u}(t) \in \Gamma \right\} \quad (\in [0, +\infty]).$$

The *Minimum Time Problem* with target Γ is the following:

(MTP) minimize $\tau_\Gamma(x, u)$ over all controls $u : [0, +\infty[\to \overline{B}_1(0)$.

Then, the *minimum time function*, defined as

$$T(x) = \inf_{u(\cdot)} \tau_\Gamma(x, u) \qquad (x \in \overline{\Omega}),$$

turns out to be the unique viscosity solution of the Dirichlet problem (1.8). It is well-known that Hörmander's bracket generating condition implies that (2.16) is small time locally controllable. Hence, T is finite and continuous.

Remark 2.7 We recall that a $u(\cdot)$ is called an *optimal control* at a point $x \in \Omega$ if $T(x) = \tau_\Gamma(x, u)$. The corresponding solution of (2.16), $y^{x,u}$, is called the *time-optimal trajectory* at x associated with u.

We denote by $H_{X_j}(x, \xi), j = 1, \ldots, N$, the vector field on $\Omega' \times \mathbb{R}^n$ given by (in local canonical coordinates)

$$H_{X_j}(x, \xi) = \sum_{i=1}^{n} \left(\frac{\partial X_j}{\partial \xi_i}(x, \xi) \frac{\partial}{\partial x_i} - \frac{\partial X_j}{\partial x_i}(x, \xi) \frac{\partial}{\partial \xi_i} \right).$$

Definition 2.2 Let $x \in \Omega$ and let $y(\cdot) = y^{x,u}(\cdot)$ be a time-optimal trajectory, with $u : [0, T(x)] \to \overline{B}_1(0)$. We say that $y(\cdot)$ is *singular* if there exists an absolutely continuous arc $\xi : [0, T(x)] \to \mathbb{R}^n \setminus \{0\}$ such that, setting $\rho(t) = (y(t), \xi(t))$, for a.e. $t \in [0, T(x)]$ we have

$$\rho'(t) = \sum_{j=1}^{N} u_j(t) H_{X_j}(\rho(t)), \quad \rho(t) \in \Sigma, \tag{2.17}$$

and

$$\xi(T(x)) = \lambda \nu(y(T(x))), \qquad \lambda > 0. \tag{2.18}$$

Remark 2.8 We recall that the adjective "singular" in Definition 2.2 is motivated by the fact that if $x_0 \in \Omega$ and $y^{x_0,u}$ is a singular time-optimal trajectory then T fails to be Lipschitz continuous at $y^{x_0,u}(t)$. (This property is the content of Corollary 3.1 in [3].)

The following characterization clarifies the nature of the singular time-optimal trajectories (see Theorem 3.1 in [3]).

Theorem 2.5 *Let $x \in \Omega$ and let $y^{x,u}$ be a time-optimal trajectory. Then, $y^{x,u}$ is singular if and only if $y^{x,u}(T(x)) \in E$.*

Finally, let recall the basic regularity result proved in [3].

Theorem 2.6 (Interior regularity) *Under assumption (H), the following properties are equivalent:*

(1) (MTP) *admits no singular time-optimal trajectory;*
(2) *T is locally semiconcave in Ω;*
(3) *T is locally Lipschitz continuous in Ω.*

2.2 Proof of Theorem 1.4

Let us suppose, by contradiction, that T is not locally Lipschitz continuous near x_0. Then, by Theorem 2.6, possibly reducing δ there exist a point

$$\bar{\rho} \in \Sigma \quad \text{with} \quad \pi(\bar{\rho}) = \bar{x} \in \Omega \cap B_\delta(x_0),$$

and a control function $u : [0, T(\bar{x})] \longrightarrow \overline{B}_1(0)$ such that, denoting by $\rho(\cdot)$ the solution of (2.17) with $\rho(0) = \bar{\rho}$, $\pi(\rho(t))$ is a singular time–optimal trajectory and

$$\rho_0 := \rho(T(\bar{x})) \in N^*\Gamma \cap \Sigma.$$

We observe that

$$\Sigma = \{X_1 = \cdots = X_N = 0\} \implies \text{span}\{dX_1(\rho), \dots, dX_N(\rho)\} \subset (T_\rho\Sigma)^\perp,$$

for every $\rho \in \Sigma$ near ρ_0. (Here $\perp$ denotes the orthogonal with respect to the standard Euclidean product in $\mathbb{R}^{2n}$.) Hence, we find

$$\text{span}\{H_{X_1}(\rho), \dots, H_{X_N}(\rho)\} \subset (T_\rho\Sigma)^\sigma, \tag{2.19}$$

for every $\rho \in \Sigma$ (near ρ_0). Let F_{ρ_0} be the Hamiltonian leaf going through ρ_0. Then, we have that

$$\rho(t) \in F_{\rho_0}, \qquad \text{for every } t \text{ near } T(\bar{x})$$

(because of (2.19) and, by definition, $\rho'(t) \in T_{\rho(t)}\Sigma$, for t near $T(\bar{x})$).

Then, we conclude that $\pi(\rho(t)) \cap \Omega = \emptyset$ for t near $T(\bar{x})$, in contradiction with the fact that $\pi(\rho(\cdot))$ is a time–optimal trajectory. This completes our proof.

We observe that if the assumptions of Theorem 1.4 are not satisfied, then T may be not locally Lipschitz continuous on Ω. Indeed, we have the following

Example 7 In $\mathbb{R}^3$ consider the vector fields

$$X_1 = \partial_x, \qquad X_2 = (1-x)\partial_y + x^2\partial_t$$

(see [8] for a detailed study of this model). Let Ω be the open set with smooth boundary defined as follows: let $a \in]0, 1[$ and set

$$\Gamma_a = \{(x, y, -(y-a)^2) \ : \ x \in \mathbb{R}, \quad y > 0\}.$$

Consider the Euclidean open ball with center at $(0, a, 0)$ and radius $a/2$, B, and extend, smoothly, Γ_a outside such a ball. We observe that $\Omega \cap B = \{(x, y, t) \in B \ : \ t > -(y-a)^2\}$. In [3, Example 1], it is shown that the solution of (1.8) is not locally Lipschitz continuous on Ω near the point $(0, a, 0) \in \Gamma$. Let us show that near this point the assumptions of Theorem 1.4 are not satisfied. Indeed,

$$X_1(0, a, 0) = \partial_x, \quad X_2(0, a, 0) = \partial_y \in T_{(0,a,0)}\Gamma,$$

i.e. $(0, a, 0) \in E$. Furthermore, we have that, for $\tau_0 \neq 0$,

$$(0, a, 0, 0, 0, \tau_0) \in N^*\Gamma \cap \Sigma.$$

and

$$F_{(0,a,0,0,0,\tau_0)} = \{(0, y, 0, 0, 0, \tau_0) \ : \ y \in \mathbb{R}\}.$$

Then, we conclude that

$$\pi(F_{(0,a,0,0,0,\tau_0)}) \cap \Omega \neq \emptyset.$$

Acknowledgements The author was supported by the INdAM Project 2017 Regolarità delle soluzioni viscose per equazioni a derivate parziali non lineari degeneri.

Appendix A

In this appendix we provide a couple of results on the existence of an Hamiltonian leaf.

Theorem A.1 *Let $\Sigma \subset T^*\Omega'$ be a submanifold, let $\rho_0 \in \Sigma$ and let suppose that* $\operatorname{rank} \sigma|_\Sigma$ *is constant* ($< \dim \Sigma$) *near ρ_0. Then, there exists the Hamiltonian leaf going through the point ρ_0.*

Remark A.9 Theorem A.1 is a simple consequence of Theorem 21.2.4 of [7] (see Theorem 21.2.7 of [7]). We recall that Theorem 21.2.4 is a (local) canonical form for a submanifold of a symplectic manifold under a constant rank assumption on the symplectic form. For the reader convenience we will provide a direct proof of

Theorem A.1 based on the integrability Frobenius theorem (see e.g. Theorem C.1.1 in [7]).

Proof of Theorem A.1 We observe that, because of the rank of σ is constant on Σ, then

$$\Sigma \ni \rho \mapsto T_\rho\Sigma \cap (T_\rho\Sigma)^\sigma \tag{A.1}$$

defines a vectorial distribution on Σ of constant rank. (Here we are also using the assumption that Σ is a manifold.) In order to apply the Frobenius Theorem to (A.1), we need to check that the commutator $[X, Y] := XY - YX$, of two arbitrary (local) sections of (A.1), X and Y, is still a local section of (A.1).

For this purpose we use the following general result on the differential of a two form (see Lemma 21.1.5 in [7]).

Lemma A.1 *Let ω be a C^1 two form on a C^2 manifold M and let X, Y, Z be three C^1 vector fields on M. Then*

$$\begin{aligned} d\omega(X, Y, Z) &= X\omega(Y, Z) + Y\omega(Z, X) + Z\omega(X, Y) \\ &\quad - \omega([X, Y], Z) - \omega([Y, Z], X) - \omega([Z, X], Y). \end{aligned} \tag{A.2}$$

In order to use the Frobenius Theorem, let us apply the lemma with $\omega = \sigma$, $M = \Sigma$, X and Y two local sections of the distribution (A.1). We need to verify that

$$[X, Y](\rho) \in T_\rho\Sigma \cap (T_\rho\Sigma)^\sigma, \tag{A.3}$$

for $\rho \in \Sigma$. Then, let $\rho \in \Sigma$ and let Z be a vector field on Σ, we want to show that

$$\sigma([X, Y](\rho), Z(\rho)) = 0. \tag{A.4}$$

(This means that $[X, Y](\rho) \in (T_\rho\Sigma)^\sigma$.) We observe that, since σ is a closed form, then the left hand side of (A.2) is zero. Furthermore, the assumptions $X(\rho), Y(\rho) \in T_\rho\Sigma \cap (T_\rho\Sigma)^\sigma$ and $Z(\rho) \in T_\rho\Sigma$ yield that

$$\sigma(Y(\rho), Z(\rho)) = \sigma(Z(\rho), X(\rho)) = \sigma(X(\rho), Y(\rho)) = 0.$$

Finally, using the fact that $[Y, Z](\rho), [Z, X](\rho) \in T_\rho\Sigma$, we deduce that

$$\sigma([Y, Z](\rho), X(\rho)) = \sigma([Z, X](\rho), Y(\rho)) = 0$$

(because of $X(\rho), Y(\rho) \in (T_\rho\Sigma)^\sigma$). Then, (A.2) implies (A.4), i.e. $[X, Y](\rho) \in (T_\rho\Sigma)^\sigma$. The same argument, taking as Z a vector field on Σ with values in $(T_\rho\Sigma)^\sigma$ yields that

$$[X, Y](\rho) \in T_\rho\Sigma,$$

and we deduce that (A.3) holds true. Then, we may apply the Frobenius integrability Theorem and the existence of F_{ρ_0}, with $\rho_0 \in F_{\rho_0}$, follows. This completes our proof. □

Let us now consider a more interesting case: rank $\sigma|_\Sigma$ is not constant. The simpler geometrical situation is that the rank of the symplectic form varies on a submanifold. More precisely we have the following

Theorem A.2 *Let $\rho_0 \in \Sigma$ and let us suppose that Σ, near ρ_0, is a manifold and that there exists a submanifold $\Sigma_1 \subset \Sigma$ defined near ρ_0, with $\rho_0 \in \Sigma_1$, such that*

(1) $\Sigma \setminus \Sigma_1$ *is a symplectic manifold;*
(2) $\operatorname{rank} \sigma|_{\Sigma_1}$ $(< \operatorname{rank} \sigma|_{\Sigma\setminus\Sigma_1})$ *is constant*[2] $< \dim \Sigma_1$.

Then, there exists the Hamiltonian leaf going through the point ρ_0.

We omit the proof of Theorem A.2 since it is a repetition of the one of Theorem A.1.

References

1. Albano, P.: On the Eikonal equation for degenerate elliptic operators. Proc. Am. Math. Soc. **140**, 1739–1747 (2012)
2. Albano, P.: On the subelliptic Eikonal equation. Bruno Pini Math. Anal. Semin. **8**, 73–89 (2017)
3. Albano, P., Cannarsa, P., Scarinci, T.: Regularity results for the minimum time function with Hörmander vector fields. J. Differ. Equ. **264**(5), 3312–3335 (2018)
4. Albano, P., Cannarsa, P., Scarinci, T.: On the smoothness of the solution of the Dirichlet problem for the subelliptic Eikonal equation. C. R. Math. Acad. Sci. Paris **356**(2), 172–176 (2018)
5. Cannarsa, P., Rifford, L.: Semiconcavity results for optimal control problems admitting no singular minimizing controls. Ann. Inst. H. Poincaré Anal. Non Linéaire **25**(4), 773–802 (2008)
6. Derridj, M.: Sur un théorème de traces. Ann. Inst. Fourier (Grenoble) **22**, 73–83 (1972)
7. Hörmander, L.: The analysis of linear partial differential operators. III. Pseudo-differential operators. Reprint of the 1994 edition. In: Classics in Mathematics, Springer, Berlin (2007)
8. Liu, W., Sussmann, H.J.: Abnormal sub-Riemannian minimizers. In: Differential Equations, Dynamical Systems, and Control Science. Lecture Notes in Pure and Applied Mathematics, vol. 152, pp. 705–716. Dekker, New York (1994)
9. Montgomery, R.: A Tour of Subriemannian Geometries, Their Geodesics and Applications. Mathematical Surveys and Monographs. American Mathematical Society (2002)
10. Strichartz, R.S.: Sub-Riemannian geometry. J. Differ. Geom. **24**(2), 221–263 (1986)
11. Strichartz, R.S.: Corrections to: "sub-Riemannian geometry". J. Differ. Geom. 24(2), 221–263 (1986). J. Differ. Geom. 30(2), 595–596 (1989)

[2] Once more we are assuming that $\operatorname{rank} \sigma|_{\Sigma_1} < \dim \Sigma_1$, otherwise Σ_1 is a symplectic submanifold of Σ and the integral manifold going through ρ_0 reduces to $\{\rho_0\}$.

Lipschitz Continuity of the Value Function for the Infinite Horizon Optimal Control Problem Under State Constraints

Vincenzo Basco and Hélène Frankowska

Abstract This paper investigates sufficient conditions for Lipschitz regularity of the value function for an infinite horizon optimal control problem subject to state constraints. We focus on problems with a cost functional that includes a discount rate factor and allow time dependent dynamics and Lagrangian. Furthermore, our state constraints may be unbounded and with nonsmooth boundary. The key technical result used in our proof is an estimate on the distance of a given trajectory from the set of all its viable (feasible) trajectories (provided the discount rate is sufficiently large). These distance estimates are derived under a uniform inward pointing condition on the state constraint and imply, in particular, that feasible trajectories depend on initial states in a Lipschitz way with an exponentially increasing in time Lipschitz constant. As a corollary, we show that the value function of the original problem coincides with the value function of the relaxed infinite horizon problem.

Keywords Infinite horizon · Value function · State constraints · Relaxation

1 Introduction

Infinite time horizon models arising in mathematical economics and engineering typically involve control systems with restrictions on both controls and states. For instance it is natural to request all the variables involved in an economic model to

V. Basco
Dipartimento di Matematica, Università degli Studi di Roma Tor Vergata,
Via della Ricerca Scientifica, 1, 00133 Rome, Italy
e-mail: basco@mat.uniroma2.it

IMJ-PRG, UMR 7586, Sorbonne Université, case 247, 4 place Jussieu,
75252 Paris, France

H. Frankowska (✉)
CNRS, IMJ-PRG, UMR 7586, Sorbonne Université, case 247, 4 place Jussieu,
75252 Paris, France
e-mail: helene.frankowska@imj-prg.fr

F. Alabau-Boussouira et al. (eds.), *Trends in Control Theory and Partial Differential Equations*, Springer INdAM Series 32,
https://doi.org/10.1007/978-3-030-17949-6_2

be non-negative. While dealing with control constraints is rather well understood, the major difficulties with state constraints arise whenever for small perturbations of the initial state (or of a feasible control) the corresponding trajectory violates the constraints at a later time. More generally, it may happen that the celebrated value function associated to an infinite horizon optimal control problem takes infinite values and is discontinuous. In particular, this prevents using such a classical tool of optimal control theory as Hamilton-Jacobi-Bellman equation and its viscosity solution. In the literature one finds some results concerning continuity of the value function for state constrained infinite horizon problems, see for instance [13]. However in this last reference the state constraints are given by a compact set with a smooth boundary. This clearly does not fit the state constraint described by the cone of positive vectors. In addition, results of [13] address only the autonomous case, which is also a serious restriction, because, as it was shown later on, arguments of its proof can not be extended to the non-autonomous case whenever the time dependence is merely continuous.

Because of their presence in various applied models, addressing non-autonomous control systems subject to unbounded and non smooth state constraints remains crucial. Let us note that (the finite horizon) state-constrained Mayer's and Bolza's problems have been successfully investigated by many authors, see for instance [6, 9, 14] and the references therein. However in the infinite horizon framework these results can not be used, because restricting optimal trajectories of the infinite horizon problem to a finite time interval, in general, does not lead to optimal trajectories of the corresponding finite horizon problem. See [7] for a further discussion of this issue.

Infinite horizon problems exhibit many phenomena not arising in the finite horizon context and for this reason their study is still going on, even in the absence of state constraints, cfr. [1, 2, 7, 8, 12].

This paper deals with the infinite horizon optimal control problem $\mathscr{B}_\infty$:

$$\text{minimize} \int_{t_0}^{\infty} e^{-\lambda t} l(t, x(t), u(t))\, dt \tag{1}$$

over all trajectory-control pairs $(x(\cdot), u(\cdot))$ of the state constrained control system

$$\begin{cases} x'(t) = f(t, x(t), u(t)) \text{ a.e. } t \in [t_0, \infty) \\ x(t_0) = x_0 \\ u(t) \in U(t) \qquad\qquad\qquad \text{a.e. } t \in [t_0, \infty) \\ x(t) \in A \qquad\qquad\qquad\quad \forall t \in [t_0, \infty), \end{cases} \tag{2}$$

where $\lambda > 0$, $f : [0, \infty) \times \mathbb{R}^n \times \mathbb{R}^m \to \mathbb{R}^n$ and $l : [0, \infty) \times \mathbb{R}^n \times \mathbb{R}^m \to \mathbb{R}$ are given functions, $U : [0, \infty) \rightrightarrows \mathbb{R}^m$ is a Lebesgue measurable set-valued map with closed nonempty images, A is a closed subset of $\mathbb{R}^n$, and $(t_0, x_0) \in [0, \infty) \times A$ is the initial datum. Every trajectory-control pair $(x(\cdot), u(\cdot))$ that satisfies the state constrained control system (2) is called *feasible*. The infimum of the cost functional in (1) over all feasible trajectory-control pairs, with the initial datum (t_0, x_0), is denoted by $V(t_0, x_0)$ (if no feasible trajectory-control pair exists at (t_0, x_0) or if the integral in (1) is not defined for every feasible pair, we set $V(t_0, x_0) = +\infty$). The function $V : [0, \infty) \times A \to \mathbb{R} \cup \{\pm\infty\}$ is called the *value function* of problem $\mathscr{B}_\infty$.

Lipschitz continuity of V for a compact set of constraints A was recently investigated in [4] for *autonomous* control systems and lagrangian functions. It was used to get a maximum principle under state constraints and also to obtain sensitivity relations. However, in [4] the maximum principle was proved for the non-autonomous case and for possibly unbounded A under the assumption that $V(t, \cdot)$ is locally Lipschitz on A for every $t \geq 0$. So the open question remained: how to guarantee the Lipschitz continuity of $V(t, \cdot)$ when the data are time dependent and without imposing the compactness of A. Then recovering Lipschitz continuity of the value function is not straightforward and calls for distinct arguments. Here we propose sufficient conditions (cfr. Sect. 3) for it, allowing both f and l to be *time dependent* and not requiring *boundedness* of A and *smoothness* of ∂A. Our proof differs substantially from the one in [4].

The outline of the paper is as follows. In Sect. 2, we provide basic definitions, terminology, and facts from nonsmooth analysis. In Sect. 3, we state a new neighboring feasible trajectory theorem under a uniform inward pointing condition. In Sect. 4, we give an example when the uniform inward pointing condition is satisfied for functional state constraints and in Sect. 5 we prove our main result on Lipschitz continuity of the value function. Section 6 is devoted to an application to the relaxation of our control problem.

2 Preliminaries

Let $B(x, \delta)$ stand for the closed ball in $\mathbb{R}^n$ with radius $\delta > 0$ centered at $x \in \mathbb{R}^n$ and set $\mathbb{B} = B(0, 1)$, $S^{n-1} = \partial\mathbb{B}$. Denote by $|\cdot|$ and $\langle \cdot, \cdot \rangle$ the Euclidean norm and scalar product, respectively. Let $C \subset \mathbb{R}^n$ be a nonempty set. We denote the interior of C by $\operatorname{int} C$, the convex hull of C by $\operatorname{co} C$, and the distance from $x \in \mathbb{R}^n$ to C by $d_C(x) := \inf\{|x - y| : y \in C\}$. If C is closed, we let $\Pi_C(x)$ be the set of all projections of $x \in \mathbb{R}^n$ onto C. For $p \in \mathbb{R}^+ \cup \{\infty\}$ and a Lebesgue measurable set $I \subset \mathbb{R}$ we denote by $L^p(I; \mathbb{R}^n)$ the space of $\mathbb{R}^n$-valued Lebesgue measurable functions on I endowed with the norm $\|\cdot\|_{p,I}$. We say that $f \in L^p_{\mathrm{loc}}(I; \mathbb{R}^n)$ if $f \in L^p(J; \mathbb{R}^n)$ for any compact subset $J \subset I$. In what follows μ stands for the Lebesgue measure on $\mathbb{R}$.

Let I be an open interval in $\mathbb{R}$. For $f \in L^1_{\mathrm{loc}}(\overline{I}; \mathbb{R}^n)$ and all $\sigma \in [0, \mu(I))$ define

$$\theta_f(\sigma) = \sup\left\{ \int_J |f(\tau)|\, d\tau \ :\ J \subset \overline{I},\ \mu(J) \leqslant \sigma \right\}.$$

We denote by $\mathscr{L}_{\mathrm{loc}}$ the set of all $f \in L^1_{\mathrm{loc}}([0, \infty); \mathbb{R}^+)$ such that $\lim_{\sigma\to 0} \theta_f(\sigma) = 0$. Notice that $L^\infty([0, \infty); \mathbb{R}^+) \subset \mathscr{L}_{\mathrm{loc}}$ and, for any $f \in \mathscr{L}_{\mathrm{loc}}$, $\theta_f(\sigma) < \infty$ for every $\sigma > 0$.

A set-valued map $F : \mathbb{R}^n \rightrightarrows \mathbb{R}^n$ taking nonempty images is said to be *L-Lipschitz continuous* for some $L \geqslant 0$, if $F(x) \subset F(\tilde{x}) + L|x - \tilde{x}|\mathbb{B}$ for all $x, \tilde{x} \in \mathbb{R}^n$.

Let $I \subset \mathbb{R}$ be an open interval and $G : \overline{I} \times \mathbb{R}^n \rightrightarrows \mathbb{R}^n$ be a multifunction taking nonempty values. We say that G has a sub-linear growth (in x) if, for some $c \in L^1_{\text{loc}}(\overline{I}; \mathbb{R}^+)$, $\sup_{v \in G(t,x)} |v| \leqslant c(t)(1 + |x|)$ for a.e. $t \in \overline{I}$ and all $x \in \mathbb{R}^n$.

Let $\Lambda \subset \mathbb{R}^n$. We say that $G(\cdot, x)$ is *γ-left absolutely continuous, uniformly for* $x \in \Lambda$, where $\gamma \in L^1_{\text{loc}}(\overline{I}; \mathbb{R}^+)$, if

$$G(s, x) \subset G(t, x) + \int_s^t \gamma(\tau)\, d\tau \mathbb{B} \qquad \forall s, t \in \overline{I} : s < t,\ \forall x \in \Lambda. \tag{3}$$

If $\overline{I} = [S, T]$, then we have the following characterization of uniform absolute continuity from the left: $G(\cdot, x)$ is left absolutely continuous uniformly for $x \in \Lambda$, for some $\gamma \in L^1_{\text{loc}}(\overline{I}; \mathbb{R}^+)$, if and only if for every $\varepsilon > 0$ there exists $\delta > 0$ such that for any finite partition $S \leqslant t_1 < \tau_1 \leqslant t_2 < \tau_2 \leqslant \cdots \leqslant t_m < \tau_m \leqslant T$ of $[S, T]$,

$$\sum_{i=1}^{m} (\tau_i - t_i) < \delta \quad \Longrightarrow \quad \sum_{i=1}^{m} d_{G(\tau_i, x)}(G(t_i, x)) < \varepsilon \quad \forall\, x \in \Lambda,$$

where $d_E(\tilde{E}) := \inf\{\beta > 0 : \tilde{E} \subset E + \beta\mathbb{B}\}$ for any $E, \tilde{E} \subset \mathbb{R}^n$ with $\inf \emptyset = +\infty$.

Consider a closed set $E \subset \mathbb{R}^n$ and $x \in E$. The *Clarke tangent cone* $T^C_E(x)$ to E at x is defined by

$$T^C_E(x) := \{\xi \in \mathbb{R}^n : \forall x_i \to_E x,\ \forall t_i \downarrow 0,\ \exists v_i \to \xi \text{ such that } x_i + t_i v_i \in E \quad \forall i\ \},$$

where $x_i \to_E x$ means $x_i \in E$ for all i. We denote by $N^C_E(x) := (T^C_E(x))^-$ the *Clarke normal cone* to E at x, where "$^-$" stands for the negative polar of a set.

3 Uniform Distance Estimates

We provide here sufficient conditions for uniform linear L^∞ estimates on intervals of the form $I = [t_0, t_1]$, with $0 \leqslant t_0 < t_1$, for the state constrained differential inclusion

$$\begin{cases} x'(t) \in F(t, x(t)) \text{ a.e. } t \in I \\ x(t) \in A \qquad\qquad \forall t \in I, \end{cases}$$

where $F : [0, \infty) \times \mathbb{R}^n \rightrightarrows \mathbb{R}^n$ is a given set-valued map and $A \subset \mathbb{R}^n$ is a closed set.

A function $x : [t_0, t_1] \to \mathbb{R}^n$ is said to be an *F-trajectory* if it is absolutely continuous and $x'(t) \in F(t, x(t))$ for a.e. $t \in [t_0, t_1]$, and a *feasible F-trajectory* if $x(\cdot)$ is an F-trajectory and $x([t_0, t_1]) \subset A$.

We denote by (H) the following hypothesis on $F(\cdot, \cdot)$:

(i) F has closed, nonempty values, a sub-linear growth, and $F(\cdot, x)$ is Lebesgue measurable for all $x \in \mathbb{R}^n$;
(ii) there exist $M \geqslant 0$ and $\alpha > 0$ such that

$$\sup\{|v| \,:\, v \in F(t, x),\ (t, x) \in [0, \infty) \times (\partial A + \alpha\mathbb{B})\} \leq M; \qquad (4)$$

(iii) there exists $\varphi \in \mathscr{L}_{\text{loc}}$ such that $F(t, \cdot)$ is $\varphi(t)$-Lipschitz continuous for a.e. $t \in \mathbb{R}^+$.

We shall also need the following two assumptions:

(AC) there exist $\tilde{\eta} > 0$ and $\gamma \in \mathscr{L}_{\text{loc}}$ such that $F(\cdot, x)$ is γ-left absolutely continuous, uniformly for $x \in \partial A + \tilde{\eta}\mathbb{B}$;
(IPC) for some $\varepsilon > 0$, $\eta > 0$ and every $(t, x) \in [0, \infty) \times (\partial A + \eta\mathbb{B}) \cap A$ there exists $v \in \text{co}\, F(t, x)$ satisfying

$$\{y + [0, \varepsilon](v + \varepsilon\mathbb{B}) \,:\, y \in (x + \varepsilon\mathbb{B}) \cap A\} \subset A. \qquad (5)$$

We state next a uniform neighboring feasible trajectory theorem for left absolutely continuous with respect to time set-valued maps.

Theorem 1 *Assume* (H), (AC), *and* (IPC). *Then for every $\delta > 0$ there exists a constant $\beta > 0$ such that for any $[t_0, t_1] \subset [0, \infty)$ with $t_1 - t_0 = \delta$, any F-trajectory $\hat{x}(\cdot)$ defined on $[t_0, t_1]$ with $\hat{x}(t_0) \in A$, and any $\rho > 0$ satisfying*

$$\rho \geqslant \sup_{t \in [t_0, t_1]} d_A(\hat{x}(t)),$$

we can find an F-trajectory $x(\cdot)$ on $[t_0, t_1]$ such that $x(t_0) = \hat{x}(t_0)$,

$$\|\hat{x} - x\|_{\infty, [t_0, t_1]} \leqslant \beta\rho \quad \& \quad x(t) \in \text{int}\, A \quad \forall t \in (t_0, t_1].$$

The following Proposition can be proved using the same arguments as in [5, pp. 1922–1923].

Proposition 1 *Assume* (H), (AC), (IPC), *and that the assertion of Theorem 1 is valid under the additional hypothesis: $F(t, x)$ is convex for all $(t, x) \in [0, \infty) \times \mathbb{R}^n$. Then the assertion of Theorem 1 is valid under* (H), (AC), *and* (IPC) *alone.*

Proof (of Theorem 1*)* Fix $\delta > 0$ and let us relabel by η the constant given by $\min\{\eta, \tilde{\eta}, \alpha\}$. Let

$$k > 0,\ \Delta > 0,\ \bar{\rho} > 0, \text{ and } m \in \mathbb{N}^+ \qquad (6)$$

be such that $k > 1/\varepsilon$,

$$(i)\ \ \Delta \leqslant \varepsilon; \quad (ii)\ \ \bar{\rho} + M\Delta < \varepsilon, \quad k\bar{\rho} < \varepsilon; \quad (iii)\ \ 4\Delta M \leqslant \eta, \qquad (7)$$

$$\begin{aligned} &(i) \quad e^{\theta_\varphi(\Delta)}(\theta_\gamma(\Delta) + \theta_\varphi(\Delta)M) < \varepsilon; \\ &(ii) \quad 2e^{\theta_\varphi(\Delta)}(\theta_\gamma(\Delta) + \theta_\varphi(\Delta)M)k < (k\varepsilon - 1), \end{aligned} \tag{8}$$

and

$$\frac{\delta}{m} \leqslant \Delta. \tag{9}$$

We remark that all the constants appearing in (6) do not depend on the time interval $[t_0, t_1]$, the trajectory $\hat{x}(\cdot)$, and ρ.

By Proposition 1, we may assume that $F(\cdot, \cdot) = \text{co}\, F(\cdot, \cdot)$. We consider three cases.

Case 1: $\rho \leqslant \bar{\rho}$ and $\delta \leqslant \Delta$.

By (7)–(iii), if $\hat{x}(t_0) \in A \backslash (\partial A + \frac{\eta}{2}\mathbb{B})$, then $x(\cdot) = \hat{x}(\cdot)$ is as desired. Suppose next that $\hat{x}(t_0) \in (\partial A + \frac{\eta}{2}\mathbb{B}) \cap A$. Let $v \in F(t_0, \hat{x}(t_0))$ be as in (IPC) and define $y : [t_0, t_1] \to \mathbb{R}^n$ by $y(t_0) = \hat{x}(t_0)$ and

$$y'(t) = \begin{cases} v & t \in [t_0, (t_0 + k\rho) \wedge t_1] \\ \hat{x}'(t - k\rho) & t \in (t_0 + k\rho, t_1] \cap J, \end{cases} \tag{10}$$

where $J = \{s \in (t_0 + k\rho, t_1] \,:\, \hat{x}'(s - k\rho) \text{ exists}\}$. Hence

$$\|\hat{x} - y\|_{\infty,[t_0,t_1]} \leqslant 2Mk\rho. \tag{11}$$

By Filippov's theorem (cfr. [3]) there exists an F-trajectory $x(\cdot)$ on $[t_0, t_1]$ such that $x(t_0) = y(t_0)$ and

$$\|y - x\|_{\infty,[t_0,t]} \leqslant e^{\int_{t_0}^t \varphi(\tau)\,d\tau} \int_{t_0}^t d_{F(s,y(s))}(y'(s))\,ds \tag{12}$$

for all $t \in [t_0, t_1]$. Then, using (H)-(iii), (3), and (10), it follows that

$$d_{F(s,y(s))}(y'(s)) \leqslant \begin{cases} \theta_\gamma(\Delta) + \varphi(s)M(s - t_0) & \text{a.e. } s \in [t_0, (t_0 + k\rho) \wedge t_1] \\ \varphi(s)Mk\rho + \int_{s-k\rho}^s \gamma(\tau)\,d\tau & \text{a.e. } s \in (t_0 + k\rho, t_1]. \end{cases} \tag{13}$$

Hence, we obtain for any $t \in [t_0, (t_0 + k\rho) \wedge t_1]$

$$\int_{t_0}^t d_{F(s,y(s))}(y'(s))\,ds \leqslant (\theta_\gamma(\Delta) + \theta_\varphi(\Delta)M)(t - t_0),$$

and, using the Fubini theorem, for any $t \in (t_0 + k\rho, t_1]$,

$$\int_{t_0+k\rho}^t d_{F(s,y(s))}(y'(s))\,ds \leqslant (\theta_\varphi(\Delta)M + \theta_\gamma(\Delta))k\rho.$$

Thus, by (12), for all $t \in [t_0, (t_0 + k\rho) \wedge t_1]$

$$\|y - x\|_{\infty,[t_0,t]} \leqslant e^{\theta_\varphi(\Delta)}(\theta_\gamma(\Delta) + \theta_\varphi(\Delta)M)(t - t_0) \tag{14}$$

and

$$\|y - x\|_{\infty,[t_0,t_1]} \leqslant 2e^{\theta_\varphi(\Delta)}(\theta_\gamma(\Delta) + \theta_\varphi(\Delta)M)k\rho. \tag{15}$$

Finally, taking note of (11), it follows that $\|\hat{x} - x\|_{\infty,[t_0,t_1]} \leqslant \beta_1\rho$, where $\beta_1 = 2(M + e^{\theta_\varphi(\Delta)}(\theta_\gamma(\Delta) + \theta_\varphi(\Delta)M))k$.

We claim next that $x(t) \in \operatorname{int} A$ for all $t \in (t_0, t_1]$. Indeed, if $t \in (t_0, (t_0 + k\rho) \wedge t_1]$, then from (IPC), (7)-(i) and (10) it follows that

$$y(t) + (t - t_0)\varepsilon\mathbb{B} = \hat{x}(t_0) + (t - t_0)(v + \varepsilon\mathbb{B}) \subset A,$$

and it is enough to use (14) and (8)-(i).

On the other hand, if $t \in (t_0 + k\rho, t_1]$, then for $\pi(t) \in \Pi_A(\hat{x}(t - k\rho))$ we have $|\hat{x}(t - k\rho) - \pi(t)| = d_A(\hat{x}(t - k\rho)) \leqslant \rho$, and, from (10), it follows that

$$y(t) \in \pi(t) + k\rho v + \rho\mathbb{B}. \tag{16}$$

Now, since $|\pi(t) - \hat{x}(t_0)| \leqslant |\hat{x}(t - k\rho) - \pi(t)| + |\hat{x}(t - k\rho) - \hat{x}(t_0)| \leqslant \bar{\rho} + M\Delta$, from (5) and (7)-$(ii)$ we have

$$\pi(t) + k\rho v + k\rho\varepsilon\mathbb{B} = \pi(t) + k\rho(v + \varepsilon\mathbb{B}) \subset A. \tag{17}$$

Finally, (16) and (17) imply that $y(t) + (k\varepsilon - 1)\rho\mathbb{B} \subset A$. So, the claim follows from (8)-(ii) and (15).

<u>Case 2</u>: $\rho > \bar{\rho}$ and $\delta \leqslant \Delta$.

By the viability theorem from [10], we know that there exists a feasible F-trajectory $\bar{x}(\cdot)$ on $[t_0, t_1]$ starting from $\hat{x}(t_0)$. Note that $d_A(\bar{x}(t)) = 0$ for all $t \in [t_0, t_1]$. By the Case 1, replacing $\hat{x}(\cdot)$ with $\bar{x}(\cdot)$, it follows that there exists a feasible F-trajectory $x(\cdot)$ on $[t_0, t_1]$ such that $x(t_0) = \hat{x}(t_0)$ and $x((t_0, t_1]) \subset \operatorname{int} A$. Hence, by (4), we have $\|\hat{x} - x\|_{\infty,[t_0,t_1]} \leqslant 2M\Delta \leqslant \beta_2\rho$, with $\beta_2 = \frac{2M\Delta}{\bar{\rho}}$.

<u>Case 3</u>: $\delta > \Delta$.

The above proof implies that in Cases 1 and 2, β_1, β_2 can be taken the same if δ is replaced by any $0 < \delta_1 < \delta$. Define $\tilde{\beta} = \beta_1 \vee \beta_2$ and let $\{[\tau_-^i, \tau_+^i]\}_{i=1}^m$ be a partition of $[t_0, t_1]$ by the intervals with the length at most δ/m.

Put $x_0(\cdot) := \hat{x}(\cdot)$. From Cases 1 and 2, replacing $[t_0, t_1]$ by $[\tau_-^1, \tau_+^1]$ and setting

$$\rho_0 = \max\{\rho, \sup_{t\in[t_0,t_1]} d_A(x_0(t))\} = \rho,$$

we conclude that there exists an F-trajectory $x_1(\cdot)$ on $[\tau_-^1, \tau_+^1] = [t_0, \tau_+^1]$ such that $x_1(t_0) = \hat{x}(t_0)$, $x_1((t_0, \tau_+^1]) \subset \operatorname{int} A$, and

$$\|x_1 - x_0\|_{\infty,[\tau_-^1,\tau_+^1]} \leqslant \tilde{\beta}\rho_0.$$

Using Filippov's theorem, we can extend the trajectory $x_1(\cdot)$ on whole interval $[t_0, t_1]$ so that

$$\|x_1 - x_0\|_{\infty,[t_0,t_1]} \leqslant e^{\int_{t_0}^{t_1} \varphi(\tau)\,d\tau} \tilde{\beta}\rho_0 \leqslant K\tilde{\beta}\rho_0,$$

where $K := e^{\theta_\varphi(\delta)}$.

Repeating recursively the above argument on each time interval $[\tau_-^i, \tau_+^i]$, we conclude that there exists a sequence of F-trajectories $\{x_i(\cdot)\}_{i=1}^m$ on $[t_0, t_1]$, such that $x_i(t_0) = \hat{x}(t_0)$, $x_i((t_0, \tau_+^i]) \subset \operatorname{int} A$ for all $i = 1, \ldots, m$, $x_j(\cdot)|_{[t_0,\tau_+^{j-1}]} = x_{j-1}(\cdot)$ for all $j = 2, \ldots, m$, and

$$\|x_i - x_{i-1}\|_{\infty,[t_0,t_1]} \leqslant K\tilde{\beta}\rho_{i-1} \qquad \forall i = 1, \ldots, m, \tag{18}$$

where $\rho_{i-1} = \max\{\rho, \sup_{t\in[t_0,t_1]} d_A(x_{i-1}(t))\}$. Notice that

$$\rho_i \leqslant \rho_{i-1} + \|x_i - x_{i-1}\|_{\infty,[t_0,t_1]} \qquad \forall i = 1, \ldots, m. \tag{19}$$

Taking note of (18) and (19) we get for all $i = 1, \ldots, m$

$$\begin{aligned}\|x_i - x_{i-1}\|_{\infty,[t_0,t_1]} &\leqslant K\tilde{\beta}(\rho_{i-2} + \|x_{i-1} - x_{i-2}\|_{\infty,[t_0,t_1]})\\ &\leqslant K\tilde{\beta}(1 + K\tilde{\beta})\rho_{i-2} \leq \cdots \leqslant K\tilde{\beta}(1 + K\tilde{\beta})^{i-1}\rho_0.\end{aligned}$$

Then, letting $x(\cdot) := x_m(\cdot)$ and recalling that $\rho_0 = \rho$, we obtain

$$\|x - \hat{x}\|_{\infty,[t_0,t_1]} \leqslant \sum_{i=1}^m \|x_i - x_{i-1}\|_{\infty,[t_0,t_1]} \leqslant K\tilde{\beta}\rho_0 \sum_{i=1}^m (1 + K\tilde{\beta})^{i-1} \leqslant \beta_3\rho,$$

where $\beta_3 = (1 + K\tilde{\beta})^m - 1$.

Then all conclusions of the theorem follow with $\beta = \tilde{\beta} \vee \beta_3$. Observe that β depends only on ε, η, M, δ, and on functions $\gamma(\cdot)$ and $\varphi(\cdot)$.

When F is merely measurable with respect to time, then a stronger inward pointing condition has to be imposed:

(IPC)$'$ there exist $\eta > 0$, $r > 0$, $M \geqslant 0$ such that for a.e. $t \in [0, \infty)$, any $y \in \partial A + \eta\mathbb{B}$, and any $v \in F(t, y)$, with $\sup_{n\in N_{y,\eta}^1} \langle n, v\rangle \geq 0$, there exists $w \in F(t, y) \cap B(v, M)$ such that

$$\sup_{n\in N_{y,\eta}^1} \{\langle n, w\rangle,\ \langle n, w - v\rangle\} \leqslant -r,$$

where $N_{y,\eta}^1 := \{n \in S^{n-1} : n \in N_A^C(x),\ x \in \partial A \cap B(y, \eta)\}$.

Let us denote by (H)$'$ the assumption (H) with (H)-(ii) replaced by a weaker requirement:

(H)$'$ (ii) $\exists q \in \mathscr{L}_{\text{loc}}$ such that $F(t,x) \subset q(t)\mathbb{B}, \ \forall x \in \partial A,$ for a.e. $t \in [0,\infty)$.

Remark 1 We notice that from (H)$'$-(ii) and (iii) it follows that for any $\alpha > 0$ there exists $q_\alpha \in \mathscr{L}_{\text{loc}}$ such that $F(t,x) \subset q_\alpha(t)\mathbb{B}$ for a.e. $t \in [0,\infty)$ and all $x \in \partial A + \alpha\mathbb{B}$.

Theorem 2 *Let us assume* (H)$'$ *and* (IPC)$'$. *Then the assertion of Theorem 1 is valid.*

Proof (IPC)$'$ corresponds to the conclusion of [9, Proposition 7] with r, η, and M defined uniformly over A. Thanks to this observation and Remark 1 exactly the same arguments as those in [9, proof of Theorem 5] can be used to prove the theorem.

We provide next a condition that simplifies (IPC)$'$.

Proposition 2 *Assume that for some $\eta > 0$, $r > 0$, $M \geqslant 0$, and $\Gamma \subset [0,\infty)$, with $\mu(\Gamma) = 0$, and for any $t \in [0,\infty)\backslash\Gamma$, $y \in \partial A + \eta\mathbb{B}$, and $v \in F(t,y)$, with $\sup_{n\in N^1_{y,\eta}} \langle n, v\rangle > -r$, there exists $w \in F(t,y) \cap B(v,M)$ satisfying $\sup_{n\in N^1_{y,\eta}} \langle n, w - v\rangle \leqslant -r$. Then,* (IPC)$'$ *holds true for all $t \in [0,\infty)\backslash\Gamma$.*

Proof Indeed, otherwise there exist $t \in [0,\infty)\backslash\Gamma$, $y \in \partial A + \eta\mathbb{B}$, and $v \in F(t,y)$, with $\sup_{n\in N^1_{y,\eta}} \langle n, v\rangle > -r$, such that for any $w \in F(t,y) \cap B(v,M)$ satisfying $\sup_{n\in N^1_{y,\eta}} \langle n, w - v\rangle \leqslant -r$ we have $\sup_{n\in N^1_{y,\eta}} \langle n, w\rangle > -r$. Now, by our assumptions, there exists $w_1 \in F(t,y) \cap B(v,M)$ such that $\sup_{n\in N^1_{y,\eta}} \langle n, w_1 - v\rangle \leqslant -r$. Since ad absurdum we supposed that $\sup_{n\in N^1_{y,\eta}} \langle n, w_1\rangle > -r$, it follows that there exists $w_2 \in F(t,y) \cap B(v,M)$ satisfying $\sup_{n\in N^1_{y,\eta}} \langle n, w_2 - w_1\rangle \leqslant -r$. Then for any $n \in N^1_{y,\eta}$,

$$\langle n, w_2 - v\rangle = \langle n, w_2 - w_1\rangle + \langle n, w_1 - v\rangle \leqslant -2r.$$

Iterating the same argument, we conclude that there exists a sequence $\{w_i\}_{i\in\mathbb{N}^+}$ in $F(t,y) \cap B(v,M)$ such that $\sup_{n\in N^1_{y,\eta}} \langle n, w_i - v\rangle \leqslant -ir$ for all $i \in \mathbb{N}^+$. This contradicts the boundedness of $F(t,y) \cap B(v,M)$ and ends the proof.

Now, consider the following state constrained differential inclusion

$$\begin{cases} x'(t) \in F(t,x(t)) & \text{a.e. } t \in [t_0,\infty) \\ x(t) \in A & \forall t \in [t_0,\infty), \end{cases}$$

where $t_0 \geqslant 0$. A function $x : [t_0,\infty) \to \mathbb{R}^n$ is said to be an F_∞-*trajectory* or a *feasible F_∞-trajectory* if $x|_{[t_0,t_1]}(\cdot)$ is an *F-trajectory* or a *feasible F-trajectory*, respectively, for all $t_1 > t_0$.

Theorem 3 *Assume that either* (H), (AC), *and* (IPC) *or* (H)$'$ *and* (IPC)$'$ *hold true. Furthermore, suppose that*

$$\limsup_{t\to\infty} \frac{1}{t}\int_0^t \varphi(\tau)\,d\tau < \infty.$$

Then there exist $C > 1$, $K > 0$ such that for any $t_0 \geqslant 0$, any $x^0, x^1 \in A$, and any feasible F_∞-trajectory $x : [t_0, \infty) \to \mathbb{R}^n$, with $x(t_0) = x^0$, we can find a feasible F_∞-trajectory $\tilde{x} : [t_0, \infty) \to \mathbb{R}^n$, with $\tilde{x}(t_0) = x^1$, such that

$$|\tilde{x}(t) - x(t)| \leqslant Ce^{Kt}|x^1 - x^0| \qquad \forall t \geqslant t_0.$$

Proof Let $\delta = 1$ and $\beta > 0$ be as in Theorem 1 (or Theorem 2). Consider $K_1 > 0$, $K_2 > 0$, $\tilde{k} > 0$ such that

$$2\beta + 1 < e^{K_1} \quad \text{and} \quad \int_0^{t+1} \varphi(s)\,ds \leqslant K_2 t + \tilde{k} \qquad \forall t \geqslant 0. \tag{20}$$

Fix $x^0, x^1 \in A$, with $x^1 \neq x^0$, and a feasible F_∞-trajectory $x : [t_0, \infty) \to \mathbb{R}^n$ with $x(t_0) = x^0$. By Filippov's theorem, there exists an F-trajectory $y_0 : [t_0, t_0 + 1] \to \mathbb{R}^n$ such that $y_0(t_0) = x^1$ and

$$\|y_0 - x\|_{\infty,[t_0,t_0+1]} \leqslant e^{\int_{t_0}^{t_0+1} \varphi(s)\,ds}|x^1 - x^0|.$$

Denote by $x_0 : [t_0, t_0 + 1] \to \mathbb{R}^n$ the feasible F-trajectory, with $x_0(t_0) = x^1$, satisfying the conclusions of Theorem 1 with $\hat{x}(\cdot) = y_0(\cdot)$. Thus

$$\begin{aligned}\|x_0 - y_0\|_{\infty,[t_0,t_0+1]} &\leqslant \beta(\max_{t\in[t_0,t_0+1]} d_A(y_0(t)) + |x^1 - x^0|)\\ &\leqslant \beta(\|y_0 - x\|_{\infty,[t_0,t_0+1]} + |x^1 - x^0|) \leqslant 2\beta e^{\int_{t_0}^{t_0+1} \varphi(s)\,ds}|x^1 - x^0|\end{aligned}$$

and therefore

$$\begin{aligned}\|x_0 - x\|_{\infty,[t_0,t_0+1]} &\leqslant \|x_0 - y_0\|_{\infty,[t_0,t_0+1]} + \|y_0 - x\|_{\infty,[t_0,t_0+1]}\\ &\leqslant (2\beta + 1)e^{\int_{t_0}^{t_0+1} \varphi(s)\,ds}|x^1 - x^0|.\end{aligned} \tag{21}$$

Now, applying again Filippov's theorem on $[t_0 + 1, t_0 + 2]$, there exists an F-trajectory $y_1 : [t_0 + 1, t_0 + 2] \to \mathbb{R}^n$, with $y_1(t_0 + 1) = x_0(t_0 + 1)$, such that, thanks to (21),

$$\|y_1 - x\|_{\infty,[t_0+1,t_0+2]} \leqslant (2\beta + 1)e^{\int_{t_0}^{t_0+2} \varphi(s)\,ds}|x^1 - x^0|. \tag{22}$$

Denoting by $x_1 : [t_0 + 1, t_0 + 2] \to \mathbb{R}^n$ the feasible F-trajectory, with $x_1(t_0 + 1) = x_0(t_0 + 1)$, satisfying the conclusions of Theorem 1, for $\hat{x}(\cdot) = y_1(\cdot)$, we deduce from (22), that

$$\|x_1 - y_1\|_{\infty,[t_0+1,t_0+2]} \leqslant \beta(2\beta+1)e^{\int_{t_0}^{t_0+2}\varphi(s)\,ds}|x^1 - x^0|. \tag{23}$$

Hence, taking note of (22) and (23),

$$\|x_1 - x\|_{\infty,[t_0+1,t_0+2]} \leqslant (2\beta+1)^2 e^{\int_{t_0}^{t_0+2}\varphi(s)\,ds}|x^1 - x^0|.$$

Continuing this construction, we obtain a sequence of feasible F-trajectories $x_i : [t_0+i, t_0+i+1] \to \mathbb{R}^n$ such that $x_j(t_0+j) = x_{j-1}(t_0+j)$ for all $j \geqslant 1$, and

$$\|x_i - x\|_{\infty,[t_0+i,t_0+i+1]} \leqslant (2\beta+1)^{i+1} e^{\int_{t_0}^{t_0+i+1}\varphi(s)\,ds}|x^1 - x^0| \;\; \forall i \in \mathbb{N}. \tag{24}$$

Define the feasible F_∞-trajectory $\tilde{x} : [t_0, \infty) \to \mathbb{R}^n$ by $\tilde{x}(t) := x_i(t)$ if $t \in [t_0 + i, t_0 + i + 1]$ and observe that $\tilde{x}(t_0) = x^1$.

Let $t \geqslant t_0$. Then there exists $i \in \mathbb{N}$ such that $t \in [t_0+i, t_0+i+1]$. So, from (24) and (20), it follows that

$$\begin{aligned}|\tilde{x}(t) - x(t)| &\leqslant (2\beta+1)^{i+1} e^{\int_{t_0}^{t_0+i+1}\varphi(s)\,ds}|x^1 - x^0| \\ &\leqslant e^{\tilde{k}}(2\beta+1)e^{(K_1+K_2)(t_0+i)}|x^1 - x^0| \leqslant Ce^{Kt}|x^1 - x^0|,\end{aligned}$$

where $K = K_1 + K_2$ and $C = e^{\tilde{k}}(2\beta+1)$.

4 Uniform IPC for Functional Set Constraints

Consider the state constraints of the form

$$A = \bigcap_{i=1}^{m} A_i, \qquad A_i = \{x \in \mathbb{R}^n \;:\; g_i(x) \leqslant 0\} \;\; i = 1, \ldots, m,$$

where $g_i : \mathbb{R}^n \to \mathbb{R}$ is a $C^{1,1}$ function with bounded $\nabla g_i(\cdot)$ for all $i \in I := \{1, \ldots, m\}$. Furthermore, we assume in this section that there exist $M \geqslant 0$ and $\varphi > 0$ such that $\sup\{|v| \;:\; v \in F(t,x),\; (t,x) \in [0,\infty) \times \partial A\} \leqslant M$ and $F(t,\cdot)$ is φ-Lipschitz continuous for any $t \geqslant 0$.

Proposition 3 *Assume that for some $\delta > 0$, $r > 0$ and for all $(t,x) \in [0,\infty) \times \partial A$ there exists $v \in \operatorname{co} F(t,x)$ satisfying*

$$\langle \nabla g_i(x), v\rangle \leqslant -r \qquad \forall i \in \bigcup_{z \in B(x,\delta)} I(z),$$

where $I(z) = \{i \in I \;:\; z \in \partial A_i\}$. Then (IPC) *holds true.*

Proof Let us set $J(x) := \bigcup_{z\in B(x,\delta)} I(z)$ for all $x \in \partial A$. Fix $(t, x) \in [0, \infty) \times \partial A$ and $v \in \text{co}\, F(t, x)$ satisfying $\langle \nabla g_i(x), v\rangle \leqslant -r$ for all $i \in J(x)$. Pick

$$k > \max_{i\in I} \sup_{x\neq y} \frac{|\nabla g_i(x) - \nabla g_i(y)|}{|x - y|} \quad \& \quad L > \max_{i\in I} \sup_{x\in\mathbb{R}^n} |\nabla g_i(x)|.$$

We divide the proof into three steps.

Step 1: We claim that there exists $\eta' > 0$, not depending on (t, x), such that for all $y \in B(x, \eta')$ we can find $w \in \text{co}\, F(t, y)$, with $|w - v| \leqslant r/4L$, satisfying for all $i \in J(x)$,

$$\langle \nabla g_i(y), w\rangle \leqslant -r/2.$$

Indeed, for all $i \in J(x)$ and $y \in B(x, r/4kM)$ we have

$$\langle \nabla g_i(y), v\rangle = \langle \nabla g_i(y) - \nabla g_i(x), v\rangle + \langle \nabla g_i(x), v\rangle \leqslant kM|y - x| - r \leqslant -\frac{3r}{4}$$

and for all $w \in \mathbb{R}^n$ such that $|w - v| \leqslant r/4L$

$$\langle \nabla g_i(y), w\rangle = \langle \nabla g_i(y), w - v\rangle + \langle \nabla g_i(y), v\rangle \leqslant L|w - v| - 3r/4 \leqslant -\frac{r}{2}.$$

Since $F(t, \cdot)$ is φ-Lipschitz continuous, there exists $w \in \text{co}\, F(t, y)$ such that $|w - v| \leqslant r/4L$ whenever $|y - x| \leqslant r/4\varphi L$. So the claim follows with $\eta' = \min\{r/4\varphi L, r/4kM\}$.

Step 2: We claim that there exists $\varepsilon' > 0$, not depending on (t, x), such that for all $y \in B(x, \eta')$ we can find $w \in \text{co}\, F(t, y)$ such that

$$\langle \nabla g_i(z), \tilde{w}\rangle \leqslant -r/4 \qquad \forall z \in B(y, \varepsilon'),\ \forall \tilde{w} \in B(w, \varepsilon'),\ \forall i \in J(x).$$

Indeed, let $y \in B(x, \eta')$ and $w \in \text{co}\, F(t, y)$ be as in Step 1. Then for any $\tilde{w} \in \mathbb{R}^n$ such that $|\tilde{w} - w| \leqslant r/8L$ and for all $i \in J(x)$ and $z \in \mathbb{R}^n$,

$$\begin{aligned}\langle \nabla g_i(z), \tilde{w}\rangle &= \langle \nabla g_i(z) - \nabla g_i(y), \tilde{w}\rangle + \langle \nabla g_i(y), \tilde{w} - w\rangle + \langle \nabla g_i(y), w\rangle \\ &\leqslant k(M + r/4L + r/8L)|z - y| + r/8 - r/2.\end{aligned}$$

So the claim follows with $\varepsilon' = \min\{k^{-1}(M + r/2L)^{-1} r/8,\ r/8L\}$.

Step 3: We prove that there exist $\eta > 0$, $\varepsilon > 0$, not depending on (t, x), such that for all $y \in B(x, \eta) \cap A$ we can find $w \in \text{co}\, F(t, y)$ satisfying

$$z + \tau\tilde{w} \in A \qquad \forall z \in B(y, \varepsilon) \cap A,\ \forall \tilde{w} \in B(w, \varepsilon),\ \forall 0 \leqslant \tau \leqslant \varepsilon. \tag{25}$$

Let $y \in B(x, \eta') \cap A$ and $w \in \text{co}\, F(t, y)$ be as in Step 2. Then, by the mean value theorem, for any $\tau \geqslant 0$, any $z \in B(y, \varepsilon') \cap A$, any $\tilde{w} \in B(w, \varepsilon')$, and any $i \in J(x)$ there exists $\sigma_\tau \in [0, 1]$ such that

$$\begin{aligned} g_i(z+\tau\tilde{w}) &= g_i(z)+\tau\langle\nabla g_i(z+\sigma_\tau\tau\tilde{w}),\tilde{w}\rangle \\ &\leqslant \tau\langle\nabla g_i(z),\tilde{w}\rangle + k(M+r/4L+\varepsilon')^2\tau^2 \\ &\leqslant -\tfrac{r\tau}{4} + k(M+r/4L+\varepsilon')^2\tau^2. \end{aligned}$$

Choosing $\eta \in (0,\eta']$ and $\varepsilon \in (0,\varepsilon']$ such that $\eta+\varepsilon(M+r/4L+\varepsilon) \leqslant \delta$ and $\varepsilon \leqslant k^{-1}(M+r/4L+\varepsilon')^{-2}r/4$, it follows that for all $z \in B(y,\varepsilon)\cap A,\ \tilde{w} \in B(w,\varepsilon)$, and all $0 \leqslant \tau \leqslant \varepsilon$

$$z+\tau\tilde{w} \in B(x,\delta) \tag{26}$$

and

$$g_i(z+\tau\tilde{w}) \leqslant 0 \qquad \forall i \in J(x). \tag{27}$$

Furthermore, by (26) and since $B(x,\delta) \subset A_j$ for all $j \in I\backslash J(x)$, we have for all $z \in B(y,\varepsilon)\cap A,\ \tilde{w} \in B(w,\varepsilon)$, and all $0 \leqslant \tau \leqslant \varepsilon$

$$g_i(z+\tau\tilde{w}) \leqslant 0 \qquad \forall i \in I\backslash J(x). \tag{28}$$

The conclusion follows from (27) and (28).

5 Lipschitz Continuity for a Class of Value Functions

Now we give an application of the results of Sect. 3 to the Lipschitz regularity of the value function for a class of infinite horizon optimal control problems subject to state constraints.

Let us consider the problem $\mathscr{B}_\infty$ stated in the Introduction. Recall that for a function $q \in L^1_{\rm loc}([t_0,\infty);\mathbb{R})$ the integral $\int_{t_0}^{\infty} q(t)\,dt := \lim_{T\to\infty}\int_{t_0}^{T} q(t)\,dt$, provided this limit exists. We denote by (h) the following assumptions on f and l:

(i) there exists $\alpha > 0$ such that f and l are bounded functions on

$$\{(t,x,u)\ :\ t \geqslant 0,\ x \in (\partial A+\alpha\mathbb{B}),\ u \in U(t)\};$$

(ii) for all $(t,x) \in [0,\infty)\times\mathbb{R}^n$ the set

$$\{(f(t,x,u),l(t,x,u))\ :\ u \in U(t)\}$$

is closed;

(iii) there exist $c \in L^1_{\rm loc}([0,\infty);\mathbb{R}^+)$ and $k \in \mathscr{L}_{\rm loc}$ such that for a.e. $t \in \mathbb{R}^+$ and for all $x,\ y \in \mathbb{R}^n$, $u \in U(t)$,

$$|f(t,x,u)-f(t,y,u)|+|l(t,x,u)-l(t,y,u)| \leqslant k(t)|x-y|,$$

$$|f(t,x,u)| + |l(t,x,u)| \leqslant c(t)(1+|x|);$$

(iv) $\limsup_{t\to\infty} \frac{1}{t}\int_0^t (c(s)+k(s))\,ds < \infty$;
(v) for all $x \in \mathbb{R}^n$ the mappings $f(\cdot,x,\cdot)$, $l(\cdot,x,\cdot)$ are Lebesgue-Borel measurable.

Furthermore, we denote by (h)′ the assumptions (h) with (h)-(i) replaced by:

(h)′ (i) $\exists q \in \mathcal{L}_{\text{loc}}$ such that for a.e. $t \in [0,\infty)$
$\sup_{u\in U(t)}(|f(t,x,u)| + |l(t,x,u)|) \leqslant q(t), \quad \forall x \in \partial A.$

In what follows $G : [0,\infty) \times \mathbb{R}^n \to \mathbb{R}^{n+1}$ is the measurable with respect to t set-valued map defined by

$$G(t,x) = \{(f(t,x,u), l(t,x,u)) \, : \, u \in U(t)\}.$$

For control systems, the conditions (IPC), (AC), and (IPC)′ take the following form:

(ipc) for some $\varepsilon > 0$, $\eta > 0$ and every $(t,x) \in [0,\infty) \times (\partial A + \eta\mathbb{B}) \cap A$ there exist $\{\alpha_i\}_{i=0}^n \subset [0,1]$, with $\sum_{i=0}^n \alpha_i = 1$, and $\{u_i\}_{i=0}^n \subset U(t)$ satisfying

$$\left\{ y + [0,\varepsilon]\left(\sum_{i=0}^n \alpha_i f(t,x,u_i) + \varepsilon\mathbb{B}\right) \, : \, y \in (x+\varepsilon\mathbb{B}) \cap A \right\} \subset A;$$

(ac) there exist $\tilde{\eta} > 0$ and $\gamma \in \mathcal{L}_{\text{loc}}$ such that $G(\cdot,x)$ is γ-left absolutely continuous, uniformly for $x \in \partial A + \tilde{\eta}\mathbb{B}$;
(ipc)′ there exist $\eta > 0$, $r > 0$, $M \geqslant 0$ such that for a.e. $t \in [0,\infty)$, any $y \in \partial A + \eta\mathbb{B}$, and any $u \in U(t)$, with $\sup_{n\in N^1_{y,\eta}} \langle n, f(t,y,u)\rangle \geqslant 0$, there exists $w \in \{w' \in U(t) \, : \, |f(t,y,w') - f(t,y,u)| \leqslant M\}$ such that

$$\sup_{n\in N^1_{y,\eta}} \langle n, f(t,y,w)\rangle, \langle n, f(t,y,w) - f(t,y,u)\rangle\} \leqslant -r.$$

Remark 2 If there exist $\tilde{\eta} > 0$, $\gamma, \tilde{\gamma} \in \mathcal{L}_{\text{loc}}$, and $k \geqslant 0$ such that $(f(\cdot,x,u), l(\cdot,x,u))$ is γ-left absolutely continuous, uniformly for $(x,u) \in (\partial A + \tilde{\eta}\mathbb{B}) \times \mathbb{R}^m$, $U(\cdot)$ is $\tilde{\gamma}$-left absolutely continuous, and $f(t,x,\cdot)$ is k-Lipschitz continuous for all $(t,x) \in [0,\infty) \times (\partial A + \tilde{\eta}\mathbb{B})$, then (ac) holds true.

Theorem 4 *Assume that either* (h), (ac), *and* (ipc) *or* (h)′ *and* (ipc)′ *hold true. Then there exist $b > 1$, $K > 0$ such that for all $\lambda > K$ and every $t \geqslant 0$ the function $V(t,\cdot)$ is $L(t)$-Lipschitz continuous on A with $L(t) = be^{-(\lambda-K)t}$. Furthermore, for all $\lambda > K$ and for every feasible trajectory $x(\cdot)$, we have* $\lim_{t\to\infty} V(t,x(t)) = 0$.

Proof We notice that, by the inward pointing conditions (ipc) or (ipc)′ and the viability theorem from [10], the problem $\mathcal{B}_\infty$ admits feasible trajectory-control pairs for any initial condition. Pick $(t_0,x_0) \in [0,\infty) \times A$. Using the sub-linear growth of f, l, and the Gronwall lemma, we have $1+|x(t)| \leqslant (1+|x_0|)e^{\int_{t_0}^t c(s)\,ds}$ for all $t \geqslant t_0$ and for any trajectory-control pair $(x(\cdot), u(\cdot))$ at (t_0, x_0).

Let $a_1 > 0,\ a_2 > 0$ be such that

$$\int_0^t c(s)\,ds \leqslant a_1 t + a_2 \qquad \forall t \geqslant 0. \tag{29}$$

For all $T > t_0$, we have

$$\begin{aligned} \textstyle\int_{t_0}^T e^{-\lambda t}|l(t,x(t),u(t))|\,dt &\leqslant \textstyle\int_{t_0}^T e^{-\lambda t} c(t)(1+|x_0|)e^{\int_{t_0}^t c(s)\,ds}\,dt \\ &\leqslant (1+|x_0|)e^{a_2}\textstyle\int_{t_0}^T e^{-(\lambda-a_1)t}c(t)\,dt. \end{aligned} \tag{30}$$

Then, by (29) and denoting $\psi(t) = \int_{t_0}^t c(s)\,ds$, for any $\lambda > a_1$

$$\begin{aligned} &\textstyle\int_{t_0}^T e^{-\lambda t}|l(t,x(t),u(t))|\,dt \\ &\leqslant (1+|x_0|)|e^{a_2}\left(\left[e^{-(\lambda-a_1)t}\psi(t)\right]_{t_0}^T + (\lambda-a_1)\textstyle\int_{t_0}^T e^{-(\lambda-a_1)t}\psi(t)\,dt\right) \\ &\leqslant (1+|x_0|)e^{a_2}\left(e^{-(\lambda-a_1)T}(a_1T+a_2) + \left(a_1t_0 + \frac{a_1}{\lambda-a_1} + a_2\right)e^{-(\lambda-a_1)t_0}\right) \end{aligned} \tag{31}$$

Passing to the limit when $T \to \infty$, we deduce that for every feasible trajectory-control pair $(x(\cdot), u(\cdot))$ at (t_0, x_0)

$$\int_{t_0}^\infty e^{-\lambda t}|l(t,x(t),u(t))|\,dt < +\infty \qquad \forall \lambda > a_1.$$

From now on, assume that $\lambda > a_1$. Fix $t \geqslant 0$ and $x^1, x^0 \in A$ with $x^1 \neq x^0$. Then, for any $\delta > 0$ there exists a feasible trajectory-control pair $(x_\delta(\cdot), u_\delta(\cdot))$ at (t, x^0) such that

$$V(t,x^0) + e^{-\delta t}|x^1 - x^0| > \int_t^\infty e^{-\lambda s} l(s, x_\delta(s), u_\delta(s))\,ds.$$

Hence

$$\begin{aligned} &V(t,x^1) - V(t,x^0) \leqslant e^{-\delta t}|x^1-x^0| + \\ &\lim_{\tau\to\infty}\left|\textstyle\int_t^\tau e^{-\lambda s}l(s,x(s),u(s))\,ds - \int_t^\tau e^{-\lambda s} l(s,x_\delta(s),u_\delta(s))\,ds\right| \end{aligned} \tag{32}$$

for any feasible trajectory-control pair $(x(\cdot), u(\cdot))$ satisfying $x(t) = x^1$.

Define $\tilde{G}(t,x,z) = G(t,x)$ for all $(t,x,z) \in [0,\infty) \times \mathbb{R}^n \times \mathbb{R}$ and consider the following state constrained differential inclusion in $\mathbb{R}^{n+1}$

$$\begin{cases} (x,z)'(s) \in \tilde{G}(s,x(s),z(s)) \text{ a.e. } s \in [t,\infty) \\ x(s) \in A \qquad\qquad\qquad\quad \forall\, s \in [t,\infty). \end{cases}$$

Putting $z_\delta(s) = \int_t^s l(\xi, x_\delta(\xi), u_\delta(\xi))\, d\xi$, by Theorem 3 applied on $A \times \mathbb{R}$ and the measurable selection theorem, there exist $C > 1$, $K > 0$ such that for all $\delta > 0$ we can find a $\tilde{G}_\infty$-trajectory $(\tilde{x}_\delta(\cdot), \tilde{z}_\delta(\cdot))$ on $[t, \infty)$, and a measurable selection $\tilde{u}_\delta(s) \in U(s)$ a.e. $s \geqslant t$, satisfying

$$(\tilde{x}_\delta, \tilde{z}_\delta)'(s) = (f(s, \tilde{x}_\delta(s), \tilde{u}_\delta(s)), l(s, \tilde{x}_\delta(s), \tilde{u}_\delta(s))) \qquad \text{a.e. } s \geqslant t,$$

$(\tilde{x}_\delta(t), \tilde{z}_\delta(t)) = (x^1, 0)$, $\tilde{x}_\delta([t, \infty)) \subset A$, and for any $s \geqslant t$

$$|\tilde{x}_\delta(s) - x_\delta(s)| + |\tilde{z}_\delta(s) - z_\delta(s)| \leqslant C e^{Ks} |x^1 - x^0|. \tag{33}$$

Now, relabelling by K the constant $K \vee a_1$, by (33) and integrating by parts, for all $\lambda > K$, all $\tau \geqslant t$, and all $\delta > 0$

$$\begin{aligned}
&\left|\int_t^\tau e^{-\lambda s} l(s, \tilde{x}_\delta(s), \tilde{u}_\delta(s))\, ds - \int_t^\tau e^{-\lambda s} l(s, x_\delta(s), u_\delta(s))\, ds\right| \\
&\leqslant \left|\left[e^{-\lambda s} \left(\int_t^s l(\xi, \tilde{x}_\delta(\xi), \tilde{u}_\delta(\xi))\, d\xi - \int_t^s l(\xi, x_\delta(\xi), u_\delta(\xi))\, d\xi\right)\right]_t^\tau\right| \\
&\quad + \lambda \left|\int_t^\tau e^{-\lambda s} \left(\int_t^s l(\xi, \tilde{x}_\delta(\xi), \tilde{u}_\delta(\xi))\, d\xi - \int_t^s l(\xi, x_\delta(\xi), u_\delta(\xi))\, d\xi\right) ds\right| \\
&\leqslant e^{-\lambda\tau} |\tilde{z}_\delta(\tau) - z_\delta(\tau)| + \lambda \int_t^\tau e^{-\lambda s} |\tilde{z}_\delta(s) - z_\delta(s)| ds \\
&\leqslant C e^{-\lambda\tau} e^{K\tau} |x^1 - x^0| + \lambda C \int_t^\tau e^{-(\lambda - K)s} |x^1 - x^0|\, ds \\
&= \left(C e^{-(\lambda-K)\tau} + \lambda C \left[-\tfrac{e^{-(\lambda-K)s}}{\lambda - K}\right]_t^\tau\right) |x^1 - x^0| \\
&= \left(-\tfrac{CK}{\lambda-K} e^{-(\lambda-K)\tau} + \tfrac{\lambda C}{\lambda-K} e^{-(\lambda-K)t}\right) |x^1 - x^0| \leqslant \tfrac{\lambda C}{\lambda - K} e^{-(\lambda-K)t} |x^1 - x^0|.
\end{aligned} \tag{34}$$

Taking note of (32), (34), and putting $\delta = \lambda - K$, we get

$$V(t, x^1) - V(t, x^0) \leqslant \left(\frac{\lambda C}{\lambda - K} + 1\right) e^{-(\lambda-K)t} |x^1 - x^0|.$$

By the symmetry of the previous inequality with respect to x^1 and x^0, and since λ, C, and K do not depend on t, x^1, and x^0, the first conclusion follows.

Now, let $(t_0, x_0) \in [0, \infty) \times A$ and consider a feasible trajectory $X(\cdot)$ at (t_0, x_0). Let $t > t_0$ and $(x(\cdot), u(\cdot))$ be a feasible trajectory-control pair at $(t, X(t))$ such that $V(t, X(t)) > \int_t^\infty e^{-\lambda s} l(s, x(s), u(s))\, ds - \frac{1}{t}$. Then

$$|V(t, X(t))| \leqslant \int_t^\infty e^{-\lambda s} |l(s, x(s), u(s))|\, ds + \frac{1}{t}.$$

From (29) and (30), we have for all $T > t$

$$\begin{aligned}
&\int_t^T e^{-\lambda s} |l(s, x(s), u(s))|\, ds \leqslant \int_t^T e^{-\lambda s} (1 + |X(t)|) e^{\int_t^s c(s')\, ds'} c(s)\, ds. \\
&\leqslant (1 + |x_0|) \int_t^T e^{-\lambda s} e^{\int_{t_0}^t c(s')\, ds'} e^{\int_t^s c(s')\, ds'} c(s)\, ds \\
&\leqslant (1 + |x_0|) \int_t^T e^{-\lambda s} e^{\int_0^s c(s')\, ds'} c(s)\, ds \leqslant (1 + |x_0|) e^{a_2} \int_t^T e^{-(\lambda - a_1)s} c(s)\, ds.
\end{aligned}$$

Then, arguing as in (31) with t_0 replaced by t and taking the limit when $T \to \infty$, we deduce that

$$|V(t, X(t))| \leqslant (1 + |x_0|)e^{a_2}\left(a_1 t + \frac{a_1}{\lambda - a_1} + a_2\right)e^{-(\lambda - a_1)t} + \frac{1}{t}.$$

Since $K \geqslant a_1$, the last conclusion follows passing to the limit when $t \to \infty$.

Corollary 1 *Assume that either* (h), (ac), *and* (ipc) *or* (h)$'$ *and* (ipc)$'$ *hold true and that* $f,\ l$ *are bounded. Consider any* $N > 0$ *with*

$$N \geq \sup\{|f(t, x, u)| + |l(t, x, u)| \ : \ t \geqslant 0,\ x \in \mathbb{R}^n,\ u \in U(t)\}.$$

Then, for any $\lambda > 0$ *sufficiently large, for any* $x \in A$, *and any* $t \geqslant 0$ *the function* $V(\cdot, x)$ *is Lipschitz continuous on* $[t, \infty)$ *with constant* $\left(L(t) + 2e^{-\lambda t}\right)N$.

Proof By Theorem 4, when $\lambda > 0$ is large enough, $V(t, \cdot)$ is $L(t)$-Lipschitz continuous on A. Fix $x \in A$ and $t \geqslant 0$. Let $s, \tilde{s} \in [t, \infty)$.

Suppose that $s \geqslant \tilde{s}$. Then, by the dynamic programming principle, there exists a feasible trajectory-control pair $(\bar{x}(\cdot), \bar{u}(\cdot))$ at $(\tilde{s}, x)$ such that

$$\begin{aligned} V(s, x) - V(\tilde{s}, x) &\leqslant |V(s, x) - V(s, \bar{x}(s))| + \textstyle\int_{\tilde{s}}^{s} e^{-\lambda\xi}|l(\xi, \bar{x}(\xi), \bar{u}(\xi))|\, d\xi \\ &\quad + N|s - \tilde{s}|e^{-\lambda t} \\ &\leqslant L(s)N|s - \tilde{s}| + N|s - \tilde{s}|e^{-\lambda\tilde{s}} + N|s - \tilde{s}|e^{-\lambda t} \\ &\leqslant \left(L(t) + 2e^{-\lambda t}\right)N|s - \tilde{s}|. \end{aligned} \tag{35}$$

Arguing in a similar way, we get (35) when $s < \tilde{s}$. Hence, by the symmetry with respect to s and $\tilde{s}$ in (35), the conclusion follows.

6 Applications to the Relaxation Problem

Let $f(\cdot)$, $l(\cdot)$, and $U(\cdot)$ be as in $\mathscr{B}_\infty$. Consider the relaxed infinite horizon state constrained problem $\mathscr{B}_\infty^{rel}$:

$$\tilde{V}(t_0, x_0) = \inf \int_{t_0}^{\infty} e^{-\lambda t}\tilde{l}(t, x(t), w(t))\, dt,$$

where the infimum is taken over all trajectory-control pairs $(x(\cdot), w(\cdot))$ subject to the state constrained control system

$$\begin{cases} x'(t) = \tilde{f}(t, x(t), w(t)) \text{ a.e. } t \in [t_0, \infty) \\ x(t_0) = x_0 \\ w(t) \in W(t) & \text{a.e. } t \in [t_0, \infty) \\ x(t) \in A & \forall t \in [t_0, \infty), \end{cases}$$

where $\lambda > 0$, $W : [0, \infty) \rightrightarrows \mathbb{R}^{(n+1)m} \times \mathbb{R}^{n+1}$ is the measurable set-valued map defined by

$$W(t) := (\times_{i=0}^{n} U(t)) \times \{(\alpha_0, \dots, \alpha_n) \in \mathbb{R}^{n+1} : \sum_{i=0}^{n} \alpha_i = 1, \ \alpha_i \geqslant 0 \ \forall i\},$$

and the functions $\tilde{f} : [0, \infty) \times \mathbb{R}^n \times \mathbb{R}^{(n+1)m} \times \mathbb{R}^{n+1} \to \mathbb{R}^n$ and $\tilde{l} : [0, \infty) \times \mathbb{R}^n \times \mathbb{R}^{(n+1)m} \times \mathbb{R}^{n+1} \to \mathbb{R}$ are defined by: for all $t \geqslant 0$, $x \in \mathbb{R}^n$, and $w = (u_0, \dots, u_n, \alpha_0, \dots, \alpha_n) \in \mathbb{R}^{(n+1)m} \times \mathbb{R}^{n+1}$

$$\tilde{f}(t, x, w) = \sum_{i=0}^{n} \alpha_i f(t, x, u_i) \quad \& \quad \tilde{l}(t, x, w) = \sum_{i=0}^{n} \alpha_i l(t, x, u_i).$$

Theorem 5 *Assume that either* (h), (ac), *and* (ipc) *or* (h)′ *and* (ipc)′ *hold true. Then, for all large* $\lambda > 0$, $\tilde{V}(\cdot, \cdot) = V(\cdot, \cdot)$ *on* $[0, \infty) \times A$.

Proof Notice that $\tilde{V}(t, x) \leqslant V(t, x)$ for any $(t, x) \in [0, \infty) \times A$, and that Theorem 4 implies that $\tilde{V}(t, \cdot)$ and $V(t, \cdot)$ are Lipschitz continuous on A for all $t \geqslant 0$ whenever $\lambda > 0$ is sufficiently large. That is, in particular, they are continuous and finite.

Fix $(t_0, x_0) \in [0, \infty) \times A$ and $\varepsilon > 0$. We claim that: for every $j \in \mathbb{N}^+$ there exists a finite set of trajectory-control pairs $\{(x_k(\cdot), u_k(\cdot))\}_{k=1,\dots,j}$ satisfying the following: $x_k'(s) = f(s, x_k(s), u_k(s))$ a.e. $s \in [t_0, t_0 + k]$ and $x_k([t_0, t_0 + k]) \subset A$ for all $k = 1, \dots, j$; if $j \geqslant 2$, $x_k|_{[t_0, t_0+k-1]}(\cdot) = x_{k-1}(\cdot)$ for all $k = 2, \dots, j$; and for any $k = 1, \dots, j$

$$\tilde{V}(t_0, x_0) \geqslant \tilde{V}(t_0 + k, x_k(t_0 + k)) + \int_{t_0}^{t_0+k} e^{-\lambda t} l(t, x_k(t), u_k(t))\, dt - \varepsilon \sum_{i=1}^{k} \frac{1}{2^i}. \tag{36}$$

We prove the claim by the induction argument with respect to $j \in \mathbb{N}^+$. By the dynamic programming principle, there exists a trajectory-control pair $(\tilde{x}(\cdot), \tilde{w}(\cdot))$ on $[t_0, t_0 + 1]$, feasible for the problem $\mathscr{B}_\infty^{rel}$ at (t_0, x_0), such that

$$\tilde{V}(t_0, x_0) + \frac{\varepsilon}{4} > \tilde{V}(t_0 + 1, \tilde{x}(t_0 + 1)) + \int_{t_0}^{t_0+1} e^{-\lambda t} \tilde{l}(t, \tilde{x}(t), \tilde{w}(t))\, dt. \tag{37}$$

By the relaxation theorem for finite horizon problems (cfr. [14]), for any $h > 0$ there exists a measurable control $\hat{u}^h(t) \in U(t)$ a.e. $t \in [t_0, t_0 + 1]$ such that the solution of the equation $(\hat{x}^h)'(t) = f(t, \hat{x}^h(t), \hat{u}^h(t))$ a.e. $t \in [t_0, t_0 + 1]$, with $\hat{x}^h(t_0) = x_0$, satisfies

$$\|\hat{x}^h - \tilde{x}\|_{\infty,[t_0,t_0+1]} < h$$

and

$$\left| \int_{t_0}^{t_0+1} e^{-\lambda t} \tilde{l}(t, \tilde{x}(t), \tilde{w}(t))\, dt - \int_{t_0}^{t_0+1} e^{-\lambda t} l(t, \hat{x}^h(t), \hat{u}^h(t))\, dt \right| < h.$$

Now, consider the following state constrained differential inclusion in $\mathbb{R}^{n+1}$

$$\begin{cases} (x, z)'(s) \in \tilde{G}(s, x(s), z(s)) & \text{a.e. } s \in [t_0, t_0 + 1] \\ x(s) \in A & \forall s \in [t_0, t_0 + 1], \end{cases}$$

where

$$\tilde{G}(t, x, z) = \{(f(t, x, u), e^{-\lambda t} l(t, x, u)) \,:\, u \in U(t)\}.$$

Letting $\hat{X}^h(\cdot) = (\hat{x}^h(\cdot), \hat{z}^h(\cdot))$, with $\hat{z}^h(t) = \int_{t_0}^{t} e^{-\lambda s} l(s, \hat{x}^h(s), \hat{u}^h(s))\, ds$, by Theorem 1, or Theorem 2, and the measurable selection theorem, there exists $\beta > 0$ (not depending on (t_0, x_0)) such that for any $h > 0$ we can find a feasible $\tilde{G}$-trajectory $X^h(\cdot) = (x^h(\cdot), z^h(\cdot))$ on $[t_0, t_0 + 1]$, with $X^h(t_0) = (x_0, 0)$, and a measurable control $u^h(s) \in U(s)$ a.e. $s \in [t_0, t_0 + 1]$, such that

$$(x^h, z^h)'(s) = (f(s, x^h(s), u^h(s)), e^{-\lambda s} l(s, x^h(s), u^h(s))) \quad \text{a.e. } s \in [t_0, t_0 + 1]$$

and

$$\|X^h - \hat{X}^h\|_{\infty,[t_0,t_0+1]} \leqslant \beta\Big(\sup_{s\in[t_0,t_0+1]} d_{A\times\mathbb{R}}(\hat{X}^h(s)) + h\Big).$$

Since $\sup_{s\in[t_0,t_0+1]} d_{A\times\mathbb{R}}(\hat{X}^h(s)) \leqslant \|\tilde{x} - \hat{x}^h\|_{\infty,[t_0,t_0+1]}$, we have

$$\begin{aligned} &\left| \int_{t_0}^{t_0+1} e^{-\lambda t} l(t, x^h(t), u^h(t))\, dt - \int_{t_0}^{t_0+1} e^{-\lambda t} \tilde{l}(t, \tilde{x}(t), \tilde{w}(t))\, dt \right| \\ &\leqslant \left| \int_{t_0}^{t_0+1} e^{-\lambda t} \tilde{l}(t, \tilde{x}(t), \tilde{w}(t))\, dt - \int_{t_0}^{t_0+1} e^{-\lambda t} l(t, \hat{x}^h(t), \hat{u}^h(t))\, dt \right| \\ &\quad + \left| \int_{t_0}^{t_0+1} e^{-\lambda t} l(t, x^h(t), u^h(t))\, dt - \int_{t_0}^{t_0+1} e^{-\lambda t} l(t, \hat{x}^h(t), \hat{u}^h(t))\, dt \right| \\ &< h(2\beta + 1) \end{aligned}$$

and

$$\|x^h - \tilde{x}\|_{\infty,[t_0,t_0+1]} \leqslant \|\tilde{x} - \hat{x}^h\|_{\infty,[t_0,t_0+1]} + \|x^h - \hat{x}^h\|_{\infty,[t_0,t_0+1]} < h(2\beta + 1).$$

Hence, choosing $0 < h < \varepsilon/4(2\beta + 1)$ sufficiently small, we can find a trajectory-control pair $(x^h(\cdot), u^h(\cdot))$ on $[t_0, t_0 + 1]$ with $x^h([t_0, t_0 + 1]) \subset A$, $u^h(s) \in U(s)$ and

$(x^h)'(s) = f(s, x^h(s), u^h(s))$ a.e. $s \in [t_0, t_0+1]$, $x^h(t_0) = x_0$ such that, by (37) and continuity of $\tilde{V}(t_0+1, \cdot)$

$$\tilde{V}(t_0, x_0) > \tilde{V}(t_0+1, x^h(t_0+1)) + \int_{t_0}^{t_0+1} e^{-\lambda t} l(t, x^h(t), u^h(t))\, dt - \frac{\varepsilon}{2}.$$

Letting $(x_1(\cdot), u_1(\cdot)) := (x^h(\cdot), u^h(\cdot))$, the conclusion follows for $j = 1$.

Now, suppose we have shown that for some $j \geq 1$ there exist $\{(x_k(\cdot), u_k(\cdot))\}_{k=1,\dots,j}$ satisfying the claim. Let us to prove it for $j+1$. By the dynamic programming principle there exists a trajectory-control pair $(\tilde{x}(\cdot), \tilde{w}(\cdot))$ on $[t_0+j, t_0+j+1]$, feasible for the problem $\mathscr{B}_\infty^{rel}$ at $(t_0+j, x_j(t_0+j))$, such that

$$\begin{aligned} \tilde{V}(t_0+j, x_j(t_0+j)) + \tfrac{\varepsilon}{2^{j+2}} > \tilde{V}(t_0+j+1, \tilde{x}(t_0+j+1)) \\ + \int_{t_0+j}^{t_0+j+1} e^{-\lambda t} \tilde{l}(t, \tilde{x}(t), \tilde{w}(t))\, dt. \end{aligned} \tag{38}$$

As before, for every $h > 0$ there exists a feasible $\tilde{G}$-trajectory $X^h(\cdot) = (x^h(\cdot), z^h(\cdot))$ on $[t_0+j, t_0+j+1]$, with $X^h(t_0) = (x_j(t_0+j), 0)$, and a measurable control $u^h(s) \in U(s)$ a.e. $s \in [t_0+j, t_0+j+1]$, such that

$$(x^h, z^h)'(s) = (f(s, x^h(s), u^h(s)), e^{-\lambda s} l(s, x^h(s), u^h(s))) \text{ a.e. } s \in [t_0+j, t_0+j+1],$$

satisfying

$$\left| \int_{t_0+j}^{t_0+j+1} e^{-\lambda t} l(t, x^h(t), u^h(t))\, dt - \int_{t_0+j}^{t_0+j+1} e^{-\lambda t} \tilde{l}(t, \tilde{x}(t), \tilde{w}(t))\, dt \right| < h(2\beta+1)$$

and

$$\|x^h - \tilde{x}\|_{\infty, [t_0+j, t_0+j+1]} < h(2\beta+1).$$

Putting

$$(x_{j+1}(\cdot), u_{j+1}(\cdot)) := \begin{cases} (x_j(\cdot), u_j(\cdot)) \text{ on } [t_0, t_0+j] \\ (x^h(\cdot), u^h(\cdot)) \text{ on } [t_0+j, t_0+j+1], \end{cases} \tag{39}$$

and choosing $0 < h < \varepsilon/2^{j+2}(2\beta+1)$ sufficiently small, it follows from (38) that

$$\begin{aligned} \tilde{V}(t_0+j, x_j(t_0+j)) \geqslant \tilde{V}(t_0+j+1, x_{j+1}(t_0+j+1)) \\ + \int_{t_0+j}^{t_0+j+1} e^{-\lambda t} l(t, x_{j+1}(t), u_{j+1}(t))\, dt - \tfrac{2\varepsilon}{2^{j+2}}. \end{aligned} \tag{40}$$

So, taking note of (39) and (40), we obtain

$$\begin{aligned}
\tilde{V}(t_0, x_0) &\geqslant \tilde{V}(t_0 + j, x_j(t_0 + j)) + \int_{t_0}^{t_0+j} e^{-\lambda t} l(t, x_j(t), u_j(t))\, dt - \varepsilon \sum_{i=1}^{j} \tfrac{1}{2^i} \\
&\geqslant \tilde{V}(t_0 + j + 1, x_{j+1}(t_0 + j + 1)) - \varepsilon \sum_{i=1}^{j} \tfrac{1}{2^i} - \tfrac{\varepsilon}{2^{j+1}} \\
&\quad + \int_{t_0+j}^{t_0+j+1} e^{-\lambda t} l(t, x_{j+1}(t), u_{j+1}(t))\, dt + \int_{t_0}^{t_0+j} e^{-\lambda t} l(t, x_j(t), u_j(t))\, dt \\
&= \tilde{V}(t_0 + j + 1, x_{j+1}(t_0 + j + 1)) + \int_{t_0}^{t_0+j+1} e^{-\lambda t} l(t, x_{j+1}(t), u_{j+1}(t))\, dt \\
&\quad -\varepsilon \sum_{i=1}^{j+1} \tfrac{1}{2^i}.
\end{aligned}$$

Hence $\{(x_k(\cdot), u_k(\cdot))\}_{k=1,\dots,j+1}$ also satisfy our claim. Now, let us define the trajectory-control pair $(x(\cdot), u(\cdot))$ by $(x(t), u(t)) := (x_k(t), u_k(t))$ if $t \in [t_0 + k - 1, t_0 + k]$. Then $(x(\cdot), u(\cdot))$ is a feasible trajectory-control pair for the problem $\mathscr{B}_\infty$ at (t_0, x_0). Since $\tilde{V}(t, x(t)) \to 0$ when $t \to +\infty$, by (36), we have

$$\tilde{V}(t_0, x_0) \geqslant \int_{t_0}^{\infty} e^{-\lambda t} l(t, x(t), u(t))\, dt - \varepsilon.$$

Hence, we deduce that $\tilde{V}(t_0, x_0) \geqslant V(t_0, x_0) - \varepsilon$. Since ε is arbitrary, the conclusion follows.

Remark 3 The authors thank the referee for attracting their attention to [11], where relaxation of differential inclusions over infinite horizon was investigated. The framework there is substantially different from ours, because in the relaxation result of [11] on one hand small variations of the initial state are permitted, on the other hand no state constraints are involved in the setting of [11].

Acknowledgements The second author was partially supported by the Air Force Office of Scientific Research under award number FA9550-18-1-0254.

References

1. Aseev, S.M., Veliov, V.M.: Maximum principle for infinite-horizon optimal control problems with dominating discount. Dyn. Contin. Discrete Impuls. Syst. Ser. B Appl. Algorithms **19**, 43–63 (2012)
2. Aseev, S.M., Veliov, V.M.: Maximum principle for infinite-horizon optimal control problems under weak regularity assumptions. Tr. Inst. Mat. Mekh. **20**(3), 41–57 (2014)
3. Aubin, J.-P., Frankowska, H.: Set-valued analysis. In: Modern Birkhäuser Classics. Birkhäuser Boston Inc., Boston, MA (2009)
4. Basco, V., Cannarsa, P., Frankowska, H.: Necessary conditions for infinite horizon optimal control problems with state constraints. Math. Control Relat. Fields **8**, 535–555 (2018)
5. Bettiol, P., Frankowska, H., Vinter, R.B.: L^∞ estimates on trajectories confined to a closed subset. J. Differ. Equ. **252**(2), 1912–1933 (2012)
6. Bettiol, P., Frankowska, H., Vinter, R.B.: Improved sensitivity relations in state constrained optimal control. Appl. Math. Optim. **71**(2), 353–377 (2015)
7. Cannarsa, P., Frankowska, H.: Value function, relaxation, and transversality conditions in infinite horizon optimal control. J. Math. Anal. Appl. **457**(2), 1188–1217 (2018)
8. Carlson, D.A., Haurie, A.: Infinite Horizon Optimal Control: Theory and Applications. Springer, New York (1987)

9. Frankowska, H., Mazzola, M.: Discontinuous solutions of Hamilton-Jacobi-Bellman equation under state constraints. Calc. Var. Partial Differ. Equ. **46**(3–4), 725–747 (2013)
10. Frankowska, H., Plaskacz, S.: A measurable upper semicontinuous viability theorem for tubes. Nonlinear Analysis: Theory, Methods & Applications **26**(3), 565–582 (1996)
11. Ingalls, B., Sontag, E., Wang, Y.: An infinite-time relaxation theorem for differential inclusions. Proc. Am. Math. Soc. **131**(2), 487–499 (2003)
12. Pickenhain, S.: On adequate transversality conditions for infinite horizon optimal control problems—a famous example of Halkin. In: Crespo Cuaresma, J., Palokangas, T., Tarasyev, A. (eds.) Dynamic Systems, Economic Growth, and the Environment, pp. 3–22. Springer, Berlin, Heidelberg (2010)
13. Soner, H.M.: Optimal control problems with state-space constraints I. SIAM J. Control Optim. **24**, 552–562 (1986)
14. Vinter, R.B.: Optimal Control. Birkhäuser, Boston, MA (2000)

Herglotz' Generalized Variational Principle and Contact Type Hamilton-Jacobi Equations

Piermarco Cannarsa, Wei Cheng, Kaizhi Wang and Jun Yan

Abstract We develop an approach for the analysis of fundamental solutions to Hamilton-Jacobi equations of contact type based on a generalized variational principle proposed by Gustav Herglotz. We also give a quantitative Lipschitz estimate on the associated minimizers.

Keywords Hamilton-Jacobi equations · Contact transformations · Herglotz variational principle

1 Introduction

The so called *generalized variational principle* was proposed by Gustav Herglotz in 1930 (see [31, 32]). It generalizes classical variational principle by defining the functional, whose extrema are sought, by a differential equation. More precisely, the functional u is defined in an implicit way by an ordinary differential equation

$$\dot{u}(s) = F(s, \xi(s), \dot{\xi}(s), u(s)), \quad s \in [0, t], \tag{1}$$

P. Cannarsa (✉)
Dipartimento di Matematica, Università di Roma "Tor Vergata",
Via della Ricerca Scientifica 1, 00133 Rome, Italy
e-mail: cannarsa@mat.uniroma2.it

W. Cheng
Department of Mathematics, Nanjing University, Nanjing 210093, China
e-mail: chengwei@nju.edu.cn

K. Wang
School of Mathematical Sciences, Shanghai Jiao Tong University,
Shanghai 200240, China
e-mail: kzwang@sjtu.edu.cn

J. Yan
Shanghai Key Laboratory for Contemporary Applied Mathematics,
School of Mathematical Sciences, Fudan University, Shanghai 200433, China
e-mail: yanjun@fudan.edu.cn

F. Alabau-Boussouira et al. (eds.), *Trends in Control Theory and Partial Differential Equations*, Springer INdAM Series 32,
https://doi.org/10.1007/978-3-030-17949-6_3

with $u(t) = u_0 \in \mathbb{R}$, for $t > 0$, a function $F \in C^2(\mathbb{R} \times \mathbb{R}^n \times \mathbb{R}^n \times \mathbb{R}, \mathbb{R})$ and a piecewise C^1 curve $\xi : [0, t] \to \mathbb{R}^n$. Here, $u = u[\xi, s]$ can be regarded as a functional, on a space of paths $\xi(\cdot)$. The generalized variational principle of Herglotz is as follows:

Let the functional $u = u[\xi, t]$ be defined by (1) with ξ in the space of piecewise C^1 functions on $[0, t]$. Then the value of the functional $u[\xi, t]$ is an extremal for the function ξ such that the variation $\frac{d}{d\varepsilon}u[\xi + \varepsilon\eta, t] = 0$ for arbitrary piecewise C^1 function η such that $\eta(0) = \eta(t) = 0$.

Herglotz reached the idea of the generalized variational principle through his work on contact transformations and their connections with Hamiltonian systems and Poisson brackets. His work was motivated by ideas from S. Lie, C. Carathéodory and other researchers. An important reference on the generalized variational principle is the monograph [30]. The variational principle of Herglotz is important for many reasons:

- The solutions of the Eq. (1) determine a family of contact transformations, see [11, 21, 28, 30];
- The generalized variational principle gives a variational description of energy-nonconservative processes even when F in (1) is independent of t.
- If F has the form $F = -\lambda u + L(x, v)$, then the relevant problems are closely connected to the Hamilton-Jacobi equations with discount factors (see, for instance, [9, 18, 19, 29, 34–37]). As an extension to nonlinear discounted problems, various examples are discussed in [14, 43].
- Even for a energy-nonconservative process which can be described with the generalized variational principle, one can systematically derive conserved quantities as Noether's theorems such as [26, 27];
- The generalized variational principle provides a link between the mathematical structure of control and optimal control theories and contact transformation (see [25]);
- There are some interesting connections between contact transformations and equilibrium thermodynamics (see, for instance, [39]).

In this note, we will clarify more connections between the generalized variational principle of Herglotz and Hamilton-Jacobi theory motivated by recent works in [41, 42] under a set of Tonelli-like conditions. We will begin with generalized variational principle of Herglotz in the frame of Lagrangian formalism different from the methods used in [41, 42]. Throughout this paper, let $L : \mathbb{R}^n \times \mathbb{R} \times \mathbb{R}^n$ be a function of class C^2 such that the following standing assumptions are satisfied:

(L1) $L(x, r, \cdot) > 0$ is strictly convex for all $(x, r) \in \mathbb{R}^n \times \mathbb{R}$.
(L2) There exist two superlinear nondecreasing function $\overline{\theta}_0, \theta_0 : [0, +\infty) \to [0, +\infty)$, $\theta_0(0) = 0$ and $c_0 > 0$, such that

$$\overline{\theta}_0(|v|) \geqslant L(x, 0, v) \geqslant \theta_0(|v|) - c_0, \quad (x, v) \in \mathbb{R}^n \times \mathbb{R}^n.$$

(L3) There exists $K > 0$ such that

$$|L_r(x,r,v)| \leqslant K, \quad (x,r,v) \in \mathbb{R}^n \times \mathbb{R} \times \mathbb{R}^n.$$

Remark 1 For each $r \in \mathbb{R}$, from the conditions (L2) and (L3) we could take

$$\overline{\theta}_r := \overline{\theta}_0 + K|r|, \quad \theta_r := \theta_0, \quad c_r := c_0 + K|r|,$$

such that

$$\overline{\theta}_r(|v|) \geqslant L(x,r,v) \geqslant \theta_r(|v|) - c_r, \quad (x,v) \in \mathbb{R}^n \times \mathbb{R}^n. \tag{2}$$

Obviously, $\overline{\theta}_r$ and θ_r are both nonnegative, superlinear and nondecreasing functions, $c_r > 0$.

It is natural to introduce the associated Hamiltonian

$$H(x,r,p) = \sup_{v \in \mathbb{R}^n} \{\langle p, v\rangle - L(x,r,v)\}, \quad (x,r,p) \in \mathbb{R}^n \times \mathbb{R} \times (\mathbb{R}^n)^*.$$

Let $x, y \in \mathbb{R}^n$, $t > 0$ and $u_0 \in \mathbb{R}$. Set

$$\Gamma^t_{x,y} = \{\xi \in W^{1,1}([0,t], \mathbb{R}^n) : \xi(0) = x, \ \xi(t) = y\}.$$

We consider a variational problem

$$\text{Minimize} \quad u_0 + \inf \int_0^t L(\xi(s), u_\xi(s), \dot{\xi}(s))\, ds, \tag{3}$$

where the infimum is taken over all $\xi \in \Gamma^t_{x,y}$ such that the Carathéodory equation

$$\dot{u}_\xi(s) = L(\xi(s), u_\xi(s), \dot{\xi}(s)), \quad a.e.\ s \in [0,t], \tag{4}$$

admits an absolutely continuous solution u_ξ with initial condition $u_\xi(0) = u_0$. It is already known that the variational problem (3) with subsidiary conditions (4) is closely connected to the Hamilton-Jacobi equations in the form

$$H(x, u(x), Du(x)) = c. \tag{5}$$

The readers can refer to [28] for a systematic approach of Hamilton-Jacobi equations in the form (5) especially in the context of contact geometry.

In [41, 42], a weak KAM type theory on Eq. (5) was developed on compact manifolds under the aforementioned Tonelli-like conditions. Problem (3) is understood as an implicit variational principle [41] and, by introducing the positive and negative Lax-Oleinik semi-groups, an existence result for weak KAM type solutions of (5) was obtained provided c in the right side of Eq. (5) belongs to the set of critical values [42]. The same approach adapts to the evolutionary equations in the form

$$D_t u + H(x, u, D_x u) = 0. \tag{6}$$

Unlike the methods used in [41, 42], in this note, our approach of the Eqs. (5) and (6) is based on the the variational problem (3) under subsidiary conditions (4). We give all the details of such a Tonelli-like theory and its connection to viscosity solutions of (5) and (6).

In view of Proposition 1 below, the infimum in (3) can be achieved. Suppose that $\xi \in \Gamma^t_{x,y}$ is a minimizer for (3) where u_ξ is uniquely determined by (4) with initial condition $u_\xi(0) = u_0$. Then we call such ξ an *extremal*. Due to Proposition 1 below, each extremal ξ and associated u_ξ are of class C^2 and satisfy the Herglotz equation (Generalized Euler-Lagrange equation by Herglotz)

$$\begin{aligned} &\frac{d}{ds} L_v(\xi(s), u_\xi(s), \dot{\xi}(s)) \\ &= L_x(\xi(s), u_\xi(s), \dot{\xi}(s)) + L_u(\xi(s), u_\xi(s), \dot{\xi}(s)) L_v(\xi(s), u_\xi(s), \dot{\xi}(s)). \end{aligned} \tag{7}$$

Moreover, let $p(s) = L_v(\xi(s), u_\xi(s), \dot{\xi}(s))$ be the so called dual arc. Then p is also of class C^2 and we conclude that (ξ, p, u_ξ) satisfies the following Lie equation

$$\begin{cases} \dot{\xi}(s) = H_p(\xi(s), u_\xi(s), p(s)); \\ \dot{p}(s) = -H_x(\xi(s), u_\xi(s), p(s)) - H_u(\xi(s), u_\xi(s), p(s))p(s); \\ \dot{u}_\xi(s) = p(s) \cdot \dot{\xi}(s) - H(\xi(s), u_\xi(s), p(s)), \end{cases} \tag{8}$$

where the reader will recognize the classical system of characteristics for (5).

The paper is organized as follows: In Sect. 2, we afford a detailed and rigorous treatment of (3) under subsidiary conditions (4). In Sect. 3, we study the regularity of the minimizers and deduce the Herglotz equation (7) and Lie equation (8) as well. In Sect. 4, we show that the two approaches between [41, 42] and ours are equivalent. We also sketch the way to move Herglotz' variational principle to manifolds.

2 Existence of Minimizers in Herglotz' Variational Principle

Fix $x_0, x \in \mathbb{R}^n$, $t > 0$ and $u_0 \in \mathbb{R}$. Let $\xi \in \Gamma^t_{x_0,x}$, we consider the Carathéodory equation

$$\begin{cases} \dot{u}_\xi(s) = L(\xi(s), u_\xi(s), \dot{\xi}(s)), \quad a.e.\, s \in [0, t], \\ u_\xi(0) = u_0. \end{cases} \tag{9}$$

We define the action functional

$$J(\xi) := \int_0^t L(\xi(s), u_\xi(s), \dot{\xi}(s))\, ds, \tag{10}$$

where $\xi \in \Gamma^t_{x_0,x}$ and u_ξ is defined in (9) by Proposition 8 in Appendix. Notice that Carathéodory's theorem (Proposition 8) is just a local result, but the existence and uniqueness of the solution of (9) holds on $[0, t]$ since condition (L3) and that $\xi \in \mathscr{A}$. Our purpose is to minimize $J(\xi)$ over

$$\mathscr{A} = \{\xi \in \Gamma^t_{x_0,x} : (9) \text{ admits an absolutely continuous solution } u_\xi\}.$$

Notice that $\mathscr{A} \neq \varnothing$ because it contains all piecewise C^1 curves connecting x_0 to x. It is not hard to check that, for each $a \in \mathbb{R}$,

$$\mathscr{A} = \mathscr{A}' := \{\xi \in \Gamma^t_{x_0,x} : \text{the function } s \mapsto L(\xi(s), a, \dot{\xi}(s)) \text{belongs to } L^1([0, t])\}.$$

Indeed, If $\xi \in \mathscr{A}$, then $L(\xi(s), u_\xi(s), \dot{\xi}(s))$ is integrable on $[0, t]$ and u_ξ is bounded. Thus $\xi \in \mathscr{A}'$ since

$$|L(\xi, 0, \dot{\xi})| \leqslant |L(\xi, u_\xi, \dot{\xi})| + K|u_\xi|.$$

On the other hand, if $\xi \in \mathscr{A}'$, then

$$\dot{u}_\xi \leqslant L(\xi, 0, \dot{\xi}) + K|u_\xi|.$$

Therefore, $\xi \in \mathscr{A}$.

For the following estimate, we define $L_0(x, v) := L(x, 0, v)$.

Lemma 1 *Let $x_0, x \in \mathbb{R}^n$, $t > 0$, $u_0 \in \mathbb{R}$. Given $\xi \in \Gamma^t_{x_0,x}$ such that (9) admits an absolutely continuous solution, then we have that*

$$|u_\xi(s)| \leqslant \exp(Ks)(|u_0| + c_0 s) \tag{11}$$

if $u_\xi(s) < 0$. In particular, we have

$$u_\xi(s) \geqslant -\exp(Ks)(|u_0| + c_0 s), \quad s \in [0, t]. \tag{12}$$

Proof Let $x_0, x \in \mathbb{R}^n$, $t > 0$, $u_0 \in \mathbb{R}$ and $\xi \in \mathscr{A}$. Suppose that $u_\xi(s_0) < 0$, $s_0 \in (0, t]$. We define $E = \{s \in [0, s_0) : u_\xi(s) \geqslant 0\}$ and

$$a = \begin{cases} 0 & E = \varnothing, \\ \sup E & E \neq \varnothing. \end{cases}$$

Then, we have that $u_\xi(s) \leqslant 0$ for all $s \in [a, s_0]$ and $u_\xi(a) = 0$ if $E \neq \varnothing$. Now, we are assuming that $E \neq \varnothing$. For any $s \in [a, s_0]$ we have that

$$\begin{aligned}
-|u_\xi(s)| = u_\xi(s) &= u_\xi(a) + \int_a^s L(\xi(\tau), u_\xi(\tau), \dot{\xi}(\tau))\, d\tau \\
&\geqslant -|u_\xi(a)| + \int_a^s L_0(\xi(\tau), \dot{\xi}(\tau))\, d\tau - K \int_a^s |u_\xi(\tau)|\, d\tau\, d\tau \\
&\geqslant -|u_\xi(a)| + \int_a^s \theta_0(|\dot{\xi}(\tau)|)\, d\tau - c_0(s-a) - K \int_a^s |u_\xi(\tau)|\, d\tau \\
&\geqslant -|u_\xi(a)| - c_0 s - K \int_a^s |u_\xi(\tau)|\, d\tau.
\end{aligned}$$

Then, we have that

$$|u_\xi(s)| \leqslant (|u_0| + c_0 s) + K \int_a^s |u_\xi(\tau)|\, d\tau, \quad s \in [a, s_0].$$

Then Gronwall inequality implies

$$|u_\xi(s)| \leqslant \exp(K(s-a))(|u_0| + c_0 s) \leqslant \exp(Ks)(|u_0| + c_0 s), \quad s \in [a, s_0].$$

If $E = \varnothing$, then $a = 0$ and the proof is the same. This leads to (11) and (12). □

In view to Lemma 1, we conclude that $\inf_{\xi \in \mathscr{A}} J(\xi)$ is bounded below. Now, for any $\varepsilon > 0$, set

$$\mathscr{A}_\varepsilon = \{\xi \in \mathscr{A} : \inf_{\eta \in \mathscr{A}} J(\eta) + \varepsilon \geqslant u_\xi(t) - u_0\}.$$

Lemma 2 *Suppose $x_0 \in \mathbb{R}^n$, $t, R > 0$, $u_0 \in \mathbb{R}$ and $|x - x_0| \leqslant R$. Let $\varepsilon > 0$ and $\xi \in \mathscr{A}_\varepsilon$. Then we have that*

$$u_\xi(t) - u_0 \leqslant t(\kappa(R/t) + K|u_0|)\exp(Kt) + \varepsilon,$$

with $\kappa(r) = \overline{\theta}_0(r) + 2c_0$. Moreover, there exist two nondecreasing and superlinear functions $F, G : [0, +\infty) \to [0, +\infty)$ such that

$$|u_\xi(t)| \leqslant tF(R/t) + G(t)|u_0| + \varepsilon, \tag{13}$$

where $F(r) = \max\{\kappa(r), c_0 \exp(Kr)\}$ *and* $G(r) = \max\{rK\exp(Kr) + 1, \exp(Kr)\}$.

Proof Suppose $x_0 \in \mathbb{R}^n$, $t, R > 0$, $u_0 \in \mathbb{R}$ and $|x - x_0| \leqslant R$. Let $\varepsilon > 0$ and $\xi \in \mathscr{A}_\varepsilon$. First, notice that

$$|L_0(x, v)| \leqslant L_0(x, v) + 2c_0 \leqslant \overline{\theta}_0(|v|) + 2c_0, \quad (x, v) \in \mathbb{R}^n \times \mathbb{R}^n. \tag{14}$$

Set $\kappa(r) = \overline{\theta}_0(r) + 2c_0$.

Define $\xi_0(s) = x_0 + s(x - x_0)/t$ for any $s \in [0, t]$, then $\xi_0 \in \mathscr{A}$. Then, for any $s \in [0, t]$, we have that

$$
\begin{aligned}
|u_{\xi_0}(s) - u_0| &\leqslant \int_0^s |L_0(\xi_0, \dot{\xi}_0)|\, d\tau + K \int_0^s |u_{\xi_0}|\, d\tau \\
&\leqslant t\kappa(R/t) + K \int_0^s |u_{\xi_0} - u_0|\, d\tau + tK|u_0|.
\end{aligned}
$$

Due to Gronwall inequality, we obtain

$$
|u_{\xi_0}(s) - u_0| \leqslant t(\kappa(R/t) + K|u_0|)\exp(Kt), \quad s \in [0, t]. \tag{15}
$$

Together with Lemma 1, this completes the proof. □

Lemma 3 *Suppose* $x_0 \in \mathbb{R}^n$, $t, R > 0$, $u_0 \in \mathbb{R}$ *and* $|x - x_0| \leqslant R$. *Let* $\varepsilon > 0$ *and* $\xi \in \mathscr{A}_\varepsilon$. *Then there exist two continuous functions* $F_1, F_2 : [0, +\infty) \times [0, +\infty) \to [0, +\infty)$ *depending on* R, *with* $F_i(r_1, \cdot)$ *being nondecreasing and superlinear and* $F_i(\cdot, r_2)$ *being nondecreasing for any* $r_1, r_2 \geqslant 0$, $i = 1, 2$, *such that*

$$
|u_\xi(s)| \leqslant tF_1(t, R/t) + C_1(t)(\varepsilon + |u_0|), \quad s \in [0, t] \tag{16}
$$

and

$$
\int_0^t |L(\xi, u_\xi, \dot{\xi})|\, d\tau \leqslant tF_2(t, R/t) + C_2(t)(\varepsilon + |u_0|), \tag{17}
$$

where $C_i(t) > 0$ *for* $i = 1, 2$.

Proof Suppose $x_0 \in \mathbb{R}^n$, $t, R > 0$, $u_0 \in \mathbb{R}$ and $|x - x_0| \leqslant R$. Let $\varepsilon > 0$ and $\xi \in \mathscr{A}_\varepsilon$.

If $u_\xi(t) \geqslant 0$, we define $E_+ = \{s \in [0, t] : u_\xi(s) > u_\xi(t)\}$. If $E_+ = \varnothing$, then we have that $u_\xi(s) \leqslant u_\xi(t)$ for all $s \in [0, t]$. Now, we suppose that $E_+ \neq \varnothing$. It is known that E_+ is the union of a countable family of open intervals $\{(a_i, b_i)\}$ which are mutually disjoint (It is possible that $a_i = 0$ and this case can be dealt with separately but similarly). For any $\tau \in E_+$, there exists an open interval (a, b), a component of E_+ containing s, such that $u_\xi(\tau) > u_\xi(t) \geqslant 0$ for all $\tau \in (a, b)$ and $u_\xi(b) = u_\xi(t)$. Therefore, for almost all $s \in [a, b]$, we have that

$$
\dot{u}_\xi(s) = L(\xi(s), u_\xi(s), \dot{\xi}(s)) \geqslant L_0(\xi(s), \dot{\xi}(s)) - Ku_\xi(s).
$$

Invoking condition (L2), it follows that, for all $s \in [a, b]$,

$$
e^{Kb}u_\xi(b) - e^{Ks}u_\xi(s) \geqslant \int_s^b e^{K\tau} L_0(\xi(\tau), \dot{\xi}(\tau))\, d\tau \geqslant -c_0(b - s)e^{Kb}
$$

Thus we obtain that

$$
\begin{aligned}
u_\xi(s) &\leqslant c_0(b - s)e^{K(b-s)} + e^{K(b-s)}u_\xi(t) \\
&\leqslant c_0te^{Kt} + e^{Kt}[(t\kappa(R/t) + K|u_0|)e^{Kt} + \varepsilon + |u_0|] \quad s \in [0, t], \\
&= tF_1(t, R/t) + G_1(t)|u_0| + \varepsilon,
\end{aligned} \tag{18}
$$

where $F_1(r_1, r_2) := e^{Kr_1}(c_0 + \kappa(r_2))$ and $G_1(r) = e^{Kr}(Ke^{Kr} + 1)$.

If $u_\xi(t) < 0$, define $v_\xi(s) = u_\xi(s) - u_\xi(t)$, then $v_\xi(s)$ satisfies the Carathéodory equation

$$\dot{v}_\xi(s) = L(\xi(s), v_\xi(s) + u_\xi(t), \dot{\xi}(s)), \quad s \in [0, t]$$

with initial condition $v_\xi(0) = u_0 - u_\xi(t)$. Similarly, We define $F_+ = \{s \in [0, t] : v_\xi(s) > v_\xi(t)\}$. If $F_+ = \varnothing$, then we have that $v_\xi(s) \leqslant v_\xi(t) = 0$ for all $s \in [0, t]$. Now, we suppose that $F_+ \neq \varnothing$ and F_+ is the union of a countable family of open intervals $\{(c_i, d_i)\}$ which are mutually disjoint. For any $s \in F_+$, there exists an open interval, say (c, d), such that $v_\xi(s) > v_\xi(t) = 0$ for all $s \in (c, d)$ and $v_\xi(d) = v_\xi(t)$. Therefore, for almost all $s \in [c, d]$, we have that

$$\dot{v}_\xi(s) \geqslant L_0(\xi(s), \dot{\xi}(s)) - Kv_\xi(s) - K|u_\xi(t)|.$$

It follows that, for all $s \in [c, d]$,

$$\begin{aligned} e^{Kd}v_\xi(d) - e^{Ks}v_\xi(s) &\geqslant \int_s^d e^{K\tau}L_0(\xi(s), \dot{\xi}(s))\, d\tau - Kt|u_\xi(t)|e^{Kt} \\ &\geqslant -c_0te^{Kd} - Kt|u_\xi(t)|e^{Kt}, \end{aligned}$$

and this gives rise to

$$v_\xi(s) \leqslant c_0te^{K(d-s)} + K|u_\xi(t)|te^{K(t-s)} + e^{K(d-s)}v_\xi(d) \leqslant c_0te^{Kt} + Kt|u_\xi(t)|e^{Kt},$$

since $v_\xi(d) = 0$. It follows that, for all $s \in [0, t]$,

$$\begin{aligned} u_\xi(s) &\leqslant c_0te^{Kt} + Kte^{Kt}|u_\xi(t)| + u_\xi(t) \leqslant c_0te^{Kt} + (Kte^{Kt} + 1)|u_\xi(t)| \\ &\leqslant c_0te^{Kt} + (Kte^{Kt} + 1)(tF_2(R/t) + G_2(t)|u_0| + \varepsilon) \end{aligned} \tag{19}$$

with F_2 and G_2 determined by Lemma 2. By combining (27) and (29) and setting

$$F_3(r_1, r_2) = \max\{F_1(r_1, r_2), c_0e^{Kr_1} + F_2(r_2)(Kr_1e^{Kr_1} + 1)\},$$
$$C_1(r) = \max\{G_1(r), G_2(t)(Kr_1e^{Kr_1} + 1)\}, \quad C_2(r) = \max\{C_1(r), e^{Kr}c_0\},$$

we conclude that

$$u_\xi(s) \leqslant tF_3(t, R/t) + C_1(t)(|u_0| + \varepsilon), \tag{20}$$
$$|u_\xi(s)| \leqslant tF_3(t, R/t) + C_2(t)(|u_0| + \varepsilon). \tag{21}$$

This leads to the proof of (16) together with Lemma 1.

Now, by (14), Lemma 2 and (21), we have that

$$\begin{aligned}\int_0^s |L_0(\xi,\dot{\xi})|d\tau &\leqslant \int_0^s (L_0(\xi,\dot{\xi})+2c_0)\,d\tau \leqslant 2c_0 s + u_\xi(s) - u_0 + K\int_0^s |u_\xi|\,d\tau\\ &\leqslant 2c_0 t + tF_2(t,R/t) + C_2(t)(|u_0|+\varepsilon) + |u_0|\\ &\quad + t^2 K F_2(t,R/t) + tKC_2(t)(|u_0|+\varepsilon)\\ &\leqslant tF_4(t,R/t) + C_3(t)(|u_0|+\varepsilon).\end{aligned}$$

Therefore, (17) follows from the estimate below

$$\begin{aligned}&\int_0^t |L(\xi,u_\xi,\dot{\xi})|\,d\tau \leqslant \int_0^t |L_0(\xi,\dot{\xi})|\,d\tau + K\int_0^t |u_\xi|\,d\tau\\ &\leqslant tF_4(t,R/t) + C_3(t)(|u_0|+\varepsilon) + tK(tF_3(t,R/t) + C_2(t)(|u_0|+\varepsilon))\\ &= tF_5(t,R/t) + C_4(t)(|u_0|+\varepsilon).\end{aligned}$$

We relabel the function F_i and this completes our proof. □

Lemma 4 *Suppose* $x_0 \in \mathbb{R}^n$, $t, R > 0$, $u_0 \in \mathbb{R}$ *and* $|x - x_0| \leqslant R$. *Let* $\varepsilon \in (0,1)$ *and* $\xi \in \mathscr{A}_\varepsilon$. *Then there exist a continuous function* $F = F_{u_0,R} : [0,+\infty) \times [0,+\infty) \to [0,+\infty)$, $F(r_1,\cdot)$ *is nondecreasing and superlinear and* $F(\cdot, r_2)$ *is nondecreasing for any* $r_1, r_2 \geqslant 0$, *such that*

$$\int_0^t |\dot{\xi}(s)|\,ds \leqslant tF(t,R/t) + \varepsilon.$$

Moreover, the family $\{\dot{\xi}\}_{\xi\in\mathscr{A}_\varepsilon}$ *is equi-integrable.*

Proof Let $\varepsilon > 0$ and $\xi \in \mathscr{A}_\varepsilon$. Then, by (L2) we obtain

$$\begin{aligned}&u_\xi(t) - u_0\\ &= \int_0^t L(\xi(s),u_\xi(s),\dot{\xi}(s))\,ds \geqslant \int_0^t \{L(\xi(s),0,\dot{\xi}(s)) - K|u_\xi(s)|\}\,ds\\ &\geqslant \int_0^t \{\theta_0(|\dot{\xi}(s)|) - c_0 - K|u_\xi(s)|\}\,ds\\ &\geqslant \int_0^t \{|\dot{\xi}(s)| - K|u_\xi(s)| - (c_0 + \theta_0^*(1))\}\,ds.\end{aligned} \tag{22}$$

In view of Lemma 2, Lemma 3 and (22), we obtain that

$$\begin{aligned}\int_0^t |\dot{\xi}(s)|\,ds &\leqslant \int_0^t K|u_\xi(s)|\,ds + t(c_0+\theta_0^*(1)) + u_\xi(t) - u_0\\ &\leqslant tK(tF_1(t,R/t) + C_1(t)(\varepsilon + |u_0|)) + t(c_0+\theta_0^*(1))\\ &\quad + tF_2(t,R/t) + \varepsilon := tF_3(t,R/t) + \varepsilon.\end{aligned}$$

Now we turn to proof of the equi-integrability of the family $\{\dot{\xi}\}_{\xi\in\mathscr{A}_\varepsilon}$. Since θ_0 is a superlinear function, then for any $\alpha > 0$ there exists $C_\alpha > 0$ such that $r \leqslant \theta_0(r)/\alpha$ for $r > C_\alpha$. Thus, for any measurable subset $E \subset [0, t]$, invoking (L2), (L3) and Lemma 3, we have that

$$\begin{aligned}
\int_{E\cap\{|\dot{\xi}|>C_\alpha\}} |\dot{\xi}|ds &\leqslant \frac{1}{\alpha}\int_{E\cap\{|\dot{\xi}|>C_\alpha\}} \theta_0(|\dot{\xi}|)ds \leqslant \frac{1}{\alpha}\int_{E\cap\{|\dot{\xi}|>C_\alpha\}} (L_0(\xi, \dot{\xi}) + c_0)ds \\
&\leqslant \frac{1}{\alpha}\int_{E\cap\{|\dot{\xi}|>C_\alpha\}} (L(\xi, u_\xi, \dot{\xi}) + K|u_\xi| + c_0)ds \\
&\leqslant \frac{1}{\alpha}(u_\xi(t) - u_0 + K(tF_1(t, R/t) + C_1(t)(\varepsilon + |u_0|)) + tc_0) \\
&\leqslant \frac{1}{\alpha}(tF_2(t, R/t) + 1 + K(tF_1(t, R/t) + C_1(t)(1 + |u_0|)) + tc_0) \\
&:= \frac{1}{\alpha}F_4(t, R/t).
\end{aligned}$$

Therefore, we conclude that

$$\int_E |\dot{\xi}|ds \leqslant \int_{E\cap\{|\dot{\xi}|>C_\alpha\}} |\dot{\xi}|ds + \int_{E\cap\{|\dot{\xi}|\leqslant C_\alpha\}} |\dot{\xi}|ds \leqslant \frac{1}{\alpha}F_4(t, R/t) + |E|C_\alpha.$$

Then, the equi-integrability of the family $\{\dot{\xi}\}_{\xi\in\mathscr{A}_\varepsilon}$ follows since the right-hand side can be made arbitrarily small by choosing α large and $|E|$ small, and this proves our claim.

Proposition 1 *The functional*

$$\Gamma^t_{x_0,x} \ni \xi \mapsto J(\xi) = \int_0^t L(\xi(s), u_\xi(s), \dot{\xi}(s))\, ds,$$

where u_ξ is determined by (9) *with initial condition $u_\xi(0) = u_0$, admits a minimizer in $\Gamma^t_{x_0,x}$.*

Remark 2 Notice that we can rewrite the functional J as

$$J(\xi) = (e^{\int_0^t \widehat{L^\xi_u}\, dr} - 1)u_0 + \int_0^t e^{\int_\tau^t \widehat{L^\xi_u}\, dr} L(\xi(\tau), 0, \dot{\xi}(\tau))\, d\tau \tag{23}$$

since $J(\xi) = u_\xi(t) - u$, where

$$\widehat{L^\xi_u}(s) = \int_0^1 L_u(\xi(s), \lambda u_\xi(s), \dot{\xi}(s))\, d\lambda.$$

We set $\mu_\xi(s) := e^{\int_s^t \widehat{L^\xi_u}\, dr}$. Therefore $J(\xi) = J_1(\xi) + J_2(\xi)$ where

$$J_1(\xi) = (\mu_\xi(a) - 1)u_0, \quad J_2(\xi) = \int_0^t \mu_\xi(\tau)L(\xi(\tau), 0, \dot{\xi}(\tau))\,d\tau.$$

Proof Fix $x_0, x \in \mathbb{R}^n$, $t > 0$ and $u_0 \in \mathbb{R}$. Consider any minimizing sequence $\{\xi_k\}$ for J, that is, a sequence such that $J(\xi_k) \to \inf\{J(\xi) : \xi \in \mathscr{A}\}$ as $k \to \infty$. We want to show that this sequence admits a cluster point which is the required minimizer. Notice there exists an associated sequence $\{u_{\xi_k}\}$ given by (9) in the definition of $J(\xi_k)$. The idea of the proof is standard but a little bit different.

First, notice that Lemma 4 implies that the sequence of derivatives $\{\dot{\xi}_k\}$ is equi-integrable. Since the sequence $\{\dot{\xi}_k\}$ is equi-integrable, by the Dunford-Pettis Theorem there exists a subsequence, which we still denote by $\{\dot{\xi}_k\}$, and a function $\eta^* \in L^1([0,t], \mathbb{R}^n)$ such that $\dot{\xi}_k \rightharpoonup \eta^*$ in the weak-L^1 topology. The equi-integrability of $\{\dot{\xi}_k\}$ implies that the sequence $\{\xi_k\}$ is equi-continuous and uniformly bounded. Invoking Ascoli-Arzela theorem, we can also assume that the sequence $\{\xi_k\}$ converges uniformly to some absolutely continuous function $\xi_\infty \in \Gamma^t_{x_0,x}$. For any test function $\varphi \in C^1_0([0,t], \mathbb{R}^n)$,

$$\int_0^t \varphi\eta^* ds = \lim_{k\to\infty} \int_0^t \varphi\dot{\xi}_k ds = -\lim_{k\to\infty} \int_0^t \dot{\varphi}\xi_k ds = -\int_0^t \dot{\varphi}\xi_\infty ds.$$

By du Bois-Reymond lemma (see, for instance, [10, Lemma 6.1.1]), we can conclude that $\dot{\xi}_\infty = \eta^*$ almost everywhere. In View of Remark 2 and condition (L3), we also have that the sequence $\{\mu_{\xi_k}\}$ is bounded and equi-continuous. Therefore μ_{ξ_k} converges uniformly to μ_ξ as $k \to \infty$ by taking a subsequence if necessary.

We recall a classical result (see, for instance, [3, Theorem 3.6] or [2, Section 3.4]) on the sequentially lower semicontinuous property on the functional

$$L^1([0,t], \mathbb{R}^m) \times L^1([0,t], \mathbb{R}^n) \ni (\alpha, \beta) \mapsto \mathbf{F}(\alpha, \beta) := \int_0^t \mathbf{L}(\alpha(s), \beta(s))\,ds.$$

One has that if (i) $\mathbf{L}$ is lower semicontinuous; (ii) $\mathbf{L}(\alpha, \cdot)$ is convex on $\mathbb{R}^n$, then the functional $\mathbf{F}$ is sequentially lower semicontinuous on the space $L^1([0,t], \mathbb{R}^m) \times L^1([0,t], \mathbb{R}^n)$ endowed with the strong topology on $L^1([0,t], \mathbb{R}^m)$ and the weak topology on $L^1([0,t], \mathbb{R}^n)$.

Now, let

$$\mathbf{L}(\mu_{\xi_k}(s), \xi_k(s), \dot{\xi}_k(s)) := \mu_{\xi_k}(s)L(\xi_k(s), 0, \dot{\xi}_k(s))$$

with $\alpha_{\xi_k}(s) = (\mu_{\xi_k}(s), \xi_k(s))$ and $\beta_{\xi_k}(s) = \dot{\xi}_k(s)$, then J_2 is lower semi-continuous in the topology mentioned above. The lower semi-continuos of J_1 is obvious (In fact, J_1 is continuous). Therefore, $\xi_\infty \in \mathscr{A}$ is a minimizer of J and this completes the proof of the existence result.

Corollary 1 *Let $u_0 \in \mathbb{R}$ and $R > 0$ be fixed. Then there exists a continuous function $F = F_{u_0,R} : [0,+\infty) \times [0,+\infty) \to [0,+\infty)$, with $F(t,r)$ nondecreasing in both* *variables and superlinear with respect to r, such that for any given* $t > 0$ *and* $x, x_0 \in \mathbb{R}^n$, *with* $|x - x_0| \leqslant R$, *every minimizer* $\xi \in \Gamma^t_{x_0,x}$ *for* (10) *satisfies*

$$\int_0^t |\dot{\xi}(s)|\, ds \leqslant tF(t, R/t)$$

and

$$\operatorname*{ess\,inf}_{s\in[0,t]} |\dot{\xi}(s)| \leqslant F(t, R/t), \quad \sup_{s\in[0,t]} |\xi(s) - x_0| \leqslant tF(t, R/t).$$

Proof The first assertion is direct from Lemma 4. The last two inequality follows from the relations

$$\operatorname*{ess\,inf}_{s\in[0,t]} |\dot{\xi}(s)| \leqslant \frac{1}{t}\int_0^t |\dot{\xi}(s)|\, ds, \quad \text{and} \quad |\xi(s) - x_0| \leqslant \int_0^t |\dot{\xi}(s)|\, ds,$$

together with the first assertion.

3 Necessary Conditions and Regularity of Minimizers

3.1 Lipschitz Estimate of Minimizers

In order to study the regularity of a minimizer ξ of (10), we need an *a priori* Lipschitz estimate for ξ. A key point of the proof of such an estimate is the following *Erdmann condition*, which is standard for classical autonomous Tonelli Lagrangians. For the results in and after this section, we suppose the following technical condition

(L2') L satisfies condition (L2). Moreover, for some $R > 1$ and every compact set $A \subset \mathbb{R}^n$ there exists a constant $C_A > 0$ such that

$$L(x, 0, rv) \leqslant C_A(1 + L(x, 0.v)), \quad \forall r \in [1, R],\ (x, v) \in A \times \mathbb{R}^n.$$

We begin with some fundamental results from convex analysis.

Lemma 5 *Let* L *satisfy conditions* (L1)–(L3). *We conclude that*

(a) The function

$$f(\varepsilon) := L_v(x, r, v/(1+\varepsilon)) \cdot v/(1+\varepsilon) - L(x, r, v/(1+\varepsilon)) \qquad (24)$$

is decreasing for $\varepsilon > -1$. *In particular,* $f(\varepsilon) \geqslant f(+\infty) = -L(x, r, 0) \geqslant -\overline{\theta}_0$ $(0) - K|r|$.

(b) If $\varepsilon_1, \varepsilon_2 > -1$ *and* $\varepsilon_1 < \varepsilon_2$, *then we have*

$$L(x, r, v/(1+\varepsilon_2)) \leqslant (\kappa+1)^{-1}L(x, r, v/(1+\varepsilon_1)) + \kappa \cdot (\kappa+1)^{-1}(\overline{\theta}_0(0) + K|r|)$$

and

$$f(\varepsilon_2) \leqslant \kappa^{-1}L(x, r, v/(1+\varepsilon_1)) - (\kappa^{-1}+1)L(x, r, v/(1+\varepsilon_2))$$

where $\kappa = (\varepsilon_2 - \varepsilon_1)/(1+\varepsilon_1) > 0$.

Proof Let $\varepsilon_1, \varepsilon_2 \in (-1, +\infty)$ and $\varepsilon_1 < \varepsilon_2$. We set $L(v) = L(x, r, v)$. By (L1) we have that

$$L(v/(1+\varepsilon_2)) \geqslant L(v/(1+\varepsilon_1)) + L_v(v/(1+\varepsilon_1)) \cdot \{v/(1+\varepsilon_2) - v/(1+\varepsilon_1)\}.$$

It follows that

$$\begin{aligned} f(\varepsilon_1) - f(\varepsilon_2) \geqslant\, & L_v(v/(1+\varepsilon_1)) \cdot (v/(1+\varepsilon_1)) - L_v(v/(1+\varepsilon_2)) \cdot (v/(1+\varepsilon_2)) \\ & + L_v(v/(1+\varepsilon_1)) \cdot \{v/(1+\varepsilon_2) - v/(1+\varepsilon_1)\} \\ =\, & L_v(v/(1+\varepsilon_1)) \cdot (v/(1+\varepsilon_2)) - L_v(v/(1+\varepsilon_2)) \cdot (v/(1+\varepsilon_2)) \\ =\, & \{L_v(v/(1+\varepsilon_1)) - L_v(v/(1+\varepsilon_2))\} \cdot \{v/(1+\varepsilon_1) - v/(1+\varepsilon_2)\} \\ & \cdot (1/(1+\varepsilon_1) - 1/(1+\varepsilon_2))^{-1} \cdot (1/(1+\varepsilon_1)) \\ \geqslant\, & 0. \end{aligned}$$

The last statement is a direct consequence of (L2) and (L3).

Now we turn to the proof of (b). Observe that by convexity

$$\begin{aligned} L(v/(1+\varepsilon_1)) \geqslant\, & L(v/(1+\varepsilon_2)) + L_v(v/(1+\varepsilon_2)) \cdot \{v/(1+\varepsilon_1) - v/(1+\varepsilon_2)\} \\ =\, & L(v/(1+\varepsilon_2)) + \kappa \cdot L_v(v/(1+\varepsilon_2)) \cdot (v/(1+\varepsilon_2)). \end{aligned}$$

In view of (a) we obtain that

$$\begin{aligned} & L(v/(1+\varepsilon_1)) - (\kappa+1)L(v/(1+\varepsilon_2)) \\ \geqslant\, & \kappa \cdot \{-L(v/(1+\varepsilon_2)) + L_v(v/(1+\varepsilon_2)) \cdot (v/(1+\varepsilon_2))\} \\ \geqslant\, & -\kappa \cdot (\overline{\theta}_0(0) + K|r|) \end{aligned}$$

Then the first assertion follows. Moreover, we have that

$$L_v(v/(1+\varepsilon_2)) \cdot (v/(1+\varepsilon_2)) \leqslant \kappa^{-1}(L(v/(1+\varepsilon_1)) - L(v/(1+\varepsilon_2)))$$

which leads to the second assertion.

Lemma 6 (Erdmann condition) *Suppose* (L1), (L2') *and* (L3) *are satisfied. Let* $\xi \in \Gamma^t_{x_0,x}$ *be a minimizer of* (10) *with* u_ξ *determined by* (9) *and* $u_\xi(0) = u_0$. *Set* $\int_0^s L_u dr = \int_0^s L_u(\xi(r), u_\xi(r), \dot{\xi}(r))dr$ *and define*

$$E(s) = e^{-\int_0^s L_u dr} \cdot \left\{L_v(\xi(s), u_\xi(s), \dot{\xi}(s)) \cdot \dot{\xi}(s) - L(\xi(s), u_\xi(s), \dot{\xi}(s))\right\}$$

for almost all $s \in [0, t]$. *Then* E *has a continuous representation* $\bar{E}$ *such that* $\bar{E}$ *is absolutely continuous on* $[0, t]$ *and* $\bar{E}' = 0$ *for almost all* $s \in [0, t]$.

Remark 3 Condition (L2') is satisfied when L has polynomial growth with respect to v. It is a key point to ensure the finiteness of the action after reparametrization.

Proof We divide the proof into several steps.

Step I: Reparametrization. For any measurable function $\alpha : [0, t] \to [1/2, 3/2]$ satisfying $\int_0^t \alpha(s)\, ds = t$ (the set of all such functions α is denoted by Ω), we define

$$\tau(s) = \int_0^s \alpha(r)\, dr, \quad s \in [0, t].$$

Note that $\tau : [0, t] \to [0, t]$ is a bi-Lipschitz map and its inverse $s(\tau)$ satisfies

$$s'(\tau) = \frac{1}{\alpha(s(\tau))}, \quad a.e.\ \tau \in [0, t].$$

Now, given $\xi \in \Gamma^t_{x_0,x}$ as above and $\alpha \in \Omega$, define the reparameterization η of ξ by $\eta(\tau) = \xi(s(\tau))$. It follows that $\dot{\eta}(\tau) = \dot{\xi}(s(\tau))/\alpha(s(\tau))$. Let u_η be the unique solution of (9) with initial condition $u_\eta(0) = u_0$. Then we have that

$$\begin{aligned} J(\xi) \leqslant J(\eta) &= \int_0^t L(\eta(\tau), u_\eta(\tau), \dot{\eta}(\tau))\, d\tau \\ &= \int_0^t L(\xi(s), u_{\xi,\alpha}(s), \dot{\xi}(s)/\alpha(s))\alpha(s)\, ds \end{aligned}$$

where $u_{\xi,\alpha}$ solves

$$\dot{u}_{\xi,\alpha}(s) = L(\xi(s), u_{\xi,\alpha}(s), \dot{\xi}(s)/\alpha(s))\alpha(s), \quad u_{\xi,\alpha}(0) = u_0.$$

By a direct calculation, for all $\alpha \in \Omega$ and almost all $s \in [0, t]$, we obtain

$$\begin{aligned} \dot{u}_{\xi,\alpha} - \dot{u}_\xi &= L(\xi, u_{\xi,\alpha}, \dot{\xi}/\alpha)\alpha - L(\xi, u_\xi, \dot{\xi}) \\ &= L(\xi, u_{\xi,\alpha}, \dot{\xi}/\alpha)\alpha - L(\xi, u_\xi, \dot{\xi}/\alpha)\alpha + L(\xi, u_\xi, \dot{\xi}/\alpha)\alpha - L(\xi, u_\xi, \dot{\xi}) \\ &= \widehat{L^\alpha_u}\, (u_{\xi,\alpha} - u_\xi) + L(\xi, u_\xi, \dot{\xi}/\alpha)\alpha - L(\xi, u_\xi, \dot{\xi}), \end{aligned}$$

and $u_{\xi,\alpha}(0) - u_\xi(0) = 0$, where

$$\widehat{L^\alpha_u}(s) = \int_0^1 L_u\big(\xi(s), u_\xi(s) + \lambda(u_{\xi,\alpha}(s) - u_\xi(s)), \dot{\xi}(s)/\alpha(s)\big)\alpha(s)\, d\lambda.$$

By solving the Carathéodory equation above, we conclude that

$$u_{\xi,\alpha}(s) - u_\xi(s) = \int_0^s e^{\int_\tau^s \widehat{L_u^\alpha} dr}(L(\xi, u_\xi, \dot{\xi}/\alpha)\alpha - L(\xi, u_\xi, \dot{\xi}))\, d\tau, \tag{25}$$

and $u_{\xi,\alpha}(t) - u_\xi(t) \geqslant 0$ for all $\alpha \in \Omega$. We claim that

$$L(\xi, u_\xi, \dot{\xi}/\alpha) \in L^1([0,t]) \text{ for all } \alpha \in \Omega. \tag{26}$$

To show (26), by Lemma 3 we first observe that

$$L(\xi, u_\xi, \dot{\xi}/\alpha) \geqslant L(\xi, 0, \dot{\xi}/\alpha) - K|u_\xi| \geqslant \theta_0(0) - c_0 - KF_1(t, |x_0 - x|/t)$$

which gives the lower bound of $L(\xi, u_\xi, \dot{\xi}/\alpha)$. For the upper bound we will treat two cases:

1. Suppose $\alpha \in [1, 3/2]$. Then Lemma 5 (b) shows that $L(\xi, u_\xi, \dot{\xi}/\alpha) \leqslant (\kappa + 1)^{-1} L(\xi, u_\xi, \dot{\xi}) + \kappa \cdot (\kappa + 1)^{-1}(\overline{\theta}_0(0) + K|u_\xi|)$;
2. For the case $\alpha \in [1/2, 1]$, we need condition (L2'). Let $A = \bar{B}(0, tF_2(t, |x_0 - x|/t))$ where F_2 is determined by Corollary 1 such that $|\xi(s)| \leqslant tF_2(t, |x_0 - x|/t)$. Invoking condition (L2') we conclude that

$$\begin{aligned} L(\xi, u_\xi, \dot{\xi}/\alpha) &\leqslant L(\xi, 0, \dot{\xi}/\alpha) + KF_1(t, |x_0 - x|/t) \\ &\leqslant C_A(1 + L(\xi, 0, \dot{\xi})) + KF_1(t, |x_0 - x|/t) \\ &\leqslant C_A(1 + L(\xi, u_\xi, \dot{\xi}) + KF_1(t, |x_0 - x|/t)) + KF_1(t, |x_0 - x|/t). \end{aligned}$$

Step II: A necessary condition. Next, we introduce the family

$$\Omega_0 = \{\beta : [0,t] \to \mathbb{R} : \beta \in L^\infty([0,t]) \text{ with } \int_0^t \beta(s) ds = 0\}$$

and let $0 \neq \beta \in \Omega_0$. For any $\varepsilon \in \mathbb{R}$ such that $|\varepsilon| < \varepsilon_0 = \frac{1}{4\|\beta\|_\infty} < \frac{1}{2\|\beta\|_\infty}$ we have that $1 + \varepsilon\beta \in \Omega$.

Define the functional $\Lambda : \Omega \to \mathbb{R}$ by $\Lambda(\alpha) = u_{\xi,\alpha}(t)$. Since $\Lambda(1 + \varepsilon\beta) \geqslant \Lambda(1)$ for $|\varepsilon| < \frac{1}{2\|\beta\|_\infty}$, then we have that $\frac{d}{d\varepsilon}\Lambda(1 + \varepsilon\beta)|_{\varepsilon=0} = 0$ if the derivative exists. Thus, by (25),

$$\frac{\Lambda(1 + \varepsilon\beta) - \Lambda(1)}{\varepsilon} = \int_0^t e^{\int_s^t \widehat{L_u^\varepsilon} dr} \lambda_\varepsilon(s)\, ds, \tag{27}$$

where $\widehat{L_u^\varepsilon} = \widehat{L_u^{1+\varepsilon\beta}}$ and

$$\lambda_\varepsilon(s) := \frac{L(\xi, u_\xi, \dot{\xi}/(1+\varepsilon\beta))(1+\varepsilon\beta) - L(\xi, u_\xi, \dot{\xi})}{\varepsilon}$$
$$= L(\xi, u_\xi, \dot{\xi}/(1+\varepsilon\beta)) \cdot \beta + \frac{1}{\varepsilon}(L(\xi, u_\xi, \dot{\xi}/(1+\varepsilon\beta)) - L(\xi, u_\xi, \dot{\xi})).$$

We claim that

$$0 = \frac{d}{d\varepsilon}\Lambda(1+\varepsilon\beta)|_{\varepsilon=0} = \int_0^t e^{\int_s^t L_u dr}\left\{L(\xi, u_\xi, \dot{\xi}) - L_v(\xi, u_\xi, \dot{\xi}) \cdot \dot{\xi}\right\} \beta \, ds. \quad (28)$$

Step III: Summability. Set

$$l_\varepsilon(s) := L_v(\xi, u_\xi, \dot{\xi}/(1+\varepsilon\beta)) \cdot \dot{\xi}/(1+\varepsilon\beta) - L(\xi, u_\xi, \dot{\xi}/(1+\varepsilon\beta)).$$

Notice that we take out the variable s on right side of the inequalities above. In view of Lemma 5 (a) and Lemma 3 we have that $l_\varepsilon(s)$ is bounded below by $-(\overline{\theta}_0(0) + KF_1(t, |x_0 - x|/t))$. By convexity we have that

$$L(\xi, u_\xi, \dot{\xi}/(1+\varepsilon\beta)) - L(\xi, u_\xi, \dot{\xi}) \leqslant L_v(\xi, u_\xi, \dot{\xi}/(1+\varepsilon\beta)) \cdot \{\dot{\xi} - \dot{\xi}/(1+\varepsilon\beta)\}$$
$$= -\varepsilon\beta \cdot L_v(\xi, u_\xi, \dot{\xi}/(1+\varepsilon\beta)) \cdot \dot{\xi}/(1+\varepsilon\beta).$$

It follows that

$$\lambda_\varepsilon(s) \leqslant -\beta(s)\{L_v(\xi, u_\xi, \dot{\xi}/(1+\varepsilon\beta)) \cdot \dot{\xi}/(1+\varepsilon\beta) - L(\xi, u_\xi, \dot{\xi}/(1+\varepsilon\beta))\}$$
$$= -\beta(s) \cdot l_\varepsilon(s), \quad (29)$$

Let $\beta \in \Omega_0$ and $0 < \varepsilon < \varepsilon_0$. We rewrite $\lambda_\varepsilon(s)$, $l_\varepsilon(s)$ as $\lambda_\varepsilon^\beta(s)$, $l_\varepsilon^\beta(s)$ respectively. Set $\beta^+ = \beta \cdot \mathbb{1}_{\{\beta \geqslant 0\}}$ and $\beta^- = -\beta \cdot \mathbb{1}_{\{\beta<0\}}$, then

$$\beta = \beta^+ - \beta^-, \quad \text{and} \quad \beta^\pm \geqslant 0.$$

By (29) we have that

$$0 \leqslant \lambda_\varepsilon^\beta(s) + \beta^+ l(s)_\varepsilon^\beta(s) \leqslant \beta^-(s) l_\varepsilon^\beta(s).$$

Now observe that $\beta^+(s)l_\varepsilon^\beta(s) = \beta^+(s)l_\varepsilon^{\beta^+}(s)$ and $\beta^-(s)l_\varepsilon^\beta(s) = \beta^-(s)l_{-\varepsilon}^{\beta^-}(s)$. Then the inequalities above can recast as follows

$$0 \leqslant \lambda_\varepsilon^\beta(s) + \beta^+(s)l_\varepsilon^{\beta^+}(s) \leqslant \beta^-(s)l_{-\varepsilon}^{\beta^-}(s). \quad (30)$$

Lemma 5 (a) ensures that $\varepsilon \mapsto l_\varepsilon^{\beta^-}$ is decreasing on $[-\varepsilon_0, \varepsilon_0]$ and we conclude that

$$\beta^- l_{-\varepsilon}^{\beta^-} \leqslant \beta^- l_{-\varepsilon_0}^{\beta^-} \quad \forall \varepsilon \in (0, \varepsilon_0). \quad (31)$$

By Lemma 5 (b), we obtain that

$$\begin{aligned}\beta^- l^{\beta^-}_{-\varepsilon} &= \beta^- \{L_v(\xi, u_\xi, \dot{\xi}/(1-\varepsilon\beta^-)) \cdot \dot{\xi}/(1-\varepsilon\beta^-) - L(\xi, u_\xi, \dot{\xi}/(1-\varepsilon\beta^-))\}\\ &\leqslant \beta^-(\kappa^{\beta^-}_\varepsilon)^{-1} L(\xi, u_\xi, \dot{\xi}/(1-\varepsilon_0\beta^-)) - \beta^-((\kappa^{\beta^-}_\varepsilon)^{-1}+1) L(\xi, u_\xi, \dot{\xi}/(1-\varepsilon\beta^-))\end{aligned}$$

where $(\kappa^{\beta^-}_\varepsilon)^{-1} = \frac{1-\varepsilon_0\beta^-}{\varepsilon_0-\varepsilon} \cdot (\beta^-)^{-1}$. In view of (26), (30) and the fact that $\beta^-(\kappa^{\beta^-}_\varepsilon)^{-1}$ is bounded, we conclude that $\beta^- l^{\beta^-}_{-\varepsilon} \in L^1([0,t])$ for all $\varepsilon \in (0, \varepsilon_0]$.

Step IV: Erdmann condition. Thus integrating (30) and by Lebesgue's theorem we obtain that

$$0 \leqslant \int_0^t e^{\int_s^t L_u dr} l_0(s)\beta^+(s) ds \leqslant \int_0^t e^{\int_s^t L_u dr} l_0(s)\beta^-(s) ds.$$

Therefore, $\int_0^t e^{\int_s^t L_u dr} l_0(s)\beta(s) ds \leqslant 0$ and (28) follows since $\beta \in \Omega_0$ is arbitrary.

Now, observe that the primitive $\mu(s) := \int_0^s \beta(r) dr$ gives a one-to-one correspondence between Ω_0 and the set

$$\Omega_1 = \{\mu : [0,t] \to \mathbb{R} : \mu \text{ is Lipschitz continuous with } \mu(0) = \mu(t) = 0\}.$$

Thus, (28) can recast as follows

$$0 = -e^{\int_0^t L_u dr} \int_0^t E(s)\mu'(s)\, ds \quad \forall \mu \in \Omega_1.$$

Recalling $E(s) = -e^{-\int_0^s L_u dr} l_0(s) \in L^1([0,t])$ by Step III, then a basic lemma in the calculus of variations ensures that $E(s)$ is constant on $[0,t]$.

Proposition 2 *Suppose* (L1), (L2') *and* (L3) *are satisfied. Let* $u_0 \in \mathbb{R}$ *and* $R > 0$ *be fixed. Then there exists a continuous function* $F = F_{u_0,R} : [0,+\infty) \times [0,+\infty) \to [0,+\infty)$, *with* $F(t,r)$ *nondecreasing in both variables and superlinear with respect to* r, *such that for any given* $t > 0$ *and* $x, x_0 \in \mathbb{R}^n$, *with* $|x - x_0| \leqslant R$, *every minimizer* $\xi \in \Gamma^t_{x_0,x}$ *for (10) satisfies*

$$\operatorname*{ess\,sup}_{s\in[0,t]} |\dot{\xi}(s)| \leqslant F(t, R/t).$$

Proof Let $\xi \in \Gamma^t_{x_0,x}$ be as above. Invoking Lemma 6, E is constant on a subset of $[0,t]$ of full measure. Let

$$E_1(s) = L_v(\xi(s), u_\xi(s), \dot{\xi}(s)) \cdot \dot{\xi}(s) - L(\xi(s), u_\xi(s), \dot{\xi}(s)).$$

Set $l_\xi(s,\alpha) = L(\xi(s), u_\xi(s), \dot{\xi}(s)/\alpha)\alpha$ for all $s \in [0,t]$ and $\alpha > 0$. A simple computation shows that $l_\xi(s,\alpha)$ is convex in α and

$$\frac{d}{d\alpha}l_\xi(s,\alpha)|_{\alpha=1} = -E_1(s).$$

Taking $s_0 \in [0,t]$ such that $|\dot\xi(s_0)| = \text{ess inf}_{s\in[0,t]} |\dot\xi(s)|$, by convexity we have that

$$-E_1(s_0) \geqslant \sup_{\alpha<1} \frac{l_\xi(s_0,\alpha) - l_\xi(s_0,1)}{\alpha - 1}$$

Let us now take, in the above inequality, $\alpha = 3/4$. Then, by (L2), (L3), Lemma 3 and Corollary 1 we conclude that

$$\begin{aligned}
-E_1(s_0) \geqslant & 4(l_\xi(s_0,1) - l_\xi(s_0,3/4)) = 4(L(\xi(s_0),u_\xi(s_0),\dot\xi(s_0)) - l_\xi(s_0,3/4)) \\
\geqslant & -4c_0 - 4KF_1(t,R/t) - 3L(\xi(s_0),u_\xi(s_0),\frac{4}{3}\dot\xi(s_0)) \\
\geqslant & -4c_0 - 4KF_1(t,R/t) - 3KF_1(t,R/t) - 3L(\xi(s_0),0,\frac{4}{3}\dot\xi(s_0)) \\
\geqslant & -4c_0 - 7KF_1(t,R/t) - \overline{\theta}_0(\frac{4}{3}|\dot\xi(s_0)|) \\
\geqslant & -4c_0 - 7KF_1(t,R/t) - \overline{\theta}_0(F_2(t,R/t)) := -F_3(t,R/t).
\end{aligned}$$

It follows that, for almost all $s \in [0,t]$,

$$E(s) = E(s_0) = e^{-\int_0^{s_0} L_u d\tau} E_1(s_0) \leqslant e^{Kt} F_3(t,R/t),$$

and

$$E_1(s) = e^{\int_0^s L_u d\tau} E(s) \leqslant e^{2Kt} F_3(t,R/t) := F_4(t,R/t). \tag{32}$$

Now, let s be such that $\dot\xi(s)$ exists and (32) holds. By convexity, we have that

$$\begin{aligned}
& L(\xi(s),u_\xi(s),\dot\xi(s)/(1+|\dot\xi(s)|)) - L(\xi(s),u_\xi(s),\dot\xi(s)) \\
\geqslant & ((1+|\dot\xi(s)|)^{-1} - 1)\cdot\langle L_v(\xi(s),u_\xi(s),\dot\xi(s)),\dot\xi(s)\rangle \\
\geqslant & ((1+|\dot\xi(s)|)^{-1} - 1)\cdot(L(\xi(s),u_\xi(s),\dot\xi(s)) + F_4(t,R/t)).
\end{aligned}$$

It follows that

$$\begin{aligned}
& L(\xi(s),u_\xi(s),\dot\xi(s)) \\
& \leqslant L(\xi(s),u_\xi(s),\dot\xi(s)/(1+|\dot\xi(s)|))(1+|\dot\xi(s)|) + F_4(t,R/t)|\dot\xi(s)|.
\end{aligned}$$

Let $C = \sup_{s\in[0,t],|v|\leqslant 1} L(\xi(s),u_\xi(s),v)$ and by (L2). By Lemma 3 we have that

$$C \leqslant \sup_{s\in[0,t],|v|\leqslant 1} \{L(\xi(s),0,v) + K|u_\xi(s)|\} \leqslant \overline{\theta}_0(1) + KF_1(t,R/t) := F_5(t,R/t).$$

It follows that

$$L(\xi(s), u_\xi(s), \dot{\xi}(s)) \leqslant F_5(t, R/t) + (F_5(t, R/t) + F_4(t, R/t))|\dot{\xi}(s)|.$$

Therefore, invoking Lemma 3, we obtain

$$\begin{aligned}&(F_5(t, R/t) + F_4(t, R/t) + 1)|\dot{\xi}(s)| - (\theta_0^*(F_5(t, R/t) + F_4(t, R/t) + 1) + c_0)\\ &\leqslant \theta_0(|\dot{\xi}(s)|) - c_0 \leqslant L(\xi(s), 0, \dot{\xi}(s)) \leqslant L(\xi(s), u_\xi(s), \dot{\xi}(s)) + K|u_\xi(s)|\\ &\leqslant F_5(t, R/t) + (F_5(t, R/t) + F_4(t, R/t))|\dot{\xi}(s)| + KF_1(t, R/t).\end{aligned}$$

This leads to

$$\begin{aligned}|\dot{\xi}(s)| &\leqslant (\theta_0^*(F_5(t, R/t) + F_4(t, R/t) + 1) + c_0) + F_5(t, R/t) + KF_1(t, R/t)\\ &:= F_6(t, R/t),\end{aligned}$$

which completes the proof.

Proposition 3 *Suppose $L_\lambda(x, r, v) = L_0(x, v) - \lambda r$, $r \in \mathbb{R}$, where L_0 is a Tonelli Lagrangian. Then L_λ satisfies condition* (L1), (L2) *and* (L3). *Moreover, the Lipschitz estimate in Proposition 2 holds.*

Proof By solving the Carathéodory equation (4), we have that

$$u_\xi(t) = e^{-\lambda t}u_0 + e^{-\lambda t}\int_0^t e^{\lambda s}L_0(\xi, \dot{\xi})\,ds.$$

Therefore problem (3) is essentially a basic problem in the calculus of variations with a time-dependent Lagrangian $G(t, x, v) = e^{\lambda t}L_0(x, v)$. Moreover, G satisfies a restricted growth condition

$$G_t(t, x, v) = \lambda G(t, x, v).$$

Then any minimizer ξ of (3) is Lipschitz continuous (see, for instance, [3, Theorem 4.9]). Therefore, Erdmann condition in Lemma 6 holds with a slight modification of the proof and the expected Lipschitz estimate can obtained as in the proof of Proposition 2 similarly.

3.2 *Regularity of Minimizers-Herglotz Equations–Lie Equations*

Let $\xi \in \Gamma^t_{x_0,x}$ be a minimizer of (10) where u_ξ is uniquely determined by (9). For any $\lambda \in \mathbb{R}$ and any Lipschitz function $\eta \in \Gamma^t_{0,0}$, we denote $\xi_\lambda(s) = \xi(s) + \lambda\eta(s)$. It is clear that $\xi_\lambda \in \Gamma^t_{x_0,x}$ and $J(\xi) \leqslant J(\xi_\lambda)$. Let u_{ξ_λ} be the associated unique solution of (9) with respect to ξ_λ and the initial condition u_0. Notice that

$$\frac{\partial}{\partial\lambda}J(\xi_\lambda)|_{\lambda=0}=\frac{\partial}{\partial\lambda}u_{\xi_\lambda}(t)|_{\lambda=0}=0.$$

Now for any $s\in[0,t]$ we set

$$\Delta_\lambda(s)=\frac{u_{\xi_\lambda}(s)-u_\xi(s)}{\lambda}=\frac{1}{\lambda}\int_0^s L(\xi_\lambda,u_{\xi_\lambda},\dot\xi_\lambda)-L(\xi,u_\xi,\dot\xi)\,d\tau,$$

and

$$f_1^\lambda(s)=\frac{1}{\lambda}\int_0^s L(\xi_\lambda,u_{\xi_\lambda},\dot\xi_\lambda(s))-L(\xi_\lambda,u_{\xi_\lambda},\dot\xi)\,d\tau,$$
$$f_2^\lambda(s)=\frac{1}{\lambda}\int_0^s L(\xi_\lambda,u_{\xi_\lambda},\dot\xi)-L(\xi,u_{\xi_\lambda},\dot\xi)\,d\tau.$$

Then f_1^λ and f_2^λ are all absolutely continuous functions on $[0,t]$, and it follows

$$\Delta_\lambda(s)=f_1^\lambda(s)+f_2^\lambda(s)+\frac{1}{\lambda}\int_0^s \widehat{L_u^\lambda}\cdot(u_{\xi_\lambda}-u_\xi)\,d\tau,\quad s\in[0,t],$$

where

$$\widehat{L_u^\lambda}(\tau)=\int_0^1 L_u(\xi(\tau),u_\xi(\tau)+\theta(u_{\xi_\lambda}(\tau)-u_\xi(\tau)),\dot\xi(\tau))\,d\theta,\quad \tau\in[0,t].$$

Thus, we conclude that for almost all $s\in[0,t]$, the following Carathéodory equation holds:

$$\dot\Delta_\lambda(s)=\dot f_1^\lambda(s)+\dot f_2^\lambda(s)+\widehat{L_u^\lambda}(s)\cdot\Delta_\lambda(s)\tag{33}$$

with initial condition $\Delta_\lambda(t)=a_\lambda$. Notice that $\lim_{\lambda\to0}\Delta_\lambda(t)$ exists and $\lim_{\lambda\to0}\Delta_\lambda(t)$ $=\lim_{\lambda\to0}a_\lambda=\frac{\partial}{\partial\lambda}u_{\xi_\lambda}(t)|_{\lambda=0}=0$ since ξ is a minimizer of J. It is not difficult to solve (33), we obtain that

$$\Delta_\lambda(s)=a_\lambda e^{\int_t^s\widehat{L_u^\lambda}(r)\,dr}+e^{\int_t^s\widehat{L_u^\lambda}(r)\,dr}\cdot\int_t^s e^{-\int_t^r\widehat{L_u^\lambda}(\tau)\,d\tau}\cdot(\dot f_1^\lambda(r)+\dot f_2^\lambda(r))\,dr.$$

Since $(\xi_\lambda(s),\dot\xi_\lambda(s),u_{\xi_\lambda}(s))$ tends $(\xi(s),\dot\xi(s),u_\xi(s))$ as $\lambda\to0$ for almost all $s\in[0,t]$, together with Proposition 2, it follows that, for all $s\in[0,t]$, we have

$$f(s):=\frac{\partial}{\partial\lambda}u_{\xi_\lambda}(s)|_{\lambda=0}=e^{\int_t^s h(r)\,dr}\cdot\int_t^s e^{-\int_t^r h(\tau)\,d\tau}\cdot g(r)\,dr,\quad f(t)=0,\tag{34}$$

where $g=L_x\cdot\eta+L_v\cdot\dot\eta$ and $h=L_u$ which are both measurable and bounded. Notice that (33) implies that

$$f(s) = \int_0^s g(r) + h(r) f(r)\, dr, \quad s \in [0, t].$$

Then, invoking (34), we conclude that

$$\begin{aligned}
0 &= \int_0^t g(s) + h(s) \cdot e^{\int_t^s h(r)\, dr} \cdot \left\{ \int_t^s e^{-\int_t^r h(\tau)\, d\tau} \cdot g(r)\, dr \right\} ds \\
&= \int_0^t g(s)\, ds + e^{\int_t^s h(r)\, dr} \cdot \left\{ \int_t^s e^{-\int_t^r h(\tau)\, d\tau} \cdot g(r)\, dr \right\} \Bigg|_0^t \\
&\quad - \int_0^t e^{\int_t^s h(r)\, dr} \cdot e^{-\int_t^s h(r)\, dr} \cdot g(s)\, ds \\
&= e^{\int_t^0 h(r)\, dr} \cdot \left\{ \int_0^t e^{-\int_t^s h(r)\, dr} \cdot g(s)\, ds \right\}.
\end{aligned}$$

It follows that

$$0 = \int_0^t e^{-\int_t^s h(r)\, dr} \cdot g(s)\, ds = \int_0^t e^{-\int_t^s h(r)\, dr} \cdot (L_x \cdot \eta + L_v \cdot \dot{\eta})(s)\, ds.$$

Invoking the fundamental lemma in calculus of variation (see, for instance, Lemma 6.1.1 in [10]), we obtain that, for almost all $s \in [0, t]$,

$$\frac{d}{ds} e^{-\int_t^s h(r)\, dr} L_v(\xi(s), u_\xi(s), \dot{\xi}(s)) = e^{-\int_t^s h(r)\, dr} L_x(\xi(s), u_\xi(s), \dot{\xi}(s)).$$

This leads to the so called Herglotz equation (for almost all $s \in [0, t]$)

$$\begin{aligned}
&\frac{d}{ds} L_v(\xi(s), u_\xi(s), \dot{\xi}(s)) \\
&= L_x(\xi(s), u_\xi(s), \dot{\xi}(s)) + L_u(\xi(s), u_\xi(s), \dot{\xi}(s)) L_v(\xi(s), u_\xi(s), \dot{\xi}(s)).
\end{aligned} \tag{35}$$

Since L is of class C^2 and $L(x, u, \cdot)$ is strictly convex, then by the standard argument as in [10, Sect. 6.2], we conclude that:

Theorem 1 *Under our standing assumptions, we have the following regularity properties for any minimizer ξ for (10):*

(1) Both ξ and u_ξ are of class C^2 and ξ satisfies Herglotz equation (35) for all $s \in [0, t]$ where u_ξ is the unique solution of (9);
(2) Let $p(s) = L_v(\xi(s), u_\xi(s), \dot{\xi}(s))$ be the dual arc. Then p is also of class C^2 and we conclude that (ξ, p, u_ξ) satisfies Lie equation (8).

Proof We first need to show that ξ is of class C^1. Let N be the set of zero Lebesgue measure where $\dot{\xi}$ does not exist. For $\bar{t} \in [0, t]$, choose a sequence $\{t_k\} \in [0, T] \setminus N$ such that $t_k \to \bar{t}$. Then $\dot{\xi}(t_k) \to \bar{v}$ for some $\bar{v} \in \mathbb{R}^n$ (up to subsequences) and

$$L_v(\xi(\bar{t}), u_\xi(\bar{t}), \dot{\xi}(\bar{t})) = \lim_{k\to\infty} L_v(\xi(t_k), u_\xi(t_k), \dot{\xi}(t_k))$$
$$= \int_0^t \{L_x(\xi(s), u_\xi(s), \dot{\xi}(s)) + L_u(\xi(s), u_\xi(s), \dot{\xi}(s))L_v(\xi(s), u_\xi(s), \dot{\xi}(s))\}ds$$

by (35). From the strict convexity of L it follows that the map $v \mapsto L_v(\xi(s), u_\xi(s), v)$ is a diffeomorphism. This implies that $\bar{v}$ is uniquely determined, i.e.,

$$\lim_{[0,t]\setminus N \ni s \to \bar{t}} \dot{\xi}(s) = \bar{v}.$$

Now, by Lemma 6.2.6 in [10], $\dot{\xi}(\bar{t})$ exists and $\lim_{[0,t]\setminus N\ni s\to\bar{t}} \dot{\xi}(s) = \dot{\xi}(\bar{t})$. It follows ξ is of class C^1. In view of (9), u_ξ is also of class C^1.

In view of (7), by setting

$$F(s) = \int_0^s \{L_x(\xi, u_\xi, \dot{\xi}) + L_u(\xi, u_\xi, \dot{\xi})L_v(\xi, u_\xi, \dot{\xi})\}\, d\tau,$$

we have that

$$\{L_v(\xi(s), u_\xi(s), v) - F(s)\}|_{v=\dot{\xi}(s)} = L_v(\xi(0), u_\xi(0), \dot{\xi}(0)).$$

Then, the implicit function theorem implies $\dot{\xi}$ is of class C^1 since both F and L_v are of class C^1. Therefore we conclude that ξ is of class C^2 and u_ξ is of class C^2 by (9).

The rest part of the proof is standard and we omit it.

4 Concluding Remarks

4.1 Equivalence of Herglotz' Variational Principle and the Implicit Variational Principle

In the recent work [41, 42], the authors introduce an implicit variational principle on closed manifolds which is essentially equivalent to Herglotz' principle.

Proposition 4 ([41]) *Let M be a C^2 closed manifold and let $L : TM \to \mathbb{R}$ be of class C^3 and satisfy conditions (L1)–(L3) for M instead of $\mathbb{R}^n$ here. Given any $x_0 \in M$ and $u_0 \in \mathbb{R}$, there exists a (unique) continuous function $h_{x_0,u_0}(t, x)$ defined on $(0, +\infty) \times M$ satisfying*

$$h_{x_0,u_0}(t, x) = u_0 + \inf_\xi \int_0^t L(\xi(s), h_{x_0,u_0}(s, \xi(s)), \dot{\xi}(s))\, ds,$$

where ξ is taken over all the Lipschitz continuous curves on M connecting $\xi(0) = x_0$ and $\xi(t) = x$.

Moreover, let ξ be any curve achieving the infimum together with the curves p and u defined by

$$u(s) = h_{x_0,u_0}(s, \xi(s)), \quad p(s) = L_v(\xi(s), u(s), \dot{\xi}(s)).$$

Then (ξ, p, u_ξ) is a solution of (8) with conditions $\xi(0) = x_0$, $\xi(t) = x$ and $\lim_{s\to 0^+} u(s) = u_0$.

Proposition 5 *Let $x_0, x \in \mathbb{R}^n$, $t > 0$ and $u_0 \in \mathbb{R}$. For any $\xi \in \Gamma^t_{x_0,x}$ being a minimizer of (10), we denote by $u_\xi(s, u_0)$ the unique solution of (9) with $u_\xi(0, u_0) = u_0$. Then, for any $0 < t' < t$, the restriction of ξ on $[0, t']$ is a minimizer for*

$$A(t', x_0, x, u_0) := u_0 + \inf \int_0^{t'} L(\xi(s), u_\xi(s), \dot{\xi}(s))\, ds$$

with u_ξ the unique solution of (9) restricted on $[0, t']$. Moreover,

$$A(s, x_0, \xi(s), u_0) = u_\xi(s, u_0), \quad \forall s \in [0, t], \tag{36}$$

and $A(s_1 + s_2, x_0, \xi(s_1 + s_2), u_0) = A(s_2, \xi(s_1), \xi(s_1 + s_2), u_\xi(s_1))$ for any $s_1, s_2 > 0$ and $s_1 + s_2 \leqslant t$.

In particular, if h_{x_0,u_0} is from Proposition 4, then

$$u_\xi(s) = h_{x_0,u_0}(s, \xi(s)), \quad s \in [0, t]. \tag{37}$$

Remark 4 The relation (37) holds only when ξ is a minimizer of (10).

Proof Suppose $x_0, x \in \mathbb{R}^n$, $t > 0$ and $u_0 \in \mathbb{R}$. Let $\xi \in \Gamma^t_{x_0,x}$ be a minimizer of (10) and $u_\xi(s) = u_\xi(s, u_0)$ be the unique solution of (9) with $u_\xi(0) = u_0$.

Now, let $0 < t' < t$. Let $\xi_1 \in \Gamma^{t'}_{x,\xi(t')}$ and $\xi_2 \in \Gamma^{t-t'}_{\xi(t'),y}$ be the restriction of ξ on $[0, t']$ and $[t', t]$ respectively. Then, we have that

$$u_\xi(t', u_0) = u_0 + \int_0^{t'} L(\xi_1(s), u_{\xi_1}(s), \dot{\xi}_1(s))\, ds,$$

$$u_\xi(t, u_0) - u_\xi(t', u_0) = \int_{t'}^{t} L(\xi_2(s), u_{\xi_2}(s), \dot{\xi}_2(s))\, ds.$$

Then both ξ_1 and ξ_2 are minimal curve for (10) restricted on $[0, t']$ and $[t', t]$ respectively by summing up the equalities above and the assumption that ξ is a minimizer of (10). In particular, (36) follows. The next assertion is direct from the relation

$$u_\xi(s_1 + s_2, u_0) = u_\xi(s_2, u_\xi(s_1)), \quad \forall s_1, s_2 > 0, \ s_1 + s_2 \leqslant t,$$

since u_ξ solves (9). The last assertion is a direct application of Gronwall's inequality. Indeed, we know that for all $s \in [0, t]$,

$$\begin{aligned} u_\xi(s) &= u_0 + \int_0^s L(\xi(r), u_\xi(r), \dot{\xi}(r))\, dr, \\ h_{x_0,u_0}(s, \xi(s)) &= u_0 + \int_0^s L(\xi(r), h_{x_0,u_0}(r, \xi(r)), \dot{\xi}(r))\, dr. \end{aligned}$$

By condition (L3), it follows that

$$|h_{x_0,u_0}(s, \xi(s)) - u_\xi(s)| \leqslant K \int_0^s |h_{x_0,u_0}(r, \xi(r)) - u_\xi(r)|\, dr.$$

Our conclusion is a consequence of Gronwall's inequality.

4.2 Herglotz' Generalized Variational Principle on Manifolds

Now, we try to explain how to move the Herglotz' generalized variational principle to any closed manifold M.

Fix $x, y \in M, t > 0$ and $u \in \mathbb{R}$. Let $\xi \in \Gamma^t_{x,y}(M)$, we consider the Carathéodory equation

$$\begin{cases} \dot{u}_\xi(s) = L(\xi(s), u_\xi(s), \dot{\xi}(s)), & a.e.\ s \in [0, t], \\ u_\xi(0) = u. \end{cases} \tag{38}$$

We define the action functional

$$J(\xi) := \int_0^t L(\xi(s), u_\xi(s), \dot{\xi}(s))\, ds, \tag{39}$$

where $\xi \in \Gamma^t_{x,y}(M)$ and u_ξ is defined in (9). Our purpose is to minimize $J(\xi)$ over

$$\mathscr{A}(M) = \{\xi \in \Gamma^t_{x,y}(M) : \text{(9) admits an absolutely continuous solution } u_\xi\}.$$

Notice that $\mathscr{A}(M) \neq \varnothing$ because it contains all piecewise C^1 curves connecting x to y. In view of the remark before Lemma 1, for each $a \in \mathbb{R}$,

$$\mathscr{A}(M) = \{\xi \in \Gamma^t_{x,y}(M) : \text{the function } s \mapsto L(\xi(s), a, \dot{\xi}(s)) \text{ belongs to } L^1([0, t])\}.$$

We begin with the case when $M = \mathbb{R}^n$. Fix $\kappa > 0$. Suppose $0 < t \leqslant 1, x, y \in \mathbb{R}^n$ such that $|x - y| \leqslant \kappa t$. Suppose $\eta \in \mathscr{A}(\mathbb{R}^n)$ is a minimizer of the action functional $\eta \mapsto J(\eta)$. Invoking the aforementioned *a priori* estimates, η is as smooth as L. Moreover, there exist constants $C_1(\kappa) > 0$, $C_2(u, t, \kappa) > 0$ such that

$$\sup_{s\in[0,t]}|\dot{\eta}(s)|\leqslant C_1(\kappa),\quad \sup_{s\in[0,t]}|\eta(s)-x|\leqslant C_1(\kappa)t,\quad \sup_{s\in[0,t]}|u_\eta(s)|\leqslant C_2(u,t,\kappa).$$

Let $D_1=B_{\mathbb{R}^n}(x,\kappa t)$ and $D_2=B_{\mathbb{R}^n}(x,(C_1(\kappa)+1)t)$, where the subscript is used for the ball in $\mathbb{R}^n$. Then, $D_1\subset D_2$ since $\kappa\leqslant C_1(\kappa)$. By denoting

$$\mathscr{B}(\mathbb{R}^n)=\{\eta\in\mathscr{A}(\mathbb{R}^n):\eta(s)\in D_2 \text{ for all } s\in[0,t]\}.$$

Therefore we can claim that for any $x\in\mathbb{R}^n$ and $y\in D_1$ the following problems are equivalent:

$$\inf_{\mathscr{A}(\mathbb{R}^n)}J(\xi)=\inf_{\mathscr{B}(\mathbb{R}^n)}J(\xi)$$

They admit the same minimizers.

Now we move to the manifold case. Let $\{(B_i,\Phi_i)\}$ be a local chart for the C^2 closed manifold M. We can suppose that $\{B_i\}_{i=1}^N$ is a finite open cover of M and $\Phi_i:B_i\to D_2\subset\mathbb{R}^n$ is a C^2-diffeomorphism for each $i=1,\dots,N$ and $\Phi_j^{-1}\circ\Phi_i: B_i\cap B_j\to B_i\cap B_j$ is a C^2-diffeomorphism for each $i\neq j=1,\dots,N$.

Fix i, let $B=B_i$ and $\Phi=\Phi_i:B\to D_2$ be a local coordinate. Let $L:TM\times\mathbb{R}\to\mathbb{R}$ be a Lagrangian satisfying (L1)-(L3). Then

$$(\Phi,d\Phi):TB\to D_2\times\mathbb{R}^n$$

defines a local trivialization of TB. Let $L_\Phi: D_2\times\mathbb{R}^n\times\mathbb{R}\to\mathbb{R}$ be defined by

$$L_\Phi(\bar{x},u,\bar{v})=L(\Phi^{-1}(\bar{x}),u,d\Phi^{-1}(\bar{x})\bar{v}),\quad (\bar{x},\bar{v})\in D_2\times\mathbb{R}^n,\ u\in\mathbb{R}.$$

Therefore, Herglotz' generalized variational principle for L restricted to $TB\times\mathbb{R}$ is equivalent to the one for L_Φ on $D_2\times\mathbb{R}^n\times\mathbb{R}\to\mathbb{R}$ since Φ is a bi-Lipschitz homeomorphism and a C^2-diffeomorphism.

Proposition 6 *Fix $\kappa>0$, $0<t\leqslant 1$. Then there exist a local chart $\{(B_i,\Phi_i)\}_{i=1}^N$ and a constant $C_2(\kappa)>0$ such that each $B_i\subset B_M(x,C_2(\kappa)t)$, and for any $x,y\in B_i$ and $u\in\mathbb{R}$, the following points on the Herglotz's generalized variational principle hold:*

(a) The functional

$$\mathscr{A}(B_i)\ni\xi\mapsto J(\xi)=\int_0^t L(\xi(s),u_\xi(s),\dot{\xi}(s))\,ds,$$

where u_ξ is determined by (9) with initial condition $u_\xi(0)=u$, admits a minimizer in $\mathscr{A}(M)$.

(b) Suppose $x,y\in B_i$. Let $\xi\in\mathscr{A}(B_i)$ be a minimizer of J. Then there exists a function $F=F_{B_i}:[0,+\infty)\times[0,+\infty)\to[0,+\infty)$, with $F(\cdot,r)$ being non-decreasing for any $r\geqslant 0$, such that

$$|u_\xi(s)| \leqslant tF(t,\kappa) + C(t)|u|, \quad s \in [0,t]$$

where $C(t) > 0$ is also nondecreasing in t.

(c) *Suppose $x, y \in B_i$. Let $\xi \in \mathscr{A}(B_i)$ be a minimizer of J. Then, there exists a function $F = F_{u,B_i} : [0,+\infty) \times [0,+\infty) \to [0,+\infty)$, with $F(\cdot, r)$ is nondecreasing for any $r \geqslant 0$, such that*

$$\operatorname*{ess\,sup}_{s\in[0,t]} |\dot{\xi}(s)| \leqslant F(t,\kappa).$$

(d) *We have the following regularity properties for any minimizer ξ for (39):*

   (1) *Both ξ and u_ξ are of class C^2 and ξ satisfies Herglotz equation (35) in local charts for all $s \in [0,t]$ where u_ξ is the unique solution of (9);*
   (2) *Let $p(s) = L_v(\xi(s), u_\xi(s), \dot{\xi}(s))$ be the dual arc. Then p is also of class C^2 and we conclude that (ξ, p, u_ξ) satisfies Lie equation (8) in local charts for all $s \in [0,t]$.*

Thus, by using L_Φ, it is not difficult to see that there exists a finer open cover, which we also denote by $\{(B_i, \Phi_i)\}_{i=1}^N$, such that the Herglotz' generalized variational principle can be applied in the case when $x, y \in B_i$ and $0 < t \leqslant 1$ $(i = 1, \ldots, N)$ since $\{\Phi_i\}_{i=1}^N$ is equi-bi-Lipschitz.

Now, let us recall the standard "broken geodesic" argument. Pick any $x, y \in M$, $t > 0$ and $u \in \mathbb{R}$. Let $\{(B_i, \Phi_i)\}_{i=1}^N$ be the local chart in the proposition above. We suppose without loss of generality that $x \in B_1$ and $y \in B_N$. Let $\xi \in \mathscr{A}(M)$. Then there exists a partition $0 = t_0 < t_1 < t_2 < \cdots < t_{k-1} < t_k = t$ such that $z_j = \xi(t_j)$ and $z_{j+1} = \xi(t_{j+1})$ are contained in the same B_i. For each j, we define

$$h_L^j(t_{j+1} - t_j, z_j, z_{j+1}, u_j) = \inf_{\xi_j} \int_{t_j}^{t_{j+1}} L(\xi_j(s), u_{\xi_j}(s), \dot{\xi}_j(s))\, ds,$$

where ξ_j is an absolutely continuous curve constrained in B_i connecting z_j to z_{j+1} and u_{ξ_j} is uniquely determined by (9) with initial condition u_j. Now we consider the problem

$$g(t,x,y,u) := \inf \sum_{j=1}^{k} h_L^j(t_{j+1} - t_j, z_j, z_{j+1}, u_j), \tag{40}$$

where the infimum is taken over any partition $0 = t_0 < t_1 < t_2 < \cdots < t_{k-1} < t_k = t$, $z_j, z_{j+1} \in M$ contained in the same B_i and $u_j \in \mathbb{R}$. Due to Proposition 6 (b), $\{u_j\}$ can be constrained in a compact subset of $\mathbb{R}$ depending on u, x, y and t. Therefore the infimum in (40) can be attained. Thanks to the local semiconcavity of the fundamental solution h_L^j, h_L^j is differentiable at each minimizer which leads to the fact

$$h_L(t,x,y,u) = g(t,x,y,u).$$

Proposition 7 *Proposition 6 holds for any connected and closed C^2 manifold M.*

4.3 Further Remarks

Comparing to the method used in [41, 42], one can see more from our approach as follows:

- We can derive the generalized Euler-Lagrange equations in a modern and rigorous way which does not appear in both [41, 42];
- There should be an extension of the main results of this paper under much more general conditions (like Osgood type conditions) to guarantee the existence and uniqueness of the solutions of the associated Carathéodory equation (9).
- Along this line, the quantitative semiconcavity and convexity estimate of the associated fundamental solutions have been obtained in [7] recently, which is useful for the intrinsic study of the global propagation of singularities of the viscosity solutions of (5) and (6) [4–6, 8];
- When the Lagrangian has the form $L(x, v) - \lambda u$, by solving the associated Carathéodory equation (9) directly, one gets the representation formula for the associated viscosity solutions immediately [13, 19, 40, 43]. The representation formula bridges the PDE aspects of the problem with the dynamical ones;
- Consider a family of Lagrangians in the form $\{\mathbf{L}(x, v) + \sum_{i=1}^{k} a_{ij}u_i\}$, a problem of Herglotz' variational principle in the vector form is closely connected to certain stochastic model of weakly coupled Hamilton-Jacobi equations (see, for instance, [20, 23, 38]).

Acknowledgements This work is partly supported by Natural Scientific Foundation of China (Grant No.11871267, No. 11631006, No. 11790272 and No. 11771283), and the National Group for Mathematical Analysis, Probability and Applications (GNAMPA) of the Italian Istituto Nazionale di Alta Matematica "Francesco Severi". Kaizhi Wang is also supported by China Scholarship Council (Grant No. 201706235019). The authors acknowledge the MIUR Excellence Department Project awarded to the Department of Mathematics, University of Rome Tor Vergata, CUP E83C18000100006. The authors are grateful to Qinbo Chen, Cui Chen and Kai Zhao for helpful discussions. This work was motivated when the first two authors visited Fudan University in June 2017. The authors also appreciate the anonymous referee for helpful suggestion to improve the paper.

Appendix

Let $\Omega \subset \mathbb{R}^{n+1}$ be an open set. A function $f : \Omega \subset \mathbb{R} \times \mathbb{R}^n \to \mathbb{R}^n$ is said to satisfy *Carathéodory condition* if

- for any $x \in \mathbb{R}^n$, $f(\cdot, x)$ is measurable;
- for any $t \in \mathbb{R}$, $f(t, \cdot)$ is continuous;
- for each compact set U of Ω, there is an integrable function $m_U(t)$ such that

$$|f(t, x)| \leqslant m_U(t), \quad (t, x) \in U.$$

A classical problem is to find an absolutely continuous function x defined on a real interval I such that $(t, x(t)) \in \Omega$ for $t \in I$ and satisfies the following Carathéodory equation

$$\dot{x}(t) = f(t, x(t)), \quad a.e., t \in I. \tag{41}$$

Proposition 8 (Carathéodory) *If Ω is an open set in $\mathbb{R}^{n+1}$ and f satisfies the Carathéodory conditions on Ω, then, for any (t_0, x_0) in Ω, there is a solution of (41) through (t_0, x_0). Moreover, if the function $f(t, x)$ is also locally Lipschitzian in x with a measurable Lipschitz function, then the uniqueness property of the solution remains valid.*

For the proof of Proposition 8 and more results related to Carathéodory equation (41), the readers can refer to [17, 24].

References

1. Arnol'd, V.I.: Mathematical Methods of Classical Mechanics. Graduate Texts in Mathematics, vol. 60, 2nd edn. Springer, New York (1989)
2. Buttazzo, G.: Semicontinuity, Relaxation and Integral Representation in the Calculus of Variations. Pitman Research Notes in Mathematics Series, vol. 207. Longman (1989)
3. Buttazzo, G., Giaquinta, M., Hildebrandt, S.: One-Dimensional Variational Problems. An Introduction. Oxford Lecture Series in Mathematics and its Applications, vol. 15. The Clarendon Press, Oxford University Press (1998)
4. Cannarsa, P., Cheng, W.: Generalized characteristics and Lax-Oleinik operators: global theory. Calc. Var. Partial Differ. Equ. **56**, 125 (2017)
5. Cannarsa, P., Cheng, W., Fathi, A.: On the topology of the set of singularities of a solution to the Hamilton-Jacobi equation. C. R. Math. Acad. Sci. Paris **355**, 176–180 (2017)
6. Cannarsa, P., Cheng, W., Mazzola, M., Wang, K.: Global generalized characteristics for the Dirichlet problem for Hamilton-Jacobi equations at a supercritical energy level, preprint (2018). arXiv:1803.01591
7. Cannarsa, P., Cheng, W., Yan, J.: Regularity properties of the fundamental solutions of Hamilton-Jacobi equations of contact type, preprint (2017)
8. Cannarsa, P., Cheng, W., Zhang, Q.: Propagation of singularities for weak KAM solutions and barrier functions. Commun. Math. Phys. **331**, 1–20 (2014)
9. Cannarsa, P., Quincampoix, M.: Vanishing discount limit and nonexpansive optimal control and differential games. SIAM J. Control Optim. **53**, 1789–1814 (2015)
10. Cannarsa, P., Sinestrari, C.: Semiconcave functions, Hamilton-Jacobi equations, and optimal control. In: Progress in Nonlinear Differential Equations and Their Applications, vol. 58. Birkhäuser Boston, Inc., Boston, MA (2004)
11. Carathéodory, C.: Calculus of Variations and Partial Differential Equations of the First Order, 2nd edn. Chelsea Publishing, New York (1989)
12. Chen, C., Cheng, W.: Lasry-Lions, Lax-Oleinik and generalized characteristics. Sci. China Math. **59**, 1737–1752 (2016)
13. Chen, C., Cheng, W., Zhang, Q.: Lasry-Lions approximations for discounted Hamilton-Jacobi equations. J. Differ. Equ. **265**, 719–732 (2018)
14. Chen, Q., Cheng, W., Ishii, H., Zhao, K.: Vanishing contact structure problem and convergence of the viscosity solutions, preprint (2018). arXiv:1808.06046
15. Clarke, F.: A Lipschitz regularity theorem. Ergod. Theory Dyn. Syst. **27**, 1713–1718 (2007)

16. Clarke, F.: Functional Analysis, Calculus of Variations and Optimal Control. Graduate Texts in Mathematics, vol. 264. Springer (2013)
17. Coddington, E.A., Levinson, N.: Theory of Ordinary Differential Equations. McGraw-Hill (1955)
18. Davini, A., Fathi, A., Iturriaga, R., Zavidovique, M.: Convergence of the solutions of the discounted equation: the discrete case. Math. Z. **284**, 1021–1034 (2016)
19. Davini, A., Fathi, A., Iturriaga, R., Zavidovique, M.: Convergence of the solutions of the discounted Hamilton-Jacobi equation. Invent. Math. **206**, 29–55 (2016)
20. Davini, A., Zavidovique, M.: Aubry sets for weakly coupled systems of Hamilton-Jacobi equations. SIAM J. Math. Anal. **46**, 3361–3389 (2014)
21. Eisenhart, L.P.: Continuous Groups of Transformations. Dover Publications, New York (1961)
22. Fathi, A.: Weak KAM Theorem in Lagragian Dynamics. To be published by Cambridge University Press
23. Figalli, A., Gomes, D.; Marcon, D.: Weak KAM theory for a weakly coupled system of Hamilton-Jacobi equations. Calc. Var. Partial Differ. Equ. **55**, Art. 79 (2016)
24. Filippov, A.F.: Differential Equations with Discontinuous Righthand Sides. Mathematics and its Applications (Soviet Series), vol. 18. Kluwer Academic Publishers Group (1988)
25. Furuta, K., Sano, A., Atherton, D.: State Variable Methods in Automatic Control. Wiley, New York (1988)
26. Georgieva, B., Guenther, R.: First Noether-type theorem for the generalized variational principle of Herglotz. Topol. Methods Nonlinear Anal. **20**, 261–273 (2002)
27. Georgieva, B., Guenther, R.: Second Noether-type theorem for the generalized variational principle of Herglotz. Topol. Methods Nonlinear Anal. **26**, 307–314 (2005)
28. Giaquinta, M., Hildebrandt, S.: Calculus of Variations. II. The Hamiltonian Formalism. Grundlehren der Mathematischen Wissenschaften, vol. 311. Springer (1996)
29. Gomes, D.A.: Generalized Mather problem and selection principles for viscosity solutions and Mather measures. Adv. Calc. Var. **1**, 291–307 (2008)
30. Guenther, R., Gottsch, A., Guenther, C.: The Herglotz Lectures on Contact Transformations and Hamiltonian Systems. Schauder Center For Nonlinear Studies, Copernicus University (1995)
31. Herglotz, G.: Berührungstransformationen, Lectures at the University of Göttingen, Göttingen (1930)
32. Herglotz, G.: In: Schwerdtfeger H. (ed.) Gesammelte Schriften. Vandenhoeck & Ruprecht, Göttingen (1979)
33. Hiriart-Urruty, J.-B., Lemaréchal, C.: Fundamentals of Convex Analysis. Grundlehren Text Editions. Springer, Berlin (2001)
34. Ishii, H., Mitake, H., Tran, H.V.: The vanishing discount problem and viscosity Mather measures. Part 1: the problem on a torus. J. Math. Pures Appl. **108**(9), 125–149 (2017)
35. Ishii, H., Mitake, H., Tran, H.V.: The vanishing discount problem and viscosity Mather measures. Part 2: boundary value problems. J. Math. Pures Appl. **108**(9), 261–305 (2017)
36. Iturriaga, R., Sánchez-Morgado, H.: Limit of the infinite horizon discounted Hamilton-Jacobi equation. Discrete Contin. Dyn. Syst. Ser. B **15**, 623–635 (2011)
37. Marò, S., Sorrentino, A.: Aubry-Mather theory for conformally symplectic systems. Commun. Math. Phys. **354**, 775–808 (2017)
38. Mitake, H., Siconolfi, A., Tran, H.V., Yamada, N.: A Lagrangian approach to weakly coupled Hamilton-Jacobi systems. SIAM J. Math. Anal. **48**, 821–846 (2016)
39. Mrugała, R.: Contact transformations and brackets in classical thermodynamics. Acta Phys. Pol. A **5**, 19–29 (1980)
40. Su, X., Wang, L., Yan, J.: Weak KAM theory for Hamilton-Jacobi equations depending on unknown functions. Discrete Contin. Dyn. Syst. **36**, 6487–6522 (2016)
41. Wang, K., Wang, L., Yan, J.: Implicit variational principle for contact Hamiltonian systems. Nonlinearity **30**, 492–515 (2017)
42. Wang, K., Wang, L., Yan, J.: Variational principle for contact Hamiltonian systems and its applications. J. Math. Pures Appl. **123**(9), 167–200 (2019)
43. Zhao K., Cheng, W.: On the vanishing contact structure for viscosity solutions of contact type Hamilton-Jacobi equations I: Cauchy problem, to appear in Discrete Contin. Dyn. Syst

Observability Inequalities for Transport Equations through Carleman Estimates

Piermarco Cannarsa, Giuseppe Floridia and Masahiro Yamamoto

Abstract We consider the transport equation $\partial_t u(x,t) + H(t) \cdot \nabla u(x,t) = 0$ in $\Omega \times (0,T)$, where $T > 0$ and $\Omega \subset \mathbb{R}^d$ is a bounded domain with smooth boundary $\partial\Omega$. First, we prove a Carleman estimate for solutions of finite energy with piecewise continuous weight functions. Then, under a further condition which guarantees that the orbits of H intersect $\partial\Omega$, we prove an energy estimate which in turn yields an observability inequality. Our results are motivated by applications to inverse problems.

Keywords Carleman estimates · Transport equation · Observability inequality

1 Introduction

Let $d \in \mathbb{N}$ and $\Omega \subset \mathbb{R}^d$ be a bounded domain with smooth boundary $\partial\Omega$, $\nu = \nu(x)$ be the unit outward normal vector at x to $\partial\Omega$, and let $x \cdot y$ and $|x|$ denote the scalar product of $x, y \in \mathbb{R}^d$ and the norm of $x \in \mathbb{R}^d$, respectively. We set $Q := \Omega \times (0,T)$, and we consider

$$Pu(x,t) := \partial_t u + H(t) \cdot \nabla u = 0 \quad \text{in } Q, \tag{1}$$

P. Cannarsa
Department of Mathematics, University of Rome "Tor Vergata", 00133 Rome, Italy
e-mail: cannarsa@mat.uniroma2.it

G. Floridia (✉)
Department of Mathematics and Applications "R. Caccioppoli",
University of Naples Federico II, 80126 Naples, Italy
e-mail: giuseppe.floridia@unina.it

M. Yamamoto
Department of Mathematical Sciences, The University of Tokyo,
Komaba, Meguro, Tokyo 153, Japan
e-mail: myama@ms.u-tokyo.ac.jp

Honorary Member of Academy of Romanian Scientists,Splaiul Independentei Street, no 54, 050094 Bucharest, Romania

F. Alabau-Boussouira et al. (eds.), *Trends in Control Theory and Partial Differential Equations*, Springer INdAM Series 32,
https://doi.org/10.1007/978-3-030-17949-6_4

where $H(t) := (H_1(t), \ldots, H_d(t)) : [0, T] \to \mathbb{R}^d$, $H \in C^1([0, T]; \mathbb{R}^d)$.

Equation (1) is called a transport equation and $H(t)$ describes the velocity of the flow, which is here assumed to be independent of the spatial variable x.

Problem Formulation

We assume

$$H_0 := \min_{t \in [0,T]} |H(t)| > 0, \tag{2}$$

and, without loss of generality, we suppose that $\mathbf{0} = (0, \ldots, 0) \in \overline{\Omega}$.

Let us recall the following definition.

Definition 1.1 A partition $\{t_j\}_0^m$ of $[0, T]$ is a strictly increasing finite sequence $t_0, t_1, \ldots, t_m$ (for some $m \in \mathbb{N}$) of real numbers starting from the initial point $t_0 = 0$ and arriving at the final point $t_m = T$.

Hereafter, we will call $\{t_j\}_0^m$ a *uniform partition* of $[0, T]$ when the length of the intervals $[t_j, t_{j+1}]$ is constant for $j = 0, \ldots, m-1$, that is, $t_j = \frac{T}{m} j$, $j = 0, \ldots, m$.

Lemma 1.2 below ensures that any vector-valued function $H(t)$, satisfying (2), admits a partition $\{t_j\}_0^m$ of $[0, T]$ such that the angles of oscillations of the vector $H(t)$ are less than $\frac{\pi}{2}$ in any time interval $[t_j, t_{j+1}]$, $j = 0, \ldots, m-1$ (see Fig. 1).

Given a partition $\{t_j\}_0^m$ of $[0, T]$, let us set

$$\eta_j := \frac{H(t_j)}{|H(t_j)|}, \quad j = 0, \ldots, m-1. \tag{3}$$

Lemma 1.2 *Let $S_* \in (1/\sqrt{2}, 1)$. For any given $H \in Lip([0, T]; \mathbb{R}^d)$, satisfying condition (2), there exist $m \in \mathbb{N}$ and a partition $\{t_j\}_0^m$ of $[0, T]$ such that*

$$\frac{H(t)}{|H(t)|} \cdot \eta_j \geq S_*, \quad \forall t \in [t_j, t_{j+1}], \quad \forall j = 0, \ldots, m-1, \tag{4}$$

where η_j are defined in (3).

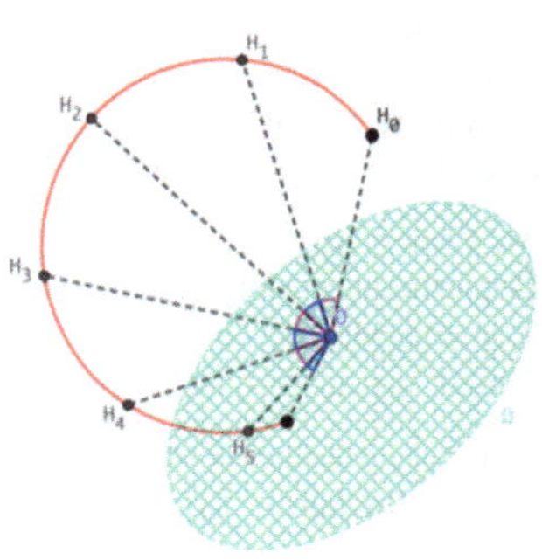

Fig. 1 In this picture $S_* = \cos\frac{\pi}{6}$, $m = 6$ and $H_j := H(t_j)$, $j = 0, \ldots, 5$.

Lemma 1.2 is proved in the Appendix.

Remark 1.3 Condition (4) means that there exist m cones in $\mathbb{R}^d$ such that the axis of every cone, that is, the straight line passing through the apex about which the whole cone has a circular symmetry, is the line between $0 = (0, \ldots, 0)$ and η_j, $j = 0, \ldots, m-1$. Moreover, a straight line passing through the apex is contained in the cone if the angle between this line and the axis of the cone is less than $\pi/4$. Indeed, the inequality (4), that is $H(t) \cdot \eta_j > \cos\vartheta^*|H(t)|$ for some $\vartheta^* \in (0, \frac{\pi}{4})$, is equivalent to the fact that the angle between $H(t)$ and η_j is less than $\pi/4$. Thus, $H(t)$ is contained in the same cone $\forall t \in [t_j, t_{j+1}]$. Let us note that it can occur that $\eta_i = \eta_j$, for $i \neq j$.

Let $\delta_\Omega = diam(\Omega) = \sup_{x,y\in\overline{\Omega}} |x-y|$. Let us fix $S_* \in (1/\sqrt{2}, 1)$, $r > 0$ and define

$$x_j := -R_j\eta_j, \quad j = 0, \ldots, m-1, \tag{5}$$

where η_j is defined in (3) and

$$\begin{cases} R_j = 2^j R_0 + (2^j - 1)(\delta_\Omega + r), \\ R_0 = \frac{1+S_*}{1-S_*}\delta_\Omega. \end{cases} \tag{6}$$

We note that from (6) it follows that

$$x_j \notin \overline{\Omega}, \qquad j = 0, \ldots, m-1.$$

For every $j = 0, \ldots, m-1$, let us define

$$M_\Omega(x_j) := \max_{x\in\overline{\Omega}} |x - x_j| \qquad \text{and} \qquad d_\Omega(x_j) := \min_{x\in\overline{\Omega}} |x - x_j|. \tag{7}$$

Remark 1.4 The choice of the R_j's in (6) (see Lemma 2.2 below and Fig. 2) guarantees that the points x_j's are located sufficiently far away from Ω and at increasing distances from the origin.

By the choice of the finite sequence $R_j = |x_j|$ in (6) (R_j sufficiently large compared with δ_Ω) we deduce in Lemma 2.1 below that

$$(x + R_j\eta_j) \cdot \eta_j \geq S_*|x + R_j\eta_j|, \quad \forall x \in \overline{\Omega}.$$

In other words, the apex angle of the minimum cone with the apex x_j which includes Ω is less than $2 \arccos S_* (<\pi/2)$ (see Fig. 3).

We now introduce the weight function $\varphi(x, t)$ to be used in our Carleman estimate, as follows. First, we define φ on $\overline{\Omega} \times [0, T)$ setting, for every $x \in \overline{\Omega}$,

$$\varphi(x,t) = \varphi_j(x,t) := -\beta(t - t_j) + |x - x_j|^2, \quad t \in [t_j, t_{j+1}), \ j = 0, \ldots, m-1, \tag{8}$$

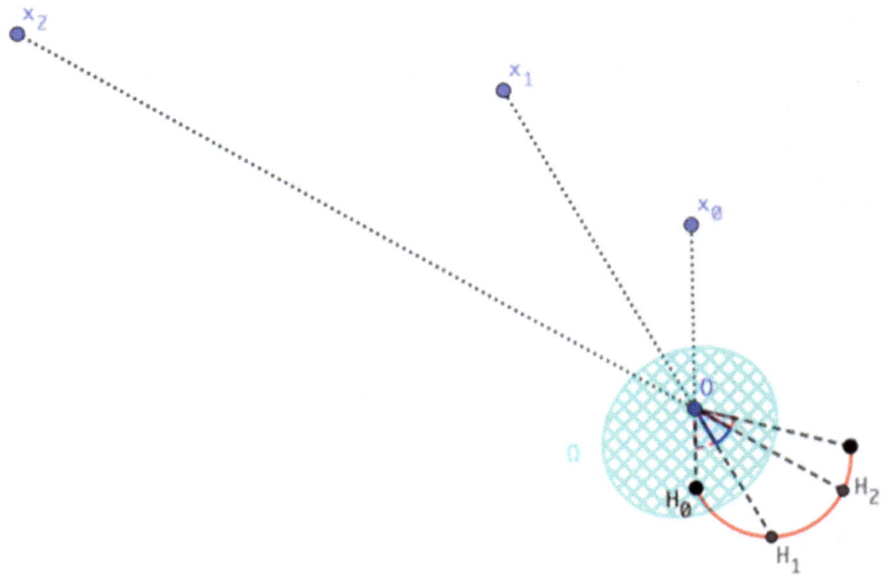

Fig. 2 In this picture $S_* = \cos\frac{\pi}{6}$, $m = 3$ and $H_j := H(t_j)$, $j = 0, 1, 2$.

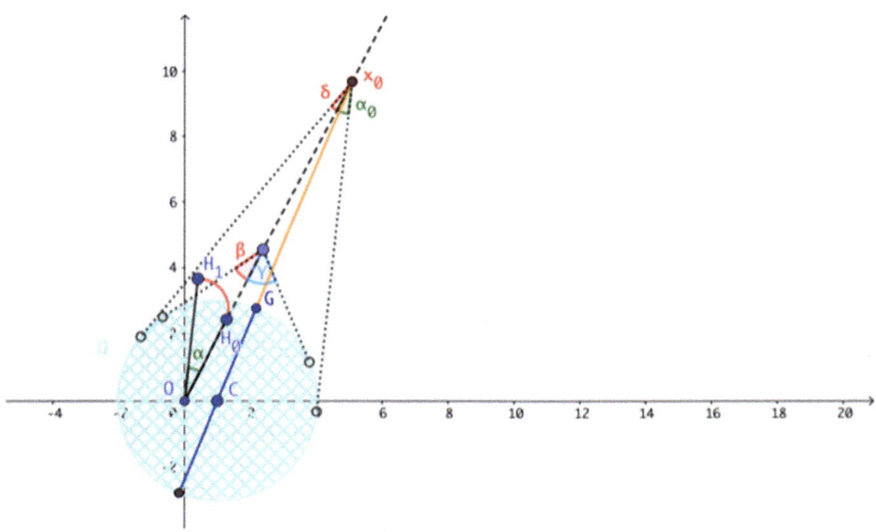

Fig. 3 In this picture: $\Omega := \{(x, y) \in R^2 : |(x, y) - (1, 0)| < 3\}$, $C = (1, 0)$, $S_* = \cos\alpha \in (\frac{1}{\sqrt{2}}, 1)$, $m = 1$, $H_j := H(t_j)$, $j = 0, 1$, and $\beta, \gamma > \alpha$, $\alpha_0 = \alpha$, $\delta \le \alpha$. We note that $d_\Omega(x_0) = \text{dist}\,(x_0, G)$ and $M_\Omega(x_0) = d_\Omega(x_0) + 6$.

where

$$\beta := (2S_*^2 - 1)H_0 d_\Omega(x_0), \tag{9}$$

with H_0 and $d_\Omega(x_0)$ defined by (2) and (7), respectively. Then we extend φ to $\overline{\Omega} \times [0, T]$ by continuity. Observe that φ is piecewise smooth in t and smooth in x.

Main Results

In this article, under condition (2), we establish an observability inequality for (1) which estimates the L^2-norm of $u(x, 0)$ by lateral boundary data $u|_{\partial\Omega\times(0,T)}$ under some conditions on $H(t)$ (see Theorem 1.6). This observability inequality is a consequence of the following Carleman estimate.

Theorem 1.5 *Let $u \in H^1(Q)$ be a solution of Eq. (1), where $H \in C^1([0, T]; \mathbb{R}^d)$ satisfies (2). Let $\{t_j\}_0^m$ be a partition of $[0, T]$ satisfying (4). Then, there exist constants $s_0, C_0, C > 0$ such that for all $s > s_0$ we have*

$$s^2 \int_Q |u|^2 e^{2s\varphi} dxdt + se^{-C_0 s} \sum_{j=0}^{m-1} \int_\Omega |u(x, t_j)|^2 dx$$
$$\leq C \int_Q |Pu|^2 e^{2s\varphi} dxdt + Cse^{Cs} \int_\Sigma |u|^2 d\gamma dt + Cse^{Cs} \int_\Omega |u(x, T)|^2 \, dx,$$

where $\varphi(x, t) : Q \longrightarrow \mathbb{R}$ is the weight function defined in (8), and

$$\Sigma = \{(x, t) \in \partial\Omega \times (0, T) : \ H(t) \cdot \nu(x) \geq 0\} \tag{10}$$

is the subboundary of all exit points for H.

We now give the observability inequality for Eq. (1).

Theorem 1.6 *Let $g \in L^2(\partial\Omega \times (0, T))$ and let us consider the following problem*

$$\begin{cases} \partial_t u + H(t) \cdot \nabla u = 0 \quad in \ Q := \Omega \times (0, T), \\ u|_{\partial\Omega\times(0,T)} = g. \end{cases} \tag{11}$$

Let us suppose that there exists a partition $\{t_j\}_0^m$ of $[0, T]$ associated to $H(t)$ satisfying (4) such that the following condition holds

$$\max_{0\leq j\leq m-1} \frac{(t_{j+1} - t_j) d_\Omega(x_j)}{M_\Omega^2(x_j)} > \frac{1}{H_0(2S_*^2 - 1)}, \tag{12}$$

where $M_\Omega(x_j)$, $d_\Omega(x_j)$ and H_0 are defined in (7) and (2), respectively. Then, there exists a constant $C > 0$ such that the following inequality holds

$$\|u(\cdot, t)\|_{L^2(\Omega)} \leq C \|g\|_{L^2(\partial\Omega\times(0,T))}, \quad 0 \leq t \leq T,$$

for any $u \in H^1(Q)$ satisfying (11).

Assumption (12) is meant to guarantee that the orbit $\{H(t) \in \mathbb{R}^d : t \in [0, T]\}$ intersects $\partial\Omega$. In the following example, we show that this or a similar condition is indeed necessary: observability fails without some extra assumption.

In the following, for $\eta > 0$ we consider $\Omega_\eta := \{z \in \mathbb{R}^2 : |z| < \eta\}$.

Example 1 Let $\sigma > 0$ and $\rho \in (0, 2\sigma/3)$. Let $\Omega := \Omega_\rho$ and let $f \in C^1(\overline{\Omega_\sigma}; \mathbb{R})$ satisfy $\text{supp}(f) \subset \Omega_{\rho/2} \subseteq \overline{\Omega}_\sigma$ and let $\alpha(t) = (\rho\cos t, \rho\sin t)$, $t \in [0, 2\pi]$. We set

$$v(x, y, t) = f(x - \rho\cos t, y - \rho\sin t).$$

Thus, v satisfies (1), where $H(t) = \alpha'(t), 0 \le t \le T$, and v vanishes at the boundary of Ω_σ. So,

$$\begin{cases} \partial_t v + \alpha'(t) \cdot \nabla v = 0 & \text{in } \Omega_\sigma \times (0, T), \\ v|_{\partial\Omega_\sigma \times (0,T)} = g, \end{cases} \tag{13}$$

with $g \equiv 0$. We note that $|\alpha'(t)| = \rho > 0$ and, for $t \in [0, T]$, the support of $v(\cdot, \cdot, t)$ is

$$\text{supp}(v(\cdot, \cdot, t)) = \left\{(x, y) \in \mathbb{R}^2 : |(x - \rho\cos t, y - \rho\sin t)| < \frac{\rho}{2}\right\}. \tag{14}$$

Then, from (13) and (14) it follows that observability fails. □

We conclude this introduction with some comments on our main results.

1. One could establish an estimate similar to the one in Theorem 1.6 with the maximum norm by the method of characteristics. Our proof is based on Carleman estimates, which naturally provide L^2-estimates for solutions over $\Omega \times \{t\}$. The method of characteristics does not yield such global L^2-estimates directly. L^2-estimates, not estimates in the maximum norm, are related to exact controllability and are more flexibly applied to other problems such as inverse problems, although we discuss no such aspects in this paper.
2. Although, due to the simplicity of Eq. (1), the method of characteristics can be easily applied to explain the validity of observability results, the one point we would like to stress is the fact that, in this paper, we intend to derive a Carleman estimate under minimal assumptions. Essentially, we want to give an explicit construction of the weight function that only depends on the lower bound (2) and the modulus of continuity of H.
3. It is worth noting that Theorem 1.6 aims at the determination of the solution u on the whole cylinder $\Omega \times [0, T]$, not only of $u(\cdot, 0)$ in Ω. For this reason, in Theorem 1.6, we have to measure data on the whole lateral boundary $\partial\Omega \times (0, T)$, not just on a subboundary as we did for the Carleman estimate in Theorem 1.5—where, however, the norm of $u(\cdot, T)$ in Ω is included. The fact that measurements on the whole boundary are necessary to majorize u on $\Omega \times [0, T]$ can be easily understood by looking at the representation solutions given by characteristics.
4. Another purpose of this paper is to single out an assumption which suffices to derive observability from a Carleman estimate. We do so with condition (12), which has a clear geometric meaning: one requires $H(t)$ not to oscillate too much

for enough time, giving an explicit evaluation of such a time. We do not pretend our method to provide the optimal evaluation of the observability time. On the other hand, Example 1 shows that some assumption is needed for observability: (12) is an example of a sufficient quantitative condition for the observability of solutions on $\Omega \times [0, T]$.

Main References and Outline of the Paper

Carleman estimates for transport equations are proved in Gaitan and Ouzzane [5], Gölgeleyen and Yamamoto [6], Cannarsa et al. [4], Klibanov and Pamyatnykh [7], Machida and Yamamoto [8] to be applied to inverse problems of determining spatially varying coefficients, where coefficients of the first-order terms in x are assumed not to depend on t. In order to improve results for inverse problems by the application of Carleman estimates, we need a better choice of the weight function in the Carleman estimate. The works [5] and [7] use one weight function which is very conventional for a second-order hyperbolic equation but seems less useful to derive analogous results for a time-dependent function $H(t)$. Our choice is more similar to the one in [8] and [6], but even these papers allow no time dependence for H. Although it is very difficult to choose the best possible weight function for the partial differential equation under consideration, our choice (8) of the weight function seems more adapted for the nature of the transport Eq. (1).

In [4] we consider the transport equation $\partial_t u(x,t) + (H(x) \cdot \nabla u(x,t)) + p(x) u(x,t) = 0$ in $\Omega \times (0,T)$ ($\Omega \subset \mathbb{R}^n$ bounded domain), and discuss two inverse problems which consist of determining a vector-valued function $H(x)$ or a real-valued function $p(x)$ by initial values and data on a subboundary of Ω. In particular in [4] we obtain conditional stability of Hölder type in a subdomain D provided that the outward normal component of $H(x)$ is positive on $\partial D \cap \partial\Omega$. The proofs are based also on a Carleman estimate where the weight function depends on H.

As it is commented above, the method of characteristics is applicable to inverse problems for first-order hyperbolic systems as well as transport equations and we refer for example to Belinskij [2] and Chap. 5 in Romanov [9], which discuss an inverse problem of determining an $N \times N$-matrix $C(x)$ in

$$\partial_t U(x,t) + \Lambda \partial_x U(x,t) + C(x)U(x,t) = F(x,t), \quad 0 < x < \ell,\ t > 0$$

with a suitably given matrix Λ and vector-valued function F. The works [2] and [9] apply the method of characteristics to prove the uniqueness and the existence of $C(x)$ realizing extra data of U provided that $\ell > 0$ is sufficiently small.

The method by Carleman estimates for establishing both energy estimates like Theorem 1.6 and inverse problems of determining spatial varying functions is well-known for hyperbolic and parabolic equations and we refer to Beilina and Klibanov [1], Bellassoued and Yamamoto [3], Yamamoto [10].

The plan of the paper is the following. In Sect. 2, we prove the Carleman estimate (Theorem 1.5). In Sect. 3, we obtain the observability inequality (Theorem 1.6). Finally, in Appendix we put the proof of Lemma 1.2.

2 Proof of the Carleman Estimate

Let $S_* \in \left(\frac{1}{\sqrt{2}}, 1\right)$ and $\{t_j\}_0^m$ a partition of $[0, T]$ associated to $H(t)$ such that (4) is satisfied.

2.1 Some Preliminary Lemmas

Lemma 2.1 *Given R_j, $j = 0, \ldots, m-1$, as in (6), then*

$$(x + R_j\eta_j)\cdot\eta_j \ge S_*|x + R_j\eta_j|, \quad \forall x \in \overline{\Omega}, \tag{15}$$

where η_j are defined in (3).

Proof For every $x \in \overline{\Omega}$, we have $|x| = |x - \mathbf{0}| \le \delta_\Omega$ since $\mathbf{0} \in \overline{\Omega}$, and

$$S_*|x + R_j\eta_j| \le S_*\left(|x| + R_j|\eta_j|\right) = S_*\left(|x| + R_j\right) \le S_*\left(\delta_\Omega + R_j\right), \tag{16}$$

and, since $-x\cdot\eta_j \le |x\cdot\eta_j| \le |x||\eta_j| = |x| \le \delta_\Omega$,

$$(x + R_j\eta_j)\cdot\eta_j = x\cdot\eta_j + R_j\eta_j\cdot\eta_j = x\cdot\eta_j + R_j \ge R_j - |x| \ge R_j - \delta_\Omega. \tag{17}$$

From (16) and (17) it follows that a sufficient condition for the inequality (15) is the following

$$R_j - \delta_\Omega \ge S_*(\delta_\Omega + R_j),$$

that is, $R_j \ge \frac{1+S_*}{1-S_*}\delta_\Omega$. For every $j = 1, \ldots, m-1$, the last condition is verified by R_j defined as in (6). □

By the definition (6) of the sequence $\{R_j\}$ the following Lemma 2.2 follows.

Lemma 2.2 *Let $x_j = -R_j\eta_j$, $j = 0, \ldots, m-1$, with R_j defined as in (6). Then*

$$M_\Omega(x_j) = \max_{x\in\overline{\Omega}}|x - x_j| < \min_{x\in\overline{\Omega}}|x - x_{j+1}| = d_\Omega(x_{j+1}), \qquad j = 0, \ldots, m-2. \tag{18}$$

By Lemma 2.2 (see also Fig. 2) we deduce

$$\max_{j=0,\ldots,m-1} M_\Omega(x_j) = M_\Omega(x_{m-1}) \qquad \text{and} \qquad \min_{j=0,\ldots,m-1} d_\Omega(x_j) = d_\Omega(x_0). \tag{19}$$

Lemma 2.3 *Let $x_j = -R_j\eta_j$, $j = 0, \ldots, m-1$, with R_j defined as in (6). Then,*

$$H(t)\cdot(x - x_j) \ge C_*\, H_0\, d_\Omega(x_0), \quad t_j \le t \le t_{j+1},\ j = 0, \ldots, m-1, \quad x \in \overline{\Omega},$$

where $C_* = 2S_*^2 - 1 > 0$ and $H_0 = \min_{t\in[0,T]}|H(t)| > 0$.

Proof Let $\vartheta^* \in (0, \pi/4)$ satisfy $\cos\vartheta^* = S_*$. For $t \in [t_j, t_{j+1}]$, $j = 0, \ldots, m-1$, from (15) and Remark 1.3 we deduce that

$$H(t) \cdot (x - x_j) \geq \cos 2\vartheta^* \, H_0 d_\Omega(x_j) \geq (2S_*^2 - 1)\, H_0 d_\Omega(x_0), \quad x \in \overline{\Omega}$$

which is our conclusion. □

2.2 Derivation of the Carleman Estimate

After introducing the previous lemmas in Sect. 2.1, we are able to prove Theorem 1.5. In this section, for simplicity of notations, for $j = 0, \ldots, m-1$ let us set

$$M_j := M_\Omega(x_j) \qquad \text{and} \qquad \mu_j := d_\Omega(x_j), \tag{20}$$

see (7) for the definitions of $M_\Omega(x_j)$ and $d_\Omega(x_j)$.

Proof (of Theorem 1.5). We derive a Carleman estimate on

$$Q_j := \Omega \times (t_j, t_{j+1}), \quad 0 \leq j \leq m-1.$$

Let $w_j := e^{s\varphi_j} u$, where φ_j is defined in (8), and

$$L_j w_j := e^{s\varphi_j} P(e^{-s\varphi_j} w_j). \tag{21}$$

By direct calculations, we obtain

$$L_j w_j = \partial_t w_j + H(t) \cdot \nabla w_j - s(P\varphi_j) w_j \quad \text{in} \quad Q_j, \tag{22}$$

where, keeping in mind (8) and the definition of the operator P contained in (1),

$$P\varphi_j(x,t) = \partial_t \varphi_j + H(t) \cdot \nabla \varphi_j = -\beta + 2H(t) \cdot (x - x_j), \quad 0 \leq j \leq m-1.$$

By Lemma 2.3 and (9), since $\beta = (2S_*^2 - 1)H_0\mu_0 \in \big(0, 2(2S_*^2 - 1)H_0\mu_0\big)$ we have

$$P\varphi_j = -\beta + 2H(t) \cdot (x - x_j) \geq C_* H_0 \mu_0, \tag{23}$$

where $C_* = 2S_*^2 - 1$. Therefore, by (23) we obtain

$$\int_{Q_j} |L_j w_j|^2 dxdt \ge -2s \int_{Q_j} (P\varphi_j) w_j (\partial_t w_j + H(t) \cdot \nabla w_j) dxdt$$
$$+ s^2 \int_{Q_j} |2H(t) \cdot (x - x_j) - \beta|^2 |w_j|^2 dxdt$$
$$\ge I_1 + I_2 + C_*^2 H_0^2 \mu_0^2 s^2 \int_{Q_j} |w_j|^2 dxdt, \qquad (24)$$

where

$$I_1 := -2s \int_{Q_j} (P\varphi_j) w_j \partial_t w_j dxdt \quad \text{and} \quad I_2 := -2s \int_{Q_j} (P\varphi_j) H(t) \cdot (w_j \nabla w_j) dxdt.$$

We have

$$I_1 = -2s \int_{Q_j} (P\varphi_j) w_j \partial_t w_j dxdt = -s \int_{t_j}^{t_{j+1}} \int_{\Omega} (P\varphi_j) \partial_t (w_j^2) dxdt$$
$$= s \int_{\Omega} \left[P\varphi_j(x,t) |w_j(x,t)|^2 \right]_{t=t_{j+1}}^{t=t_j} dx + s \int_{Q_j} \partial_t (P\varphi_j(x,t)) |w_j|^2 dxdt. \qquad (25)$$

Recalling (20), we obtain

$$\partial_t (P\varphi_j(x,t)) = 2(x - x_j) \cdot H'(t) \ge -2M_{m-1} \max_{t \in [0,T]} |H'(t)| =: -H_0'.$$

Consequently, from (25) we deduce

$$I_1 \ge s \int_{\Omega} \left[P\varphi_j(x,t) |w_j(x,t)|^2 \right]_{t=t_{j+1}}^{t=t_j} dx - s\, H_0' \int_{Q_j} |w_j|^2 dxdt. \qquad (26)$$

Then, for I_2 we deduce

$$I_2 = -2s \int_{Q_j} (P\varphi_j) H(t) \cdot (w_j \nabla w_j) dxdt = -s \int_{t_j}^{t_{j+1}} \int_{\Omega} P\varphi_j \sum_{k=1}^{d} H_k(t) \partial_k (w_j^2) dxdt$$
$$= s \int_{t_j}^{t_{j+1}} \int_{\Omega} \sum_{k=1}^{d} (\partial_k (P\varphi_j)) H_k(t) |w_j|^2 dxdt - s \int_{t_j}^{t_{j+1}} \int_{\partial\Omega} P\varphi_j (H(t) \cdot \nu(x)) |w_j|^2 d\gamma dt.$$

We note that

$$H(t) \cdot (x - x_j) \le |H(t)| |x - x_j| \le H_* M_*, \qquad (27)$$

where we set (see (19))

$$M_* = M_{m-1} \qquad \text{and} \qquad H_* := \max_{t \in [0,T]} |H(t)| > 0.$$

Therefore, since $P\varphi_j > 0$ by (23) and $\partial_k(P\varphi_j) = 2H_k(t)$, we estimate I_2 in the following way:

$$
\begin{aligned}
I_2 &= 2s\int_{t_j}^{t_{j+1}}\int_\Omega \sum_{k=1}^{d} H_k^2(t)|w_j|^2 dxdt - s\int_{t_j}^{t_{j+1}}\int_{\partial\Omega} P\varphi_j (H(t)\cdot\nu(x))|w_j|^2 d\gamma dt \\
&\geq 2s\int_{t_j}^{t_{j+1}}\int_\Omega |H(t)|^2|w_j|^2 dxdt \\
&- s\int_{\Sigma_j}(-\beta + 2H(t)\cdot(x-x_j))(H(t)\cdot\nu(x))|w_j|^2 d\gamma dt \\
&\geq 2\,H_0^2 s\int_{t_j}^{t_{j+1}}\int_\Omega |w_j|^2 dxdt - 2s\int_{\Sigma_j}(H(t)\cdot(x-x_j))(H(t)\cdot\nu(x))|w_j|^2 d\gamma dt \\
&\geq 2\,H_0^2 s\int_{Q_j}|w_j|^2 dxdt - 2H_* M_* s\int_{\Sigma_j}|H(t)||\nu(x)||w_j|^2 d\gamma dt \\
&\geq 2\,H_0^2 s\int_{Q_j}|w_j|^2 dxdt - 2H_*^2 M_* s\int_{\Sigma_j}|w_j|^2 d\gamma dt, \qquad (28)
\end{aligned}
$$

where

$$
\Sigma_j = \{(x,t)\in\partial\Omega\times(t_j,t_{j+1}) : H(t)\cdot\nu(x)\geq 0\}.
$$

Hence, by (24), (26) and (28), we obtain

$$
\begin{aligned}
\int_{Q_j}|L_j w_j|^2 dxdt &\geq s\int_\Omega \left[P\varphi_j(x,t)|w_j(x,t)|^2\right]_{t=t_{j+1}}^{t=t_j} dx \\
&- H_0' s\int_{Q_j}|w_j|^2 dxdt + C_1 s^2\int_{Q_j}|w_j|^2 dxdt \\
&- 2H_*^2 M_* s\int_{\Sigma_j}|w_j|^2 d\gamma dt\,,
\end{aligned}
$$

for some positive constant C_1. Since $w_j := e^{s\varphi_j}u$, from the previous inequality, for $j = 0,\ldots,m-1$, by (21) we deduce that there exists also a positive constant C_2 such that

$$
\begin{aligned}
\int_{t_j}^{t_{j+1}}\int_\Omega |Pu|^2 e^{2s\varphi_j} dxdt &\geq s\int_\Omega \psi_j(x)dx + (C_1 s^2 - H_0' s)\int_{Q_j} e^{2s\varphi_j}|u|^2 dxdt \\
&- C_2 s e^{C_2 s}\int_{\Sigma_j}|u|^2 d\gamma dt, \qquad (29)
\end{aligned}
$$

where $C_1,\ C_2$ are positive constants and

$$
\psi_j(x) := \left[P\varphi_j(x,t)e^{2s\varphi_j(x,t)}\,|u(x,t)|^2\right]_{t=t_{j+1}}^{t=t_j}.
$$

By (8) and (23) we obtain

$$\begin{aligned}\psi_j(x) &= \Big[\big(2H(t)\cdot(x-x_j)-\beta\big)\, e^{2s(-\beta(t-t_j)+|x-x_j|^2)}|u(x,t)|^2\Big]_{t=t_{j+1}}^{t=t_j}\\ &= (2H(t_j)\cdot(x-x_j)-\beta)e^{2s|x-x_j|^2}|u(x,t_j)|^2\\ &\quad-(2H(t_{j+1})\cdot(x-x_j)-\beta)e^{2s(-\beta(t_{j+1}-t_j)+|x-x_j|^2)}|u(x,t_{j+1})|^2. \end{aligned} \tag{30}$$

Therefore, summing in j from 0 to $m-1$ and keeping in mind that $t_0=0$ and $t_m=T$ by (9) and (27) we have

$$\begin{aligned}\sum_{j=0}^{m-1}\psi_j(x) &\geq (2H(0)\cdot(x-x_0)-\beta)e^{2s(|x-x_0|^2)}|u(x,0)|^2+\sum_{j=1}^{m-1}q_j(x)|u(x,t_j)|^2\\ &\quad-(2H(T)\cdot(x-x_{m-1})-\beta)e^{2s(-\beta(T-t_{m-1})+|x-x_{m-1}|^2)}|u(x,T)|^2\\ &\geq \mu_0H_0e^{2s\mu_0^2}|u(x,0)|^2-2M_*H_*e^{2sM_*^2}|u(x,T)|^2+\sum_{j=1}^{m-1}q_j(x)|u(x,t_j)|^2,\end{aligned}$$

where, for $j=1,\dots,m-1$, we set

$$q_j(x):=(2H(t_j)\cdot(x-x_j)-\beta)e^{2s|x-x_j|^2}-\big(2H(t_j)\cdot\big(x-x_{j-1}\big)-\beta\big)\,e^{2s|x-x_{j-1}|^2}.$$

Thus, by (7), (20), (23) and (27), we obtain the following estimate

$$q_j(x)\geq\tilde{C}\mu_0H_0e^{2s\mu_j^2}-H_*M_*e^{2sM_{j-1}^2}=\tilde{C}\mu_0H_0e^{2s\mu_j^2}\left(1-\frac{M_*H_*}{\tilde{C}\mu_0H_0}e^{-2s\left(\mu_j^2-M_{j-1}^2\right)}\right).$$

Thanks to (18) (see Lemma 2.2), the choice of the points x_j permits to have $\mu_j-M_{j-1}>0$, and we deduce that there exist $s_j>0$ enough large, that is $s_j>\frac{1}{2\left(\mu_j^2-M_{j-1}^2\right)}\log\left(\frac{2H_*M_*}{\tilde{C}\mu_0H_0}\right)$, $j=1,\dots,m-1$, such that, for every $s>s_0:=\max_{j=1,\dots,m-1}s_j$, we have

$$q_j(x)\geq\frac{\mu_0H_0}{2}e^{2s\mu_j^2}\geq\frac{\mu_0H_0}{2}e^{2s\mu_0^2}\geq C_0e^{C_0s}, \tag{31}$$

for some positive constant $C_0=C_0(s)$. Thus, by (29), (30), and (31) we have that

$$\begin{aligned}\int_Q|Pu|^2e^{2s\varphi}dxdt &= \sum_{j=0}^{m-1}\int_{t_j}^{t_{j+1}}\int_\Omega|Pu|^2e^{2s\varphi_j}dxdt\\ &\geq s\sum_{j=0}^{m-1}\int_\Omega\psi_j(x)dx+(C_1s^2-H_0's)\sum_{j=0}^{m-1}\int_{Q_j}e^{2s\varphi_j}|u|^2dxdt\end{aligned}$$

$$- C_2 s e^{C_2 s} \sum_{j=0}^{m-1} \int_{\Sigma_j} |u|^2 d\gamma dt$$

$$\geq C_3 s^2 \int_Q e^{2s\varphi_j} |u|^2 dx dt - C_2 s e^{C_2 s} \sum_{j=0}^{m-1} \int_{\Sigma_j} |u|^2 d\gamma dt$$

$$+ C_0 s e^{C_0 s} \sum_{j=0}^{m-1} \int_\Omega |u(x, t_j)|^2 dx - C_2 s e^{C_2 s} \int_\Omega |u(x, T)|^2 \, dx$$

for any $0 < C_3 < C_1$ and all s sufficiently large. The last estimate completes the proof of Theorem 1.5. □

3 Proof of the Observability Inequality

Let us give in Sect. 3.1 two lemmas and in Sect. 3.2 the proof of Theorem 1.6.

3.1 Energy Estimates

Let us give the following energy estimates.

Lemma 3.1 *Let $g \in L^2(\partial\Omega \times (0, T))$ and let us consider the problem*

$$\begin{cases} \partial_t u + H(t) \cdot \nabla u = 0 \quad in \ Q := \Omega \times (0, T), \\ u|_{\partial\Omega \times (0,T)} = g. \end{cases} \tag{11}$$

Then, for every $t \in [0, T]$, the following energy estimates hold

$$\|u(\cdot, t)\|^2_{L^2(\Omega)} \leq \|u(\cdot, 0)\|^2_{L^2(\Omega)} + H_* \|g\|^2_{L^2(\partial\Omega \times (0,T))}, \tag{32}$$

$$\|u(\cdot, 0)\|^2_{L^2(\Omega)} \leq \|u(\cdot, t)\|^2_{L^2(\Omega)} + H_* \|g\|^2_{L^2(\partial\Omega \times (0,T))}, \tag{33}$$

for any $u \in H^1(Q)$ satisfying (11), where $H_* := \max_{\xi \in [0,T]} |H(\xi)|$.

Proof Let $H(t) = (H_1(t), \ldots, H_d(t)),\ t \in [0, T]$. Multiplying the equation in (11) by $2u$ and integrating over Ω, we have

$$\int_\Omega 2u \partial_t u dx + \int_\Omega \sum_{k=1}^{d} H_k(t) 2u \partial_k u dx = 0,$$

then,

$$\partial_t\left(\int_\Omega |u(x,t)|^2dx\right)+\sum_{k=1}^d\int_\Omega H_k(t)\partial_k(|u(x,t)|^2)dx=0.$$

So, integrating by parts, for every $t\in[0,T]$, we obtain

$$\partial_t\left(\int_\Omega |u(x,t)|^2dx\right)=-\sum_{k=1}^d\int_{\partial\Omega} H_k|u|^2\nu_k d\gamma=-\int_{\partial\Omega}(H\cdot\nu)|g|^2d\gamma, \tag{34}$$

where $\nu=(\nu_1,\dots,\nu_d)$ is the unit normal vector outward to the boundary $\partial\Omega$. Setting

$$E(t):=\int_\Omega |u(x,t)|^2dx, \qquad t\in[0,T],$$

by (34), integrating on $[0,t]$ we deduce

$$|E(t)-E(0)|=\left|-\int_0^t\int_{\partial\Omega}(H(\xi)\cdot\nu(x))|g(x,\xi)|^2d\gamma d\xi\right|\le H_*\|g\|^2_{L^2(\partial\Omega\times(0,T))}$$

where $H_*=\max_{\xi\in[0,T]}|H(\xi)|$. Thus, for all $t\in[0,T]$, we have

$$E(t)\le E(0)+H_*\|g\|^2_{L^2(\partial\Omega\times(0,T))},$$

and

$$E(0)\le E(t)+H_*\|g\|^2_{L^2(\partial\Omega\times(0,T))}.$$

□

Lemma 3.2 *Let $0\le s_1<s_2\le T$, $g\in L^2(\partial\Omega\times(0,T))$. Let us assume that there exists a positive constant $C=C(s_1,s_2)$ such that for every $t\in[s_1,s_2]$ the following observability inequality holds*

$$\|u(\cdot,t)\|_{L^2(\Omega)}\le C\|g\|_{L^2(\partial\Omega\times(0,T))}, \quad \textit{for all } u\in H^1(Q) \textit{ solution to } (11). \tag{35}$$

Then, there exists a positive constant $C=C(s_1,s_2,T)$ such that the inequality (35) holds for every $t\in[0,T]$.

Proof Let $E(t)=\|u(\cdot,t)\|^2_{L^2(\Omega)}$, $t\in[0,T]$. For every $t\in[0,s_1]$, keeping in mind Lemma 3.1, by (32), (33) and (35) we obtain

$$\begin{aligned}\|u(\cdot,t)\|^2_{L^2(\Omega)}=E(t)&\le E(0)+H_*\|g\|^2_{L^2(\partial\Omega\times(0,T))}\le E(s_1)+2H_*\|g\|^2_{L^2(\partial\Omega\times(0,T))}\\&\le(C^2+2H_*)\|g\|^2_{L^2(\partial\Omega\times(0,T))}\,.\end{aligned} \tag{36}$$

For every $t\in[s_2,T]$, using again Lemma 3.1, by (32) and (35) we deduce

$$\|u(\cdot,t)\|^2_{L^2(\Omega)} = E(t) \le E(s_2) + H_* \|g\|^2_{L^2(\partial\Omega\times(0,T))} \le (C^2 + H_*)\|g\|^2_{L^2(\partial\Omega\times(0,T))}. \tag{37}$$

From (36) and (37) the conclusion follows. □

3.2 The Proof

Proof (of Theorem 1.6).
Let φ be the weight function given in (8). By the assumption (12) it follows that there exists $j^* \in \{0, \dots m-1\}$ such that

$$\frac{(t_{j^*+1} - t_{j^*})d_\Omega(x_{j^*})}{M^2_\Omega(x_{j^*})} > \frac{1}{H_0(2S^2_* - 1)}. \tag{38}$$

By the definition of the weight function $\varphi(x,t)$ (see (8)), it follows that, for every $x \in \overline{\Omega}$, we have

$$\varphi(x, t_{j^*}) = \varphi_{j^*}(x, t_{j^*}) = |x - x_{j^*}|^2 > 0$$

and, since (38) holds, keeping in mind that $\beta = (2S^2_* - 1)H_0 d_\Omega(x_0)$,

$$\lim_{t\to(t_{j^*+1})^-} \varphi_{j^*}(x,t) = |x - x_{j^*}|^2 - \beta(t_{j^*+1} - t_{j^*}) < 0.$$

Therefore, there exist $\varepsilon \in \left(0, \dfrac{t_{j^*+1} - t_{j^*}}{2}\right)$ and $\delta > 0$ such that

$$\begin{cases} \varphi(x,t) = \varphi_{j^*}(x,t) > \delta, & t \in [t_{j^*}, t_{j^*} + \varepsilon],\ x \in \overline{\Omega}, \\ \varphi(x,t) = \varphi_{j^*}(x,t) < -\delta, & t \in [t_{j^*+1} - 2\varepsilon, t_{j^*+1}),\ x \in \overline{\Omega}. \end{cases} \tag{39}$$

Let $u \in H^1(Q)$ satisfy (11) on $Q = \Omega \times (0,T)$. Let us consider $Q^* := \Omega \times (t_{j^*}, t_{j^*+1}) \subseteq Q$. Now we define a cut-off function $\chi \in C^\infty([t_{j^*}, t_{j^*+1}])$ such that $0 \le \chi \le 1$ and

$$\chi(t) = \begin{cases} 1, & t \in [t_{j^*}, t_{j^*+1} - 2\varepsilon], \\ 0, & t \in [t_{j^*+1} - \varepsilon, t_{j^*+1}]. \end{cases}$$

We set

$$v(x,t) = \chi(t)u(x,t), \quad (x,t) \in Q^*, \tag{40}$$

and, keeping in mind (11) and (40), we deduce

$$\begin{cases} \partial_t v + H(t)\cdot\nabla v = u(\partial_t \chi) & \text{in } Q^*, \\ v|_{\partial\Omega\times(t_{j^*}, t_{j^*+1})} = \chi g, & \\ v(x, t_{j^*+1}) = 0, & x \in \Omega. \end{cases} \tag{41}$$

Applying Theorem 1.5 to the problem (41), since $|v(x,t)| \le |u(x,t)|$ for every $(x,t) \in Q^*$ (see (40)), we obtain

$$s^2 \int_{Q^*} |v|^2 e^{2s\varphi} dxdt \le C \int_{Q^*} |u|^2 |\partial_t \chi|^2 e^{2s\varphi} dxdt + Ce^{Cs} \int_\Sigma |u|^2 d\gamma dt, \qquad (42)$$

for all large $s > 0$ and for some positive constant C.

Therefore, by (40) and (39) we have

$$s^2 \int_{Q^*} |v|^2 e^{2s\varphi} dxdt \ge s^2 \int_{t_{j^*}}^{t_{j^*}+\varepsilon} \int_\Omega |u|^2 e^{2s\varphi_0} dxdt \ge s^2 e^{2s\delta} \int_{t_{j^*}}^{t_{j^*}+\varepsilon} \int_\Omega |u|^2 dxdt \qquad (43)$$

and, since $\chi \in C^\infty([t_{j^*}, t_{j^*+1}])$, we also deduce

$$\begin{aligned} \int_{Q^*} |u|^2 |\partial_t \chi|^2 e^{2s\varphi} dxdt &= \int_{t_{j^*+1}-2\varepsilon}^{t_{j^*+1}-\varepsilon} \int_\Omega |u|^2 |\partial_t \chi|^2 e^{2s\varphi_{j^*}} dxdt \\ &\le K_1 e^{-2s\delta} \int_{t_{j^*+1}-2\varepsilon}^{t_{j^*+1}-\varepsilon} \int_\Omega |u|^2 dxdt \le K_1 \|u\|^2_{L^2(Q^*)} e^{-2s\delta}, \end{aligned} \qquad (44)$$

for all large $s > 0$ and for some positive constant K_1.

From (42), by (43) and (44) we obtain

$$s^2 e^{2s\delta} \int_{t_{j^*}}^{t_{j^*}+\varepsilon} \int_\Omega |u|^2 dxdt \le C_1 \|u\|^2_{L^2(Q^*)} e^{-2s\delta} + C_1 e^{C_1 s} \|g\|^2_{L^2(\partial\Omega\times(0,T))}, \qquad (45)$$

for all large $s > 0$ and for some positive constant C_1.

Setting

$$E(t) := \int_\Omega |u(x,t)|^2 dx, \qquad t \in [t_{j^*}, t_{j^*+1}],$$

by the energy estimate (33) of Lemma 3.1 we deduce

$$\begin{aligned} \int_{t_{j^*}}^{t_{j^*}+\varepsilon} \int_\Omega |u|^2 dxdt &= \int_{t_{j^*}}^{t_{j^*}+\varepsilon} E(t)dt \ge \int_{t_{j^*}}^{t_{j^*}+\varepsilon} (E(t_{j^*}) - H_* \|g\|^2_{L^2(\partial\Omega\times(0,T))})dt \\ &= \varepsilon \left(E(t_{j^*}) - H_* \|g\|^2_{L^2(\partial\Omega\times(0,T))} \right) \end{aligned} \qquad (46)$$

and, by the energy estimate (32) of Lemma 3.1 we obtain

$$\begin{aligned} \|u\|^2_{L^2(Q^*)} &= \int_{t_{j^*}}^{t_{j^*+1}} E(t)dt = \int_{t_{j^*}}^{t_{j^*+1}} \left(E(t_{j^*}) + H_* \|g\|^2_{L^2(\partial\Omega\times(0,T))} \right) dt \\ &\le E(t_{j^*})T + H_* T \|g\|^2_{L^2(\partial\Omega\times(0,T))}. \end{aligned} \qquad (47)$$

Substituting (46) and (47) into (45), we have

$$
\begin{aligned}
s^2e^{2s\delta}\varepsilon\left(E(t_{j^*}) - H_*\|g\|^2_{L^2(\partial\Omega\times(0,T))}\right) &\le s^2e^{2s\delta}\int_{t_{j^*}}^{t_{j^*}+\varepsilon}\int_\Omega |u|^2dxdt \\
&\le C_1\|u\|^2_{L^2(Q^*)}e^{-2s\delta} + C_1e^{C_1s}\|g\|^2_{L^2(\partial\Omega\times(0,T))} \\
&\le C_1e^{-2s\delta}\left(E(t_{j^*})T + H_*T\|g\|^2_{L^2(\partial\Omega\times(0,T))}\right) \\
&+ C_1e^{C_1s}\|g\|^2_{L^2(\partial\Omega\times(0,T))},
\end{aligned}
$$

for all large $s > 0$. Hence, for all s large enough,

$$
(s^2e^{2s\delta}\varepsilon - C_1Te^{-2s\delta})E(t_{j^*}) \le \left(C_1e^{C_1s} + s^2e^{2s\delta}\varepsilon H_* + C_1e^{-2s\delta}H_*T\right)\|g\|^2_{L^2(\partial\Omega\times(0,T))}
$$

But, for $s > 0$ enough large, $s^2e^{2s\delta}\varepsilon - C_1Te^{-2s\delta} > 0$. Thus, using again (32), for every $t \in [t_{j^*}, t_{j^*+1}]$, we obtain

$$
\|u(\cdot,t)\|_{L^2(\Omega)} = E(t) \le E(t_{j^*}) + H_*\|g\|^2_{L^2(\partial\Omega\times(0,T))} \le C_2\|g\|_{L^2(\partial\Omega\times(0,T))},
$$

for some positive constant C_2. The conclusion of the proof of Theorem 1.6 follows from the above inequality, using Lemma 3.2 to extend the above observability inequality from $[t_{j^*}, t_{j^*+1}]$ to $[0, T]$. □

Remark 3.3 By adapting the above proof, one could easily obtain an observability inequality for $u(\cdot, 0)$ on Ω, requiring measurements just on the subboundary Σ defined in (10).

Acknowledgements This work was partially supported by Grant-in-Aid for Scientific Research (S) 15H05740 and A3 Foresight Program Modeling and Computation of Applied Inverse Problems by Japan Society for the Promotion of Science. The first and second authors were visitors at The University of Tokyo in February 2018, supported by the above grant.
The third author was a Visiting Scholar at Rome in April 2018 supported by the University of Rome "Tor Vergata". The third author was also visitor in July 2017 at the University of Naples Federico II, supported by the Department of Mathematics and Applications "R. Caccioppoli" of that University. This work was supported also by the Istituto Nazionale di Alta Matematica (INdAM), through the GNAMPA Research Project 2017 "Comportamento asintotico e controllo di equazioni di evoluzione non lineari" (the coordinator: C. Pignotti). Moreover, this research was performed within the framework of the GDRE CONEDP (European Research Group on "Control of Partial Differential Equations") issued by CNRS, INdAM and Université de Provence. This work was also supported by the research project of the University of Naples Federico II: "Spectral and Geometrical Inequalities".

Appendix

In this appendix we prove Lemma 1.2.

Proof (*of Lemma* 1.2). Since $H \in Lip([0, T]; \mathbb{R}^d)$ there exists $L > 0$ such that

$$
|H(t) - H(s)| \le L|t - s|, \ \forall t, s \in [0, T].
$$

Let us consider, for simplicity, a uniform partition $\{t_j\}_0^m$ of $[0, T]$. Let us set

$$\eta_j := \frac{H(t_j)}{|H(t_j)|}, \quad j = 0 \ldots, m-1.$$

For $t \in [t_j, t_{j+1}]$, $j = 0, ..., m-1$, we deduce

$$\begin{aligned} H(t) \cdot \eta_j &= \big(H(t) - H(t_j)\big) \cdot \eta_j + H(t_j) \cdot \eta_j \geq -|H(t) - H(t_j)| + |H(t_j)| \\ &\geq -L|t - t_j| + |H(t_j)| \geq -L\frac{T}{m} + |H(t_j)|, \end{aligned} \tag{48}$$

and, since $|H(t)| \leq \big|H(t) - H(t_j)\big| + |H(t_j)|$,

$$\big|H(t_j)\big| \geq |H(t)| - |H(t) - H(t_j)| \geq |H(t)| - L|t - t_j| \geq |H(t)| - L\frac{T}{m}. \tag{49}$$

From (48) and (49), if we choose the uniform partition with $m \geq \frac{2LT}{H_0(1-S_*)}$, where we recall that $H_0 = \min_{t \in [0,T]} |H(t)|$, we obtain the conclusion, that is,

$$H(t) \cdot \frac{H(t_j)}{|H(t_j)|} \geq |H(t)| - 2L\frac{T}{m} \geq S_*|H(t)|, \quad \forall t \in [t_j, t_{j+1}], \quad \forall j = 0, \ldots, m-1.$$

□

References

1. Beilina, L., Klibanov, M.V.: Approximate Global Convergence and Adaptivity for Coefficient Inverse Problems. Springer, Berlin (2012)
2. Belinskij, S.P.: On one of the inverse problems for linear symmetrical t-hyperbolic systems with $n + 1$ independent variables. Differ. Equ. **12**, 15–23 (1976)
3. Bellassoued, M., Yamamoto, M.: Carleman Estimates and Applications to Inverse Problems for Hyperbolic Systems. Springer, Japan, Tokyo (2017)
4. Cannarsa, P., Floridia, G., Gölgeleyen, F., Yamamoto, M.: Inverse coefficient problems for a transport equation by local Carleman estimate, Inverse Problems (IOS Science), https://doi.org/10.1088/1361-6420/ab1c69. arxiv.org/abs/1902.06355 (2019)
5. Gaitan, P., Ouzzane, H.: Inverse problem for a free transport equation using Carleman estimates. Appl. Anal. **93**, 1073–1086 (2014)
6. Gölgeleyen, F., Yamamoto, M.: Stability for some inverse problems for transport equations. SIAM J. Math. Anal. **48**, 2319–2344 (2016)
7. Klibanov, M.V., Pamyatnykh, S.E.: Global uniqueness for a coefficient inverse problem for the non-stationary transport equation via Carleman estimate. J. Math. Anal. Appl. **343**, 352–365 (2008)

8. Machida, M., Yamamoto, M.: Global Lipschitz stability in determining coefficients of the radiative transport equation. Inverse Probl. **30**, 035010 (2014)
9. Romanov, V.G.: Inverse Problems of Mathematical Physics. VNU Science Press, Utrecht, the Netherlands (1987)
10. Yamamoto, M.: Carleman estimates for parabolic equations and applications. Inverse Probl. **25**, 123013 (2009)

On the Weak Maximum Principle for Degenerate Elliptic Operators

Italo Capuzzo Dolcetta

Abstract This paper provides an overview of some more or less recent results concerning the validity of the weak Maximum Principle for fully nonlinear degenerate elliptic equations. Special attention is devoted to the presentation of sufficient conditions relating the directions of degeneracy and the geometry of the possibly unbounded domain.

Keywords Fully nonlinear · Degenerate elliptic · Cylindrical domains · Principal eigenvalue · Min-max formula · Finite difference approximations

1 Introduction

I will give an overview of some more or less recent results concerning the validity of the weak Maximum Principle, that is of the following sign propagation property:

wMP if u satisfies $F(x, u, Du, D^2u) \geq 0$ in Ω, then $u \leq 0$ on $\partial\Omega$ implies $u \leq 0$ in Ω

Here $u \in USC(\overline{\Omega})$, the set of real-valued upper semicontinuous functions on $\overline{\Omega}$, F is a degenerate elliptic fully nonlinear mapping from $\Omega \times \mathbb{R} \times \mathbb{R}^n \times \mathcal{S}^n$ into $\mathbb{R}$ and $\mathcal{S}^n$ is the space of $n \times n$ symmetric matrices.

In what follows $\Omega \subset \mathbb{R}^n$ will be a general domain with possibly irregular boundary satisfying either measure-type conditions or geometric conditions related to the directions of ellipticity of F.

The results presented below apply to upper semicontinuous functions u satisfying the partial differential inequality $F(x, u, Du, D^2u) \geq 0$ in viscosity sense [1] and, a fortiori, in the classical sense. The motivation for considering non-smooth functions is of course a strong one when dealing with fully nonlinear partial differential

I. Capuzzo Dolcetta (✉)
Sapienza Università di Roma and GNAMPA-INdAM, Rome, Italy
e-mail: i.capuzzodolcetta@gmail.com

F. Alabau-Boussouira et al. (eds.), *Trends in Control Theory and Partial Differential Equations*, Springer INdAM Series 32,
https://doi.org/10.1007/978-3-030-17949-6_5

inequalities but is also a relevant one in the linear case, as pointed out by Calabi [2]. In that paper, brought to the attention of the present author by Garofalo [3], some version of the Hopf Maximum Principle is indeed proved to hold for upper semicontinuous functions u satisfying the linear partial differential inequality

$$\mathrm{Tr}(A(x)D^2u) + b(x) \cdot Du + c(x)u \geq 0$$

in an appropriately defined weak sense. It is worth to note that the weak notion considered in [2] is in fact similar although a bit stronger than the viscosity notion, see Mantegazza et al. [4] for comments in this respect.

My presentation here is based in particular on the papers [5–9]. I refer to these papers for more informations and proofs.

2 The Weak Maximum Principle in Bounded Domains: A Numerical Criterion

We consider first the case of a bounded domain $\Omega \subseteq \mathbb{R}^n$ and report on a characterization result in [5]. Let us start by recalling some well-known facts [10] in the framework of linear uniformly elliptic operators

$$L[u] = \mathrm{Tr}(A(x)D^2u) + b(x) \cdot Du + c(x)u, \;\; \alpha I \leq A(x) \leq \beta I$$

with, say, continuous and bounded coefficients $A, b, c, \alpha > 0$.

Several sufficient conditions of different nature known to imply the validity of **wMP** in a bounded domain Ω, e.g.

- (i) $c(x) \leq 0$
- (ii) exists $\phi > 0$ in $\overline{\Omega}$ such that $L[\phi] \leq 0$
- (ii) Ω is narrow (i.e. contained in a suitably small strip)

Simple examples show however that none of these conditions is however necessary for the validity of the weak Maximum Principle. What about sufficient and also necessary conditions for the validity of the Maximum Principle?

An important characterization result is due to Berestycki, Nirenberg and Varadhan [11]:

wMP holds for uniformly elliptic operators

$$L[u] = \mathrm{Tr}(A(x)D^2u) + b(x) \cdot Du + c(x)u$$

in a bounded domain Ω if and only if the number λ_1 defined by

$$\lambda_1 := \sup\{\lambda \in \mathbb{R} : \exists\, \phi > 0 \text{ in } \Omega \;\text{ such that } L[\phi] + \lambda\phi \leq 0 \;\text{ in } \Omega\}$$

is strictly positive. In the definition of λ_1, $\phi \in W^{2,p}_{loc}(\Omega)$.

Notably, this very nice numerical criterion was proved to hold under mild conditions on the coefficients and applies to a large class of domains with rough boundary $\partial\Omega$.

In the above result the matrix $A(x)$ is required to be uniformly positive definite but not necessarily symmetric. Note that even for symmetric A the operator L is not in general self-adjoint due to the presence of the drift term b.

Nonetheless, in [11] it is proved that the number λ_1 shares some of the properties of the classical principal eigenvalue for the Dirichlet problem, namely:

- there exists a principal eigenfunction $w_1 > 0$ in Ω such that $L[w_1] + \lambda_1 w_1 = 0 \, in \ \Omega, \ w_1 = 0$ on $\partial\Omega$
- w_1 is simple
- Re$\lambda \geq \lambda_1$ for any other eigenvalue λ of L

The existence of an associated positive and simple eigenfunction follows from the Krein-Rutman theorem thanks to compactness estimates guaranteed by the uniform ellipticity of L and the boundedness of Ω.

The Berestycki-Nirenberg-Varadhan definition above can be expressed by the equivalent pointwise min-max formula

$$\lambda_1 = - \inf_{\phi(x)>0} \sup_{x\in\Omega} \frac{L\phi(x)}{\phi(x)}$$

where $\phi \in W^{2,p}_{loc}(\Omega)$. The same formula, under more restrictive conditions (smooth boundary, continuous coefficients), was considered before in [12].

In that same paper different equivalent representation formulas for λ_1 were also proposed in terms of the average long-run behaviour of the positive semigroup generated by L. More precisely,

$$\lambda_1 = - \lim_{t\to+\infty} \frac{1}{t} \log \sup_{x\in\Omega} \int_{\Omega} p(t, x, y)dy$$

where $p(t, x, y)dy$ is the positive density defining the semigroup generated by $-L$.

A much older reference is [13] where the same min-max formula is proposed for the principal Dirichlet eigenvalue for the Laplace operator, see also [14].

One motivation for considering similar formulas in the nonlinear case comes from ergodic optimal control. Consider for example the viscous Hamilton–Jacobi equation, that is the Bellman equation satisfied by the value function u_α of an infinite horizon stochastic discounted optimal control problem with running cost V:

$$-\frac{1}{2}\Delta u_\alpha + \frac{1}{2}|\nabla u_\alpha|^2 - V(x) + \alpha u_\alpha = 0 \quad , x \in \mathbb{R}^n$$

where $\alpha > 0$ is the discount parameter and the eigenvalue problem for the linear Schrödinger type equation

$$-\frac{1}{2}\Delta\Phi + V(x)\Phi = \lambda\Phi$$

If $(\lambda_1, \Phi > 0)$ is a principal eigenvalue-eigenfunction pair, it is easy to check that the function $w = -\log\Phi + \int_{\mathbb{R}^n}\Phi^2 \log\Phi\, dx$ satisfies

$$-\frac{1}{2}\Delta w + \frac{1}{2}|\nabla w|^2 - V(x) + \lambda_1 = 0 \quad , \quad \int_{\mathbb{R}^n} w\,\Phi^2 = 0$$

which is the Bellman equation of ergodic optimal control.

It can be proved, under some conditions on V, that as $\alpha \to 0$

$$\alpha u_\alpha \to \lambda_1, \quad u_\alpha - \int_{\mathbb{R}^n} u_\alpha\,\Phi^2\,dx \to w$$

This PDE approach to ergodic optimal control has been introduced by Lasry [15], developed later by P.L. Lions [16], Bensoussan-Nagai [17] and many other authors in more general settings including the case of controlled degenerate diffusions and more general Hamiltonians modelled by operators of the type $F(D^2(w)) + H(x, Dw)$ with F convex, see [18].

According to this kind of motivations a quite natural question arises: does the Berestycki-Nirenberg-Varadhan characterization holds true as it is, or may be with suitable modifications, in the case of degenerate elliptic operators

$$\mathrm{Tr}(A(x)D^2u) + b(x)\cdot Du + c(x)u$$

with $A(x)$ non-negative definite and, more generally, for fully nonlinear degenerate elliptic operators ?

That is, is there a number associated to F and Ω whose positivity enforces the validity of **wMP** and conversely?

Recall that the mapping $F : \Omega \times \mathbb{R} \times \mathbb{R}^n \times \mathcal{S}^n \to \mathbb{R}$ is degenerate elliptic if the weak monotonicity condition

$$F(x, r, p, X + Y) \geq F(x, r, p, X)$$

holds true for any non-negative definite matrix $Y \in \mathcal{S}^n$. The starting point of our research in [5] was the observation that the definition of λ_1 in [11] does not work at this purpose in the case of degenerate ellipticity as shown by very simple examples whose analysis lead us to the following definition of the number

$$\mu_1(F, \Omega) := \sup\{\lambda \in \mathbb{R} : \exists\Omega' \supset \overline{\Omega},\ \exists\phi \in C(\Omega'),\ \phi > 0,\ F[\phi] + \lambda\phi^\alpha \leq 0 \text{ in } \Omega'\}$$

associated to a degenerate elliptic mapping F which is assumed to be homogeneous of degree $\alpha > 0$ with respect to the matrix slot. Given a domain Ω in $\mathbb{R}^N$ and an open set $\mathcal{O}$ such that $\overline{\Omega} \subset \mathcal{O}$ and an operator F positively homogeneous of degree $\alpha > 0$

One cannot expect, in the general case, $\mu_1(F, \Omega)$ to be a genuine principal eigenvalue. However, under uniform ellipticity for F, this is in fact the case as can be proved following ideas in [19].

The definition seems to depend on the choice of the set $\mathcal{O}$ but in fact this is not the case as confirmed by the Theorem below, proved in [5] for a general F depending on all variables. For the sake of simplicity we confine ourselves here to the simplified model $F(x, u, Du, D^2u) = F(D^2u) - f(x)$.

Theorem 2.1 *Let Ω be a bounded domain in $\mathbb{R}^n$ and $\mathcal{O}$ an open set such that $\overline{\Omega} \subset \mathcal{O}$. Assume that F is continuous, degenerate elliptic, positively homogeneous of degree $\alpha > 0$. Assume also that $f \in C(\overline{\Omega})$.Then*

F satisfies **wMP** *in Ω if and only if $\mu_1(F, \Omega) > 0$*

For general F extra assumptions are needed, including the Crandall-Ishii-Lions structural condition to guarantee the comparison property between viscosity sub and supersolutions.

As far as we know the above result is new even for smooth subsolutions of degenerate elliptic linear operators.

Here below a few examples to illustrate the result above:

- Zero order inequalities: $F(u(x)) = c(x)u(x) \geq 0$. If $c(x) < 0$ then, trivially, $u(x) \leq 0$ and, as easy to check $\mu_1(F, \Omega) > 0$.
  More generally, if $F(x, u, Du, D^2u) = F(u(x)) \geq 0$ and F is decreasing with $F(0) = 0$ then, trivially, $\mu_1 > 0$ and $u \leq F^{-1}(0) = 0$.
- Transport operators: $b(x) \cdot \nabla u \geq 0, \ x \in \Omega\ , \ u \leq 0, \ x \in \partial\Omega$
  It is not difficult to check that if b vanishes somewhere in Ω then $\mu_1 = 0$.
  On the other hand, if there exists a Lyapunov function L such that $\nabla L \neq 0$ and $b \cdot \nabla L > 0$ then $\mu_1 > 0$.
- Subelliptic operators: if the ellipticity of F is not degenerate in some direction ν, that is
  $$F(x, r, p, X + \nu \otimes \nu) - F(x, r, p, X) \geq \beta > 0$$
  and if the positive constants C satisfy $F(x, C, 0, 0) \geq 0$ in Ω', then $\mu_1(F, \Omega) > 0$. This is seen by taking $\phi(x) = 1 - \varepsilon e^{\sigma \nu \cdot x}$, with σ large and ε small.
  Above conditions satisfied for instance by the 2-dimensional Grushin operator: $\partial_{xx} + |x|^k \partial_{yy}$ with k an even positive integer.
- Harvey-Lawson Hessian operators, see [20]:
  $$\mathcal{H}_k(D^2u) := \eta_{n-k+1}(D^2u) + \ldots + \eta_n(D^2u),$$
  k an integer between 1 and n, $\eta_1(D^2u) \leq \eta_2(D^2u) \leq \ldots \leq \eta_N(D^2u)$ the ordered eigenvalues of the matrix D^2u. A test with quadratic polynomials shows that $\mu_1(H_k) > 0$.

3 Unbounded Domains and Uniform Ellipticity

It is well-known that **wMP** may not hold in general unbounded domains: just observe that

$$u(x) = 1 - \frac{1}{|x|^{n-2}}$$

with $n \geq 3$ satifies $\Delta u = 0$ in the exterior domain $\Omega = \mathbb{R}^n \setminus \overline{B}_1(0)$, $u \equiv 0$ on $\partial\Omega$ but $u > 0$ in Ω.

Some remarkable results concerning the validity of **wMP** for linear uniformly elliptic operators in unbounded domains are due to [21]. He considered domains satisfying the following measure-geometric condition:

(G) for fixed numbers $\sigma, \tau \in (0, 1)$, there exists a positive real number $R(\Omega)$ such that for any $y \in \Omega$ there exists an n-dimensional ball B_{R_y} of radius $R_y \leq R(\Omega)$ satisfying

$$y \in B_{R_y}, \quad |B_{R_y} \setminus \Omega_{y,\tau}| \geq \sigma |B_{R_y}|$$

where $\Omega_{y,\tau}$ is the connected component of $\Omega \cap B_{R_{y/\tau}}$ containing y.

The above condition, first introduced in [11] requires, roughly speaking, that there is enough boundary near every point in Ω allowing so to carry the information on the sign of u from the boundary to the interior of the domain.

Condition **(G)** is satisfied for example by unbounded domains with finite measure with $R(\Omega) = C(n)|\Omega|^{\frac{1}{n}}$ and also for a large class of unbounded domains with possibly infinite Lebesgue measure such as infinite cylinders.

On the other hand **(G)** does not hold on cones: this can be seen as a consequence of the fact that **(G)** implies $\sup_{y\in\Omega} \text{dist}(y, \partial\Omega) < +\infty$.

Another quite simple example of an unbounded domain with infinite Lebesgue measure satisfying **(G)** is provided by the perforated plane

$$\mathbb{R}^2_{per} = \mathbb{R}^2 \setminus \bigcup_{(i,j)\in\mathbb{Z}^2} B_r(i, j)$$

where $B_r(i, j)$ is the disc of radius $r < 1$ centered at (i, j). Observe that $\sup_{y\in\mathbb{R}^2_{per}}$ dist $(y, \partial R^2_{per}) < +\infty$ which is definitely not the case for the exterior domain $\mathbb{R}^n \setminus \overline{B}_1(0)$ considered before.

For domains satisfying **(G)** Cabré [21] proved an Alexandrov-Bakelman-Pucci **(ABP)** type estimate:

If u is a $W^{2,p}$ strong solution of the uniformly elliptic partial differential inequality

$$\text{Tr}\left(A(x)\, D^2 u\right) + b(x) \cdot Du + c(x)\, u \geq f(x) \ \text{ in } \Omega$$

with

$$A(x)\xi \cdot \xi \geq \alpha|\xi|^2\,, \quad \alpha > 0$$

then

$$\sup_{\Omega} u \leq \sup_{\partial\Omega} u^+ + C\, R(\Omega)\, ||f||_{L^n(\Omega)}$$

As a consequence of the **(ABP)** estimate above, if $f \equiv 0$ and $u \leq 0$ on $\partial\Omega$, the validity of **wMP** follows in the case of linear uniformly elliptic operators.

Some of the results in [21] have been later generalized to viscosity solutions of fully nonlinear uniformly elliptic inequalities in [6] under a weaker form of **(G)**, namely

(wG) *there exist constants* $\sigma, \tau \in (0, 1)$ *such that for all* $y \in \Omega$ *there is a ball* B_{R_y} *of radius* R_y *containing* y *such that*

$$|B_{R_y} \setminus \Omega_{y,\tau}| \geq \sigma |B_{R_y}|$$

where $\Omega_{y,\tau}$ *is the connected component of* $\Omega \cap B_{R_y/\tau}$ *containing* y.

If $\sup_{y\in\Omega} R_y < +\infty$ in the above definition, then Ω satisfies condition **(G)**. Typical examples of unbounded domains satisfying condition **(wG)** but not **(G)** are nondegenerate cones of $\mathbb{R}^n$ (and all their unbounded subsets). Indeed, condition **(wG)** is satisfied in this case with $R_y = O(|y|)$ as $|y| \to \infty$.

A less standard example is the plane domain described in polar cohordinates as $\Omega = \mathbb{R}^2 \setminus \{\varrho = e^{\theta}\,,\ \theta \geq 0\}$. Here **(wG)** holds with $R_y = O\left(e^{|y|}\right)$ as $|y| \to \infty$.

The main result in [6] is the following version of the **(ABP)** estimate and as a by-product in the case $f^- = 0$ in Ω, the validity of **wMP**:

Theorem 3.1 *Assume that* F *satisfies*

- $\alpha \operatorname{Tr}(Y) \leq F(x, t, p, X + Y) - F(x, t, p, X) \leq \beta \operatorname{Tr}(Y)$ *for some* $0 < \alpha \leq \beta$
- $t \mapsto F(x, t, p, X)$ *is nonincreasing*
- $F(x, 0, p, O) \leq b(x)\,|p|$

for all (x, t, p, X) *and for all* $Y \geq O$. *Assume also that* Ω *satisfies* **(wG)** *and*

$$\textbf{(C)} \qquad \sup_{y\in\Omega} R_y\, \|b\|_{L^\infty(\Omega_{y,\tau})} < \infty$$

If $u \in USC(\overline{\Omega})$ *with* $\sup_\Omega u < +\infty$ *is a solution of* $F(x, u, Du, D^2u) \geq f(x)$ *in* Ω *where* $f \in C(\Omega) \cap L^\infty(\Omega)$, *then*

$$\textbf{(ABP)} \quad \sup_{\Omega} u \leq \sup_{\partial\Omega} u^+ + C \sup_{y\in\Omega} R_y\, \|f^-\|_{L^N(\Omega_{y,\tau})}$$

for some positive constant C *depending on* $N,\ \alpha,\ \beta,\ \sigma,\ \tau$ *and* $\sup_{y\in\Omega} R_y\, \|b\|_{L^\infty(\Omega_{y,\tau})}$.

To obtain the **ABP** estimate in this more general case we will assume, besides condition (**wG**) on the domain, the following coupled requirement on the geometry of the domain and on the growth of the first order coefficients:

$$(\mathbf{C}) \qquad \sup_{y\in\Omega} R_y \, \|b\|_{L^\infty(\Omega_{y,\tau})} < \infty$$

This condition is trivially satisfied if $\sup_{y\in\Omega} R_y \le R_0 < +\infty$ in (**wG**), i.e. if Ω satisfies (**G**), or when $b \equiv 0$, namely when F does not depend on first-order derivatives. For a complete operator, condition (**wG**) alone is not enough to guarantee the validity of the Maximum Principle. Indeed, the function

$$u(x) = u(x_1, x_2) = \left(1 - \mathrm{e}^{1-x_1^\alpha}\right)\left(1 - \mathrm{e}^{1-x_2^\alpha}\right),$$

with $0 < \varepsilon < 1$, is bounded and strictly positive in the cone

$$\Omega = \left\{x = (x_1, x_2) \in \mathbb{R}^2 \, : \, x_1 > 1, \; x_2 > 1\right\}$$

and satisfies

$$u \equiv 0 \quad \text{on } \partial\Omega, \quad \Delta u + B(x)\cdot Du = 0 \quad \text{in } \Omega$$

where the vectorfield B is given by

$$B(x) = B(x_1, x_2) = \left(\frac{\varepsilon}{x_1^{1-\varepsilon}} + \frac{1-\varepsilon}{x_1}\,,\; \frac{\varepsilon}{x_2^{1-\varepsilon}} + \frac{1-\varepsilon}{x_2}\right)$$

As observed above, Ω satisfies (**wG**) with $R_y = O(|y|)$ as $|y| \to \infty$.

Since $|B|_{L^\infty(\Omega_{y,\tau})} = 1$ for every $y \in \Omega$, the interplay condition (**C**) fails in this example.

Here are some non trivial cases in which the interplay condition (**C**) is fulfilled:

- $\Omega = \left\{(x', x_N) \in \mathbb{R}^{N-1} \times \mathbb{R} \, : \, |x'| < 1\,,\; x_N > 0\right\}$. Since Ω satisfies condition (**G**), then (**C**) is satisfied if b is any nonnegative bounded and continuous function.
- $\Omega = \left\{(x', x_N) \in \mathbb{R}^{N-1} \times \mathbb{R} \, : \, x_N > |x'|^q\right\}$ with $q > 1$. Satisfies assumption (**wG**) with radii $R_y = O\left(|y|^{1/q}\right)$ as $|y| \to \infty$. In this case, (**C**) imposes to the function b a rate of decay $b(y) = O\left(1/|y|^{1/q}\right)$ as $|y| \to \infty$.
- Ω is the strictly convex cone $\left\{x \in \mathbb{R}^N \setminus \{0\} \, : \, x/|x| \in \Gamma\right\}$ where Γ is a proper subset of the unit half-sphere $S_+^{N-1} = \left\{x = (x', x_N) \in \mathbb{R}^{N-1} \times \mathbb{R} \, : \, |x| = 1\,,\; x_N > 0\right\}$. In this case, condition (**wG**) is satisfied with $R_y = O(|y|)$ for $|y| \to \infty$ and condition (**C**) requires on the coefficient b the rate of decay $b(y) = O\left(1/|y|\right)$ as $|y| \to \infty$.

It is not hard to check that the values $\sigma = \frac{1}{2}$ and $\tau < 1$ are feasible for the validity of the measure-geometric conditions in all the above examples.

4 One-Directional Elliptic Operators on Special Unbounded Domains

This section is a brief overview of some recent results in collaboration with Vitolo, [7, 8] about the validity of various versions of the **wMP** for degenerate elliptic operators F which are strictly elliptic on unbounded domains Ω of $\mathbb{R}^n$ whose geometry is related to the direction of ellipticity.

Some results of that kind for one-directional elliptic operators in bounded domains have been previously established by Caffarelli-Li-Nirenberg [22] , see also [23].

Assume that $\mathbb{R}^n$ is decomposed as the direct sum $U \bigoplus U^{\perp}$ where U is a k-dimensional subspace of $\mathbb{R}^n$ and $U^{\perp}$ is its orthogonal complement and denote by P and Q the projection matrices on U and $U^{\perp}$, respectively.

We will consider special open domains Ω satisfying the following condition

$$(\Omega_\circ) \quad \Omega \subseteq \{x \in \mathbb{R}^n : a \leq x \cdot \nu^h \leq a + d,\ h = 1, \dots, k\} \quad \text{for some } a \in \mathbb{R}, d > 0\,,$$

where $\{\nu^1, \dots, \nu^k\}$ is an orthonormal basis for the subspace U.

Domains as Ω are contained in infinite parallelepipeds whose k-dimensional orthogonal section is a cube of edge d.

They may be unbounded and of infinite Lebesgue measure but they do indeed satisfy the measure-geometric (**wG**) condition discussed in the previous section. No regularity assumption is made on the boundary $\partial\Omega$.

We assume the following monotonicity conditions on the mapping F :

- F is degenerate elliptic: $F(x, s, p, Y) \geq F(x, s, p, X)$ if $Y \geq X$,
- F is one-directional elliptic: $F(x, 0, p, X + t\nu \otimes \nu) - F(x, 0, p, X) \geq \alpha(x)t$ for some $\nu \in U$ and for all $t > 0$
- $F(x, s, p, X) \leq F(x, r, p, X)$ if $s > r$ and $F(x, 0, 0, O) = 0$ forall $x \in \Omega$,

Here O is the zero-matrix and $\alpha(x)$ is a continuous, strictly positive function such that $\liminf_{x\to\infty} \alpha(x) > 0$.

The strict one-directional ellipticity condition $(\Omega_\circ)$ on F play a crucial role in our results.

We will assume moreover that:

- there exists $\beta > 0$ such that $F(x, 0, 0, X + tQ) - F(x, 0, 0, X) \leq \beta t\,|x|$ for all $t > 0$, as $|x| \to \infty$
- $|F(x, 0, p, X) - F(x, 0, 0, X)| \leq \gamma(x)|p|$ for all $p \in \mathbb{R}^n$

Here Q is the orthogonal projection matrix over U and γ is a continuous function such that $\frac{\gamma(x)}{\alpha(x)}$ is bounded above in Ω by some constant $\Gamma \geq 0$. Observe that both the matrices $\nu \otimes \nu$ and Q belong to $\mathcal{S}^n$ and are positive semidefinite.

We will refer collectively to conditions above as the structure condition on F, labelled $(\mathbf{SC})_{\mathrm{U}}$. It is worth noting that $(\mathbf{SC})_{\mathrm{U}}$ requires a control from below only with respect to a single direction $\nu \in U$ and a control from above in the orthogonal directions, a much weaker condition on F than uniform ellipticity.

The latter one would indeed require a uniform control of the difference quotients both from below and from above with respect to all possible increments with positive semidefinite matrices.

A very basic example of an F satisfying $(\mathbf{SC})_\mathbf{U}$ is given by the linear operator

$$F(x, u, Du, D^2u) = \lambda_1(x)\,\frac{\partial^2 u}{\partial x_1^2} + \cdots + \lambda_k(x)\,\frac{\partial^2 u}{\partial x_k^2} + \sum_{i=1}^{n} b_i(x)\,\frac{\partial u}{\partial x_i} + c(x)u$$

which satisfies conditions above with $U = \{x_{k+1} = \cdots = x_n = 0\}$, provided $\lambda_i(x) \geq \alpha$, $i = 1, \ldots, k$, $\left|\sum_i b_i^2(x)\right|^{1/2} \leq \gamma$ and $c(x) \leq 0$.

Further examples are provided by fully nonlinear operators of Bellman-Isaacs type arising in the optimal control of degenerate diffusion processes:

$$F(x, u, Du, D^2u) = \sup_{\mu} \inf_{\nu} L^{\mu\nu} u,$$

where

$$L^{\mu\nu} u = \sum_{i,j=1}^{k} a_{ij}^{\mu\nu} \frac{\partial^2 u}{\partial x_i x_j} + \sum_{i=1}^{n} b_i^{\mu\nu} \frac{\partial u}{\partial x_i} + c^{\mu\nu} u$$

with constant coefficients depending μ and ν running in some sets of indexes $\mathcal{M}, \mathcal{N}$.

If the matrices $(a_{ij}^{\alpha\beta})$ are positive semidefinite for all α, β and

$$\sum_{i,j=1}^{k} a_{ij}^{\mu\nu} \nu_i^h \nu_j^h \geq \lambda, \quad |b_i^{\mu\nu}| \leq \gamma, \quad c^{\mu\nu} \leq 0, \quad h = 1, \ldots, k,$$

for an orthonormal basis $\{\nu^1, \ldots, \nu^k\}$ of some k-dimensional subspace U.

Our results in [7] concerning the validity of (**wMP**) for one-directional elliptic operators in the special class of unbounded described above are stated in the following theorems:

Theorem 4.1 *Let Ω be a domain of $\mathbb{R}^n$ satisfying condition*

$$(\mathbf{\Omega}_\nu) \quad \Omega \subseteq \{x \in \mathbb{R}^n : a \leq x \cdot \nu^h \leq a + d,\ h = 1, \ldots, k\} \ \textit{for some } a \in \mathbb{R}, d > 0,$$

and assume that F satisfies the structure condition $(\mathbf{SC})_\mathbf{U}$.

Then (**wMP**) *holds for any $u \in USC(\overline{\Omega})$ such that $u^+(x) = o(|x|)$ as $|x| \to \infty$.*

Note that some restriction on the behaviour of u at infinity is unavoidable. Observe indeed that $u(x_1, x_2, x_3) = e^{x_1} \sin x_2 \sin x_3$ solves the degenerate Dirichlet problem

$$\frac{\partial^2 u}{\partial x_1^2} + \frac{\partial^2 u}{\partial x_2^2} = 0 \text{ in } \Omega, \quad u(x_1, x_2, x_3) = 0 \text{ on } \partial\Omega$$

in the 1-infinite cylinder $\Omega = \mathbb{R} \times (0, \pi)^2 \subset \mathbb{R}^3$ and $u(x_1, x_2, x_3) > 0$ in Ω, implying the failure of (**wMP**).

The next is a quantitative form of the above result:

Theorem 4.2 *Let Ω be a domain of $\mathbb{R}^n$ satisfying condition*

$$(\boldsymbol{\Omega}_\nu) \quad \Omega \subseteq \{x \in \mathbb{R}^n : a \leq x \cdot \nu^h \leq a + d,\ h = 1, \ldots, k\} \ \textit{for some}\ a \in \mathbb{R}, d > 0\,,$$

and assume that F satisfies the structure condition $(\mathbf{SC})_{\mathrm{U}}$. *If*

$$F(x, u, Du, D^2u) \geq f(x) \quad \textit{in}\ \Omega$$

where f is continuous and bounded from below and $u^+(x) = o(|x|)$ as $|x| \to \infty$.

Then

$$\sup_{\overline{\Omega}} u \leq \sup_{\partial\Omega} u^+ \frac{e^{1+d\Gamma}}{1 + d\Gamma} \|\frac{f^-}{\alpha}\|_\infty \, d^2$$

where $f^-(x) = -\min(f(x), 0)$.

The next result allows a weaker monotonicity requirement for $r \to F(x, r, p, M)$ which must be compensated by some narrowness condition on the domain:

Theorem 4.3 *Let Ω satisfy condition $(\Omega_\bullet)$ and assume that F satisfies* $(\mathbf{SC})_{\mathrm{U}}$ *with the weaker condition*

$$F(x, s, p, X) - F(x, r, p, Xt) \leq c(x)\,(s - r) \ \text{ if } s > r$$

for some continuous function $c(x) > 0$. Assume also that $\frac{c(x)}{\alpha(x)} \leq K < +\infty$ in Ω.

Then (**wMP**) *holds for $u \in USC(\overline{\Omega})$, u bounded above, provided $d^2 K$ is small enough.*

For fixed $c > 0$ this results applies to narrow domains, that is when the thickness parameter d in condition $(\Omega_\bullet)$ is sufficiently small. Conversely, for fixed $d > 0$ (**wMP**) holds provided c is a sufficiently small positive number.

The above result can be used as an intermediate step in the proof of the Theorem below concerning the validity of (**wMP**) for unbounded solutions with exponential growth at infinity, a qualitative Phragmen-Lindelöf type result:

Theorem 4.4 *Let Ω satisfy condition $(\Omega_\bullet)$ and assume that F satisfies* $(\mathbf{SC})_{\mathrm{U}}$. *Then, for any fixed $\beta_0 > 0$ there exists a positive constant $d = d(n, \alpha, \beta, \gamma, \beta_0)$ such that if Ω has thickness d, then* (**wMP**) *holds for functions $u \in USC(\overline{\Omega})$ such that $u^+(x) = O(e^{\beta_0|x|})$ as $|x| \to \infty$.*

Conversely, for any fixed $d_0 > 0$ there exists a positive constant $\beta = \beta(n, \alpha, \beta, \gamma, d_0)$ such that (**wMP**) *holds for functions $u \in USC(\overline{\Omega})$ such that $u^+(x) = O(e^{\beta|x|})$ as $|x| \to \infty$.*

Note that the assumption

$$F(x,0,0,X+tQ)-F(x,0,0,X)\le \beta t\,|x| \quad \text{for all } t>0, \text{ as } |x|\to\infty$$

in the directions belonging to $U^\perp$ is essential in order to go beyond a polynomial growth. Indeed, the function $u(x_1,x_2)=x_2^2\sin x_1$ is a solution of

$$\frac{\partial^2 u}{\partial x_1^2}+\frac{1}{2}x_2^2\frac{\partial^2 u}{\partial x_2^2}=0$$

in the cylinder $\Omega=(0,\pi)\times\mathbb{R}\subset\mathbb{R}^2$, $u=0$ on $\partial\Omega$ but u is strictly positive in Ω.

5 An Approximation of the Principal Eigenvalue

In this final section we present a recent approximation result and a finite difference scheme for the computation of the principal Dirichlet eigenvalue for fully nonlinear operators F in the uniformly elliptic case based on the pointwise min-max formula for the principal eigenvalue presented in the first section. Consider first the self-adjoint operator $Lu(x)=\operatorname{div}(A(x)\nabla u)$ where $A(x)$ is a symmetric positive definite matrix with continuous entries, Ω a bounded open subset of $\mathbb{R}^n$.

The minimum value λ_1 in the Rayleigh-Ritz variational formula

$$\lambda_1:=-\min_{\phi\in H_0^1(\Omega),\,||\phi||_{L^2(\Omega)}=1}\int_\Omega A(x)D\phi\cdot D\phi\,dx$$

is attained at a function w_1 such that

$$\begin{cases} Lw_1(x)+\lambda_1 w_1(x)=0 & x\in\Omega,\\ w_1(x)=0 & x\in\partial\Omega\end{cases}$$

It is also well-known that λ_1 is the principal eigenvalue of L in Ω and w_1 is the corresponding principal eigenfunction of the Dirichlet problem for L.

For linear operators in divergence form there is a vast literature on computational methods for the principal eigenvalue. On the other hand, general non-divergence type elliptic operators such as

$$\operatorname{Tr}(A(x)D^2u)+b(x)\cdot Du+c(x)u$$

are not self-adjoint in general and the spectral theory is then much more involved: in particular, the Rayleigh-Ritz variational formula is not available anymore.

In [9] we developed a finite difference scheme for the computation of the principal eigenvalue and the principal eigenfunction of fully nonlinear uniformly elliptic operators based on the min-max formula discussed above:

$$\lambda_1 = -\inf_{\phi(x)>0} \sup_{x\in\Omega} \frac{F[\phi(x)]}{\phi(x)}$$

where we used the notation $F[\phi(x)] = F(x, \phi(x), D\phi(x), D^2\phi(x))$. That formula can be seen as a pointwise alternative to the Rayleigh-Ritz L^2 formula.

Our approach applies in particular to linear operators in non-divergence form, see [24, 25]. The literature in this case is not abundant, see for example [26] and our approximation results seem to be new even in the linear case.

Let $h\mathbb{Z}^n$ be the orthogonal lattice in $\mathbb{R}^n$ where $h > 0$ is a discretization parameter and $\mathcal{C}_h$ the space of the mesh functions defined on $\Omega_h = \Omega \subset \mathbb{Z}^n_h$. Consider a discrete operator F_h defined by

$$F_h[u](x) := F_h(x, u(x), [u]_x)$$

where

- $h > 0$ is the discretization parameter (h is meant to tend to 0),
- $x \in \Omega_h$ is a grid point
- $u \in \mathcal{C}_h$
- $[\cdot]_x$ represents the stencil of the scheme, i.e. the points in $\Omega_h \backslash \{x\}$ where the value of u are computed for writing the scheme at the point x (we assume that $[w]_x$ is independent of $w(y)$ for $|x - y| > Mh$ for some fixed $M \in \mathbb{N}$).

Following [27] we introduce some basic structure assumptions which are to be satisfied by the finite difference operator F_h:

(i) F_h is of positive type, i.e. for all $x \in \Omega_h$, $z, \tau \in \mathbb{R}$, $u, \eta \in \mathcal{C}_h$ satisfying $0 \leq \eta(y) \leq \tau$ for each $y \in \Omega_h$, then

$$F_h(x, z, [u + \eta]_x) \geq F_h(x, z, [u]_x) \geq F_h(x, z + \tau, [u + \eta]_x)$$

(ii) F_h is 1-positively homogeneous, i.e. for all $x \in \Omega_h$, $z \in \mathbb{R}$, $u \in \mathcal{C}_h$ and $t \geq 0$, then

$$F_h(x, tz, [tu]_x) = tF_h(x, z, [u]_x)$$

(iii) The family $\{F_h, 0 < h \leq h_0\}$, where h_0 is a positive constant, is consistent with F on the domain $\Omega \subset \mathbb{R}^n$, i.e. for each $u \in C^2(\Omega)$

$$\sup_{\Omega_h} \left| F(x, u(x), Du(x), D^2u(x)) - F_h(x, u(x), [u]_x) \right| \to 0 \quad \text{as } h \to 0,$$

uniformly on compact subsets of Ω.

The discretized equations for this kind of approximate operators satisfy some crucial pointwise estimates which are the discrete analogues of those valid for fully nonlinear, uniformly elliptic equations.

If F is uniformly elliptic, it is always possible, see [27], to find a scheme of the previous type which is of positive type and consistent with F.We don't know how to deal with this important issue in the case of degenerate ellipticity.

Mimicking the continuous case we define a discrete principal eigenvalue for F_h by means of the formula

$$\lambda_1^h := \sup\{\lambda \in \mathbb{R} : \ \exists \phi > 0 \ \text{ in } \Omega_h, \ F_h[\phi] + \lambda\phi \leq 0\,\}$$

The number λ_1^h defined in this way has the following properties:

- there is a positive solution ϕ_1^h of

$$\begin{cases} F[\phi] + \lambda_1^h\phi = 0 \text{ in } \Omega_h, \\ \phi = 0 \qquad\qquad\quad \text{on } \partial\Omega_h, \end{cases}$$

- or any $\lambda < \lambda_1$ the weak Maximum Principle holds for $F_h + \lambda$, i.e.
  if u is such that $F_h[u] + \lambda u \geq 0$ in Ω_h and $u \leq 0$ on $\partial\Omega_h$ then $u \leq 0$ in Ω_h
- λ_1^h is given by the finite dimensional optimization problem

$$\lambda_1^h = -\inf_{\phi\in C_h,\, \phi>0}\ \sup_{x\in\Omega_h} \frac{F_h[\phi(x)]}{\phi(x)}$$

Theorem 5.1 *Let (λ_1^h, ϕ_1^h) be the sequence of the discrete eigenvalues and of the corresponding eigenfunctions associated to F_h.*
Then,

$$\lambda_1^h \to \lambda_1, \qquad \phi_1^h \to \phi_1$$

uniformly in $\overline{\Omega}$ as $h \to 0$, where λ_1 and ϕ_1 are respectively the principal eigenvalue and the corresponding eigenfunction associated to F.

The proof of the convergence result cannot rely on standard stability results in viscosity solution theory such as the so-called Barles-Souganidis' method based on on the validity of a Comparison Principle since the limit problem

$$\begin{cases} F[\phi] + \lambda_1\phi = 0 \text{ in } \Omega, \\ \phi = 0 \qquad\qquad\quad \text{on } \partial\Omega \end{cases}$$

does not have such a property which would imply uniqueness which is not the case since the principal eigenfunction $\phi_1 > 0$ and $\phi \equiv 0$ are two distinct solutions of the problem under our assumption $F[0] = 0$.

Different techniques are therefore needed for the proof whose, main ingredients are:

- the semi-relaxed limits in viscosity solution sense,

- a weak Maximum Principle for the limit problem (rather than the Comparison Principle),
- the local Hölder estimate in [27]: if u_h is a solution of $F_h[u] = f$, then

$$|u_h(x) - u_h(y)| \le C\frac{|x-y|^\delta}{R}\left(\max_{B_R^h} u_h + \frac{R}{\alpha_0}\left\{\sum_{x\in\Omega_h} h^n |f(x)|^n\right\}^{\frac{1}{n}}\right)$$

  for any $x, y \in \Omega_h$where $R = \min \mathrm{dist}(x, \partial\Omega_h)$, $B_R^h = B(0, R) \cap \Omega_h$, δ, α_0 and C are positive constants independent of h.

In the case of convex operators F such as those arising in the optimal control theory of degenerate diffusion processes, that is F is the supremum of a family of linear operators:

$$F(x, u, Du, D^2u) = \sup_{i\in I} Tr(A^i(x)D^2u) + b^i(x)\cdot Du + c^i(x)u$$

our numerical approach leads to a finite dimensional convex optimization problem. Further informations and examples of simulation performed with the Optimization Toolbox of MATLAB can be found in [9].

References

1. Crandall, M.G., Ishii, H., Lions, P.-L.: Users guide to viscosity solutions of second order partial differential equations. Bull. Am. Math. Soc. N.S. **27**(1), 167 (1992)
2. Calabi, E.: An extension of E. Hopf's maximum principle with an application to Riemannian geometry. Duke Math. J. **25**, 45–56 (1958)
3. Garofalo, N.: Private Communication (2017)
4. Mantegazza, C., Mascellani, G.: Uraltsev, Gennady on the distributional Hessian of the distance function. Pac. J. Math. **270**(1), 151–166 (2014)
5. Berestycki, H., Capuzzo Dolcetta, I., Porretta, A., Rossi, L.: Maximum principle and generalized principal eigenvalue for degenerate elliptic operators. J. Math. Pures Appl. **103**(5, 9), 1276–1293 (2015)
6. Capuzzo-Dolcetta, I., Leoni, F., Vitolo, A.: The Alexandrov-Bakelman-Pucci weak maximum principle for fully nonlinear equations in unbounded domains. Commun. Part. Differ. Equ. **30**(10–12), 1863–1881 (2005)
7. Capuzzo Dolcetta, I., Vitolo, A.: The weak maximum principle for degenerate elliptic operators in unbounded domains. Int. Math. Res. Not. **2018**(2), 412–431 (Advance Access Publication December 16) (2016)
8. Capuzzo Dolcetta, I., Vitolo, A.: in preparation
9. Birindelli, I., Camilli, F., Capuzzo Dolcetta, I.: On the approximation of the principal eigenvalue for a class of nonlinear elliptic operators. Commun. Math. Sci. **15**(1), 55–75 (2017)
10. Protter, M.H., Weinberger, H.F.: Maximum Principles in Differential Equations. Corrected reprint of the 1967 original. Springer, New York, x+261 pp. (1984)
11. Berestycki, H., Nirenberg, L., Varadhan, S.R.S.: The principal eigenvalue and maximum principle for second-order elliptic operators in general domains. Comm. Pure Appl. Math. **47**(1), 47–92 (1994)

12. Donsker, M.D., Varadhan, S.R.S.: On the principal eigenvalue of second-order elliptic differential operators. Comm. Pure Appl. Math. **29**(6), 595–621 (1976)
13. Barta, J.: Sur la vibration fondamentale d'une membrane. C. R. Acad. Sci. Paris **204**, (1937)
14. Venturino, M.: The first eigenvalue of linear elliptic operators in nonvariational form. Boll. Un. Mat. Ital. B **15**(2, 5), 576–591 (1978) (Italian)
15. Lasry, J.-M.: Contrle stationnaire asymptotique. (French) Control theory, numerical methods and computer systems modelling (International Symposium, IRIA LABORIA, Rocquencourt, 1974), pp. 296–313. Lecture Notes in Economics and Mathematical Systems, vol. 107. Springer, Berlin (1975)
16. Lions, P.-L.: Optimal control and viscosity solutions. Recent Mathematical Methods in Dynamic Programming (Rome, 1984). Lecture Notes in Mathematics, vol. 1119, pp. 94–112. Springer, Berlin (1985)
17. Bensoussan, A., Nagai, H.: An ergodic control problem arising from the principal eigenfunction of an elliptic operator. J. Math. Soc. Jpn. **43**(1), 49–65 (1991)
18. Fleming, W.H., Soner, H.M.: Controlled Markov processes and viscosity solutions. Applications of Mathematics (New York), vol. 25, xvi+428 pp. Springer, New York (1993)
19. Birindelli, I., Demengel, F.: Eigenvalue, maximum principle and regularity for fully non linear homogeneous operators. Commun. Pure Appl. Anal. **6**(2), 335–366 (2007)
20. Harvey, R., Lawson Jr., B.: Existence, uniqueness and removable singularities for nonlinear partial differential equations in geometry. To appear in Surveys in Geometry. arXiv:1303.1117
21. Cabré, X.: On the Alexandroff-Bakelman-Pucci estimate and the reversed Hölder inequality for solutions of elliptic and parabolic equations. Comm. Pure Appl. Math. **48**(5), 539–570 (1995)
22. Caffarelli, L., Li, Y., Nirenberg, L.: Some remarks on singular solutions of nonlinear elliptic equations III: viscosity solutions including parabolic operators. Comm. Pure Appl. Math. **66**(1), 109–143 (2013)
23. Mannucci, P.: The Dirichlet problem for fully nonlinear elliptic equations non-degenerate in a fixed direction. Commun. Pure Appl. Anal. **13**(1), 119–133 (2014)
24. Babuska, I., Osborn, J.E.: Finite element-Galerkin approximation of the eigenvalues and eigenvectors of selfadjoint problems. Math. Comput. **52**(186), 275–297 (1989)
25. Boffi, D.: Finite element approximation of eigenvalue problems. Acta Numer. **19**, 1120 (2010)
26. Heuveline, V., Bertsch, C.: On multigrid methods for the eigenvalue computation of nonselfadjoint elliptic operators. East-West J. Numer. Math. **8**(4), 275–297 (2000)
27. Kuo, H.J., Trudinger, N.S.: Discrete methods for fully nonlinear elliptic equations. SIAM J. Numer. Anal. **29**(1), 123–135 (1992)

On the Convergence of Open Loop Nash Equilibria in Mean Field Games with a Local Coupling

Pierre Cardaliaguet

Abstract The paper studies the convergence, as N tends to infinity, of a system of N weakly coupled Hamilton–Jacobi equations (the open loop Nash system) when the coupling between the players becomes increasingly singular. The limit equation is a mean field game system with local coupling.

Keywords Mean field games · Mean field limit · Convergence rate

In this paper we continue the investigation of the mean field limit in differential games, i.e., the limit of the Nash equilibria in N-players differential games as N tends to infinity. We focus on equilibria in "open loop control", where players observe only their own position, but not the position of the other players. This question has been first discussed by Lasry and Lions in [10, 12] for the stationary problem in a Markovian setting. In the non Markovian setting (for finite horizon problems), Fischer [6] proved the convergence under suitable independence conditions and Lacker [9] identified all possible limits of the system. The key assumption in all these papers is that the coupling between the player is nonlocal and regularizing. Here we study the time dependent problem, with a coupling which becomes increasingly singular as the number of players tends to infinity. A similar question was addressed in [2] in the more complex framework of Nash equilibria in "closed loop controls", in which the players observe each other completely. This later paper strongly relied on [3] (see also [4]), where the convergence problem for closed loop Nash equilibria with regularizing coupling functions was established and several key techniques of proof introduced.

There are two reasons to study the "open loop" case: the first one is that in term of modeling, it is as natural to assume that players do not observe each other as to assume that they observe each other completely: the reality is probably in between. Second, the question of convergence is mathematically intriguing because the previous works concerned with open-loop problems were using compactness

P. Cardaliaguet (✉)
Université Paris-Dauphine, PSL Research University, CNRS, Ceremade, 75016 Paris, France
e-mail: cardaliaguet@ceremade.dauphine.fr

F. Alabau-Boussouira et al. (eds.), *Trends in Control Theory and Partial Differential Equations*, Springer INdAM Series 32,
https://doi.org/10.1007/978-3-030-17949-6_6

techniques, which seems to be ill-suited with the fact that the coupling becomes more and more singular.

To be more specific, let us write the open-loop Nash equilibrium system (in Markovian form). It reads, for any $i \in \{1, \dots, N\}$:

$$\begin{cases} -\partial_t v^{N,i} - \Delta v^{N,i} + H(x_i, Dv^{N,i}) = \int_{(\mathbb{T}^d)^{N-1}} F^N(x_i, m_{\boldsymbol{x}}^{N,i}) \prod_{j\neq i} m^j(t, x_j) dx_j \text{ in } (0,T)\times\mathbb{T}^d, \\ \partial_t m^i - \Delta m^i - \operatorname{div}(m^i D_p H(x_i, Dv^{N,i})) = 0 \quad \text{in } (0,T)\times\mathbb{T}^d, \\ m^i(0,\cdot) = \mathcal{L}(Z^i), \quad v^{N,i}(T, x_i) = G(x_i) \quad \text{in } \mathbb{T}^d, \end{cases} \tag{1}$$

where we set, for $\boldsymbol{x} = (x_1, \dots, x_N) \in (\mathbb{T}^d)^N$, $m_{\boldsymbol{x}}^{N,i} = \frac{1}{N-1}\sum_{j\neq i} \delta_{x_j}$. In the above system, the data are the horizon $T > 0$, the Hamiltonian $H : \mathbb{T}^d \times \mathbb{R}^d \to \mathbb{R}$, the terminal condition $G : \mathbb{T}^d \to \mathbb{R}$, the initial distribution of players $m_0 \in \mathcal{P}(\mathbb{T}^d)$ ($\mathcal{P}(\mathbb{T}^d)$ being the set of probability measures on $\mathbb{T}^d$) and the maps $F^N : \mathbb{T}^d \times \mathcal{P}(\mathbb{T}^d) \to \mathbb{R}$. The maps F^N are called the coupling because they are responsible of all the interactions between the equations. To avoid issues related to the boundary conditions, we work with periodic (in space) boundary data: $\mathbb{T}^d = \mathbb{R}^d/\mathbb{Z}^d$.

We are also interested in the associated system of N coupled stochastic differential equations (SDE), which corresponds to the optimal trajectories of the players:

$$dY_{i,t} = -D_p H\big(Y_{i,t}, Dv^{N,i}(t, Y_{i,t})\big)dt + \sqrt{2}dB_t^i, \qquad t \in [0,T],\ i \in \{1,\dots,N\}, \tag{2}$$

where the $((B_t^i)_{t\in[0,T]})_{i=1,\dots,N}$ are d-dimensional independent Brownian motions.

Our main assumption is that the maps F^N become increasingly singular, in the sense that there exists a smooth (local) map $F : \mathbb{R}^d \times [0, +\infty) \to \mathbb{R}$ such that

$$\lim_{N\to+\infty} F^N(x_i, m\mathrm{d}x) = F(x_i, m(x_i)), \tag{3}$$

for any sufficiently smooth probability density $m\mathrm{d}x = m(x)\mathrm{d}x$. Our aim is to show that, under suitable assumption on the rate of convergence in (3), the limit of the $v^{N,i}$ is given by the MFG system with a local coupling, given as the forward-backward system of PDEs:

$$\begin{cases} -\partial_t u - \Delta u + H(x, Du) = F(x, m(t,x)) \quad \text{in } (0,T)\times\mathbb{T}^d, \\ \partial_t m - \Delta m - \operatorname{div}(m D_p H(x, Du)) = 0 \quad \text{in } (0,T)\times\mathbb{T}^d, \\ u(T,x) = G(x),\ m(0,\cdot) = m_0 \quad \text{in } \mathbb{T}^d \end{cases} \tag{4}$$

This system has been thoroughly studied in the literature (cf. [7, 8, 12–14] and the references therein). The typical assumptions ensuring the above MFG system to be well-posed are that F is monotone with respect to its second variable while H is convex in its second variable (plus growth conditions on F and H).

Our main result (Theorem 3.4) states that the solution of (1) converges to the solution of (4) as N tends to infinity. Moreover, we describe in Theorem 3.5 the limit of the solution to (2) in terms of the optimal solution associated to (4). The technique of proof differs in a substantial way from the ones in [2, 3] (as the Nash system is open-loop, there is no need to use the so-called master equation) and [6, 9, 10, 12] (because compactness techniques do not seem relevant). The idea consists in using a well-known monotonicity technique—first developed in [10–12] to show the uniqueness of the solution of the MFG systems—combined with the development of the map $v^{N,i}$ along the optimal trajectories, as explained in [3].

The paper is organized in the following way: we first state our main notation and assumptions and recall how to approximate the solution to the MFG system with the local coupling F by the solution of the MFG system with the nonlocal coupling F^N. In Sect. 2 we explain how to interpret system (1) in terms of open-loop Nash equilibria. The main results are given in Sect. 3, where we prove the convergence of System (1) to System (4) and describe the propagation of chaos for (2).

1 Preliminaries

1.1 Notation

For the sake of simplicity, the paper is written under the assumption that all maps are periodic in space. So the underlying state space is the torus $\mathbb{T}^d = \mathbb{R}^d/\mathbb{Z}^d$. This simplifying assumption allows to discard possible problems at infinity (or at the boundary of a domain). We denote by $|\cdot|$ the euclidean norm in $\mathbb{R}^d$ and—by abuse of notation—the corresponding distance in $\mathbb{T}^d$. The ball centered at $x \in \mathbb{T}^d$ and of radius R is denoted by $B_R(x)$.

For $k \in \mathbb{N}$ and $\alpha \in (0, 1)$, we denote by $C^{k+\alpha}$ the set of maps $u = u(x)$ which are of class C^k and $D^k u$ is α-Holder continuous. When $u = u(t, x)$ is time dependent and $\alpha \in (0, 1)$, we say that u is in $C^{0,\alpha}$ if

$$\|u\|_{C^{0,\alpha}} := \|u\|_\infty + \sup_{(t,x),(t',x')} \frac{|u(t,x) - u(t',x')}{|x - x'|^\alpha + |t - t'|^{\alpha/2}} < +\infty.$$

We say that u is in $C^{1,\alpha}$ if u and Du belong to $C^{0,\alpha}$. Finally $C^{2,\alpha}$ consists in the maps u such that $D^2 u$ and $\partial_t u$ belong to $C^{0,\alpha}$. It is known that, if u is in $C^{2,\alpha}$, then u is also in $C^{1,\alpha}$.

We denote by $\mathcal{P}(\mathbb{T}^d)$ the set of Borel probability measures on the torus $\mathbb{T}^d := \mathbb{R}^d/\mathbb{Z}^d$. It is endowed with the Monge–Kantorovitch distance:

$$\mathbf{d}_1(m, m') = \sup_\phi \int_{\mathbb{T}^d} \phi(y)\, d(m - m')(y),$$

where the supremum is taken over all 1-Lipschitz continuous maps $\phi : \mathbb{T}^d \to \mathbb{R}$.

1.2 Assumption

Throughout the paper, we suppose that the following conditions are in force.

(H1) The Hamiltonian $H : \mathbb{T}^d \times \mathbb{R}^d \to \mathbb{R}$ is smooth, H and D_pH are globally Lipschitz continuous in both variables and H is locally uniformly convex with respect to the second variable:

$$D^2_{pp}H(x,p) > 0 \qquad \forall (x,p) \in \mathbb{T}^d \times \mathbb{R}^d. \tag{5}$$

(H2) $F : \mathbb{T}^d \times [0,+\infty) \to \mathbb{R}$ is smooth, globally Lipschitz continuous in both variables and increasing with respect to the second variable with $\partial_m F \geqslant \delta > 0$ for some $\delta > 0$.

(H3) The terminal cost $G : \mathbb{T}^d \to \mathbb{R}$ is a smooth map.

(H4) For any $N \in \mathbb{N}$, $F^N : \mathbb{T}^d \times \mathcal{P}(\mathbb{T}^d) \to \mathbb{R}$ is monotone:

$$\int_{\mathbb{T}^d} (F^N(x,m) - F^N(x,m')d(m-m')(x) \geqslant 0 \qquad \forall m, m' \in \mathcal{P}(\mathbb{T}^d).$$

(H5) (difference between F^N and F) For any $R > 0$ and $\alpha \in (0,1)$, there exists $k_N^{R,\alpha} \to 0$ as $N \to +\infty$ such that

$$\|F^N(\cdot, m\mathrm{d}x) - F(\cdot, m(\cdot))\|_\infty \le k_N^{R,\alpha} \tag{6}$$

for any density m such that $\|m\|_{C^\alpha} \le R$.

(H6) (uniform regularity of F^N for regular densities) For any $R > 0$ and $\alpha \in (0,1)$, there exists $\kappa^{R,\alpha} > 0$ such that, for any $N \in \mathbb{N}$,

$$\left|F^N(x, mdx) - F^N(y, m'dx)\right| \le \kappa_{R,\alpha}\left(|x-y|^\alpha + \|m - m'\|_\infty\right) \tag{7}$$

for any density m, m' with $\|m\|_{C^\alpha}, \|m'\|_{C^\alpha} \le R$.

(H7) (regularity assumptions on F^N for general probability measures) For any $N \in \mathbb{N}$, there exists a constant $K_N \geqslant 1$ such that

$$\left|F^N(x,m) - F^N(x,m')\right| \le K_N \mathbf{d}_1(m,m') \qquad \forall x \in \mathbb{T}^d,\ \forall m, m' \in \mathcal{P}(\mathbb{T}^d). \tag{8}$$

Let us briefly comment upon the assumptions. First we note that, by **(H5)**, $F^N(x,m)$ becomes closer and closer to $F(x,m)$ for any smooth density m while its regularity at general probability measures deteriorates (i.e., $K_N \to +\infty$ as $N \to +\infty$). The monotonicity assumptions **(H2)** and **(H4)** on F and F^N and the convexity of H are known to ensure the uniqueness of the solution in the MFG systems: they are therefore natural in our study. The global Lipschitz regularity assumption on H in **(H1)** is not natural in the context of MFG, but it simplifies a lot the well-posedness of the MFG system (4) and is used in every key step of the paper. In the same way, the map G is assumed to be independent the measure in order to simplify the analysis of Sect. 1.3: indeed we do not know how to extend the estimates of this subsection when G depends on m in a local way. Note however that we could allow G to depend on m in a smooth, nonlocal and monotone way, as in [3]; however since we concentrate here on local couplings, we do not present this easy generalization. We explain in Remark 3.2 that the strong monotonicity condition on F can be avoided (F non decreasing suffices), but the convergence rates in Theorems 3.1 and 3.4 then deteriorates a little.

As explained in [2], given a map F satisfying the above conditions, a typical example for the regularization F^N is the following:

Proposition 1.1 *Assume that* $F^N = F^{\epsilon_N}$ *with*

$$F^\epsilon(x, m) := F(\cdot, \xi^\epsilon \star m(\cdot)) \star \xi^\epsilon(x) \tag{9}$$

where $\epsilon_N \to 0$ *as* $N \to +\infty$ *and* $\xi^\epsilon(x) = \epsilon^{-d}\xi(x/\epsilon)$, ξ *being a symmetric smooth nonnegative kernel with compact support. Then, for any* N, F^N *is monotone and satisfies* (7).

Moreover the constants $k_N^{R,\alpha}$ *and* K_N *in* **(H5)** *and* **(H7)** *respectively can be estimated by*

$$k_N^{R,\alpha} \le C(1+R)\epsilon_N^\alpha, \qquad K_N \le C\epsilon_N^{-2d-12-3\alpha}, \tag{10}$$

where C *depends on the regularity of* F *and of* ξ.

1.3 Regularity Estimates

Let $m_0 \in \mathcal{P}(\mathbb{T}^d)$ and let (u^N, m^N) and (u, m) be respectively the unique solution to the MFG systems

$$\begin{cases} -\partial_t u^N - \Delta u^N + H(x, Du^N) = F^N(x, m(t)) & \text{in } (0,T) \times \mathbb{T}^d, \\ \partial_t m^N - \Delta m^N - \mathrm{div}(m^N D_p H(x, Du^N)) = 0 & \text{in } (0,T) \times \mathbb{T}^d, \\ u^N(T, x) = G(x), \ m^N(0, \cdot) = m_0 & \text{in } \mathbb{T}^d, \end{cases} \tag{11}$$

and

$$\begin{cases} -\partial_t u - \Delta u + H(x, Du) = F(x, m(t, x)) & \text{in } (0,T) \times \mathbb{T}^d, \\ \partial_t m - \Delta m - \mathrm{div}(m D_p H(x, Du)) = 0 & \text{in } (0,T) \times \mathbb{T}^d, \\ u(T, x) = G(x), \ m(0, \cdot) = m_0 & \text{in } \mathbb{T}^d. \end{cases} \tag{12}$$

Following [12, 13], these systems are known to be well-posed. We recall the following estimates on the regularity of (u^N, m^N) and on the difference between (u^N, m^N) and (u, m).

Proposition 1.2 ([2]) *Assume that m_0 has a positive density of class C^2. Then the (u^N, m^N) are bounded in $C^{2,\alpha} \times C^{0,\alpha}$ independently of N. Moreover,*

$$\sup_{t\in[0,T]} \|u^N(t,\cdot) - u(t,\cdot)\|_{H^1(\mathbb{T}^d)} + \|m^N - m\|_{L^2} \leq Ck_N^{R,\alpha},$$

$$\sup_{t\in[0,T]} \|Du^N(t,\cdot) - Du(t,\cdot)\|_\infty \leq C\left(k_N^{R,\alpha}\right)^{\frac{2}{d+2}}.$$

where α, R and C depend on on the data and m_0, but not on N.

A straightforward consequence of Proposition 1.2 is the following estimate on optimal trajectories related with the MFG systems (11) and (12).

Corollary 1.3 *Let $m_0 \in \mathcal{P}(\mathbb{T}^d)$, (u^N, m^N) and (u, m) be the solution to the MFG system* (11) *and* (12) *respectively. Let Z be a random variable with law m_0 which is independent of a Brownian motion (B_t). If $(\tilde{X}_t)$ and (X_t) are the solution to*

$$\begin{cases} d\tilde{X}_t = -D_pH(\tilde{X}_t, Du(t,\tilde{X}_t))dt + \sqrt{2}dB_t, & t \in [0,T], \\ \tilde{X}_0 = Z, \end{cases}$$

and

$$\begin{cases} dX_t = -D_pH(X_t, Du^N(t,X_t))dt + \sqrt{2}dB_t, & t \in [0,T], \\ X_0 = Z, \end{cases}$$

respectively, then

$$\mathbb{E}\left[\sup_{t\in[0,T]} \left|\tilde{X}_t - X_t\right|\right] \leq C\left(k_N^{R,\alpha}\right)^{\frac{2}{d+2}},$$

where C, R and α are as in Proposition 1.2.

2 Open Loop Nash Equilibria

Fix $\mathbf{Z} = (Z_1, \ldots, Z_N)$ a family of i.i.d. random variables on $\mathbb{T}^d$ with law $m_0 \in \mathcal{P}(\mathbb{T}^d)$.

Let $\mathcal{A}$ be the set of maps $\alpha : [0,T] \times \mathbb{T}^d \to \mathbb{R}^d$ which are Borel measurable and bounded. We look at $\mathcal{A}$ as the set of strategies of the players. We call—improperly—a strategy $\alpha \in \mathcal{A}$ an *open loop strategy* because it will depend only on player i, and not of the other players, through the state equation of player i:

$$dX_t^{N,i} = \alpha(t, X_t^{N,i})dt + \sqrt{2}dB_t^{N,i}, \qquad X_0 = Z_i. \tag{13}$$

In the above equation the $(B^{N,i})$ are independent Brownian motions, independent of the (Z_i) and the equation is understood in a weak sense. Given an N-tuple $\alpha = (\alpha^1, \ldots, \alpha^N) \in \mathcal{A}^N$, the cost of player i, for $i \in \{1, \ldots, N\}$, is given by

$$J^{N,i}(m_0, \alpha_i, (\alpha_j)_{j\neq i}) := \mathbb{E}\Big[\int_0^T \int_{\mathbb{T}^d} L(X_t^{N,i}, \alpha^i(t, X_t^{N,i})) + F^N(X_t^{N,i}, m_{X_t^N}^{N,i})dt + G(X_T^{N,i})\Big]$$

where, $(X_t^{N,i})$ is the solution to (13) with initial condition Z_i and $m_{X_t^N}^{N,i} = \frac{1}{N-1}\sum_{j\neq i}\delta_{x_j}$. Note for later use that, if we set

$$F^{N,i}(t, x_i; m_0, \alpha_i, (\alpha_j)_{j\neq i}) := \mathbb{E}\Big[F(x_i, m_{X_t^N}^{N,i})\Big], \tag{14}$$

then, as the $(X^{N,i})_{i=1,\dots,N}$ are independent, we have

$$F^{N,i}(t, x_i; m_0, \alpha_i, (\alpha_j)_{j\neq i}) = \int_{(\mathbb{T}^d)^{N-1}} F^N(x_i, m_{\boldsymbol{x}}^{N,i}) \prod_{j\neq i} m^j(t, x_j)dx_j,$$

where, for any $j \in \{1, \dots, N\}$, $m^j(t)$ is the law of $X_t^{N,j}$, i.e., the solution of the Kolmogorov equation

$$\begin{cases} \partial_t m^j - \Delta m^j + \operatorname{div}(m^j \alpha^j) = 0 & \text{in } (0, T) \times \mathbb{T}^d, \\ m^i(0, \cdot) = \mathcal{L}(Z_i) = m_0 & \text{in } \mathbb{T}^d. \end{cases} \tag{15}$$

Still because $(X^{N,i})_{i=1,\dots,N}$ are independent, we also have

$$\begin{aligned} J^{N,i}(m_0, \alpha_i, (\alpha_j)_{j\neq i}) = \mathbb{E}\Big[\int_0^T \int_{\mathbb{T}^d} & L(X_t^{N,i}, \alpha^i(t, X_t^{N,i})) \\ & + F^{N,i}(t, X_t^{N,i}; m_0, \alpha_i, (\alpha_j)_{j\neq i})dt + G(X_T^{N,i})\Big]. \end{aligned}$$

Definition 2.1 Fix $N \in \mathbb{N}$ with $N \geqslant 1$. We say that an N-tuple $\bar{\alpha} = (\bar{\alpha}^1, \dots, \bar{\alpha}^N) \in \mathcal{A}^N$ is an open loop Nash equilibrium if

$$J^{N,i}(m_0, \bar{\alpha}_i, (\bar{\alpha}_j)_{j\neq i}) \leq J^{N,i}(m_0, \alpha^i, (\bar{\alpha}_j)_{j\neq i})$$

for any $\alpha^i \in \mathcal{A}$.

The following Proposition shows that our definition corresponds to the construction of Nash equilibria in [10, 12] in our time dependent setting.

Proposition 2.2 *There exists at least one open-loop Nash equilibrium. Moreover, for any open-loop equilibrium $\bar{\alpha} = (\bar{\alpha}^1, \dots, \bar{\alpha}^N)$, there exists a solution $(v^i, m^i)_{i=1,\dots,N}$ of* (1) *such that*

$$\bar{\alpha}^i(t, x) := -D_pH(x, Dv^i(t, x)) \quad \forall (t, x) \in [0, T] \times \mathbb{T}^d, \quad \forall i \in \{1, \dots, N\}.$$

Finally there exists $\bar{\alpha}_0 \in \mathcal{A}$ such that $\bar{\alpha} := (\bar{\alpha}_0, \ldots, \bar{\alpha}_0) \in \mathcal{A}^N$ is an open-loop Nash equilibrium with associated maps $(v^i, m^i)_{i=1,\ldots,N}$ of (1) *independent of i. We call this later equilibrium a symmetric open-loop Nash equilibrium.*

Proof We consider $\mathcal{A}_b$ the subset of $\alpha \in \mathcal{A}$ such that $\|\alpha\|_\infty \leq \|D_pH\|_\infty$. We endow $\mathcal{A}_b$ with the L^∞ norm. On the closed convex set $(\mathcal{A}_b)^N$, we consider the map Φ defined as follows. For $\alpha = (\alpha^1, \ldots, \alpha^N) \in (\mathcal{A}_b)^N$ and $i \in \{1, \ldots, N\}$, let $X^{N,i}$ be the weak solution to (13), $F^{N,i}$ be defined by (14). We consider the solution v^i to

$$\begin{cases} -\partial_t v^i - \Delta v^i + H(x, Dv^i) = F^{N,i}(t, x; m_0, \alpha_i, (\alpha_j)_{j\neq i}) & \text{in } (0, T) \times \mathbb{T}^d \\ v^i(T, x) = G(x) \quad \text{in } \mathbb{T}^d, \end{cases} \tag{16}$$

where the map $F^{N,i}$ is defined in (14). Note that, by our standing assumptions on F^N, the map $(t, x) \to F^{N,i}(t, x; m_0, \alpha) := F^{N,i}(t, x; m_0, \alpha_i, (\alpha_j)_{j\neq i})$ is Lipschitz continuous in space (uniformly with respect to α) and satisfies in time the inequality

$$\begin{aligned} \left|F^{N,i}(t, x; m_0, \alpha) - F^{N,i}(t', x; m_0, \alpha)\right| &\leq C\mathbb{E}\left[\mathbf{d}_1(m^{N,i}_{X^N_t}, m^{N,i}_{X^N_{t'}})\right] \leq \frac{C}{N}\sum_{j\neq i}\mathbb{E}\left[|X^{N,j}_t - X^{N,j}_{t'}|\right] \\ &\leq C(1 + \sup_j \|\alpha^j\|_\infty)|t - t'|^{1/2} \leq C(1 + \|D_pH\|_\infty)|t - t'|^{1/2}. \end{aligned}$$

(Here the constant C depends on N, which is not an issue since$N$ is fixed in this part). Recall also that, by assumption **(H3)**, the map G is in $C^{2+\alpha}$. Therefore v^i belongs to $C^{2,\alpha}$ with a norm depending only on the data and on N. Let us set $\tilde{\alpha}^i := -D_pH(\cdot, Dv^i(\cdot, \cdot))$ and $\Phi(\alpha) = (\tilde{\alpha}^1, \ldots, \tilde{\alpha}^N)$. By our estimates, the map Φ is compact and continuous on $(\mathcal{A}_b)^N$ and hence has a fixed point $\bar{\alpha}$. Note that, by the classical verification Theorem, this fixed point is an open loop Nash equilibrium in the sense of Definition 2.1.

Let us check that any open loop feedback Nash equilibrium has the claimed form. Let $\bar{\alpha} = (\bar{\alpha}^1, \ldots, \bar{\alpha}^N)$ be such an open loop Nash equilibrium. We denote by $(X^{N,i})$ the solution to (13) with $\bar{\alpha}^i$ instead of α^i. For any j, the law $m^j(t)$ of $X^{N,j}_t$ satisfies the Kolmogorov equation (15). Let v^i be the solution to (16). By the optimality of $\bar{\alpha}^i$, we have

$$\mathbb{E}[v^i(0, Z_i)] = \mathbb{E}\left[\int_0^T \int_{\mathbb{T}^d} L(X^{N,i}_t, \bar{\alpha}^i(t, X^{N,i}_t)) + F^{N,i}(t, X^{N,i}_t; m_0, \alpha)dt + G(X^{N,i}_T)\right]. \tag{17}$$

On the other hand, by Itô's formula, we have

$$\mathbb{E}[v^i(0, Z_i)] = \mathbb{E}\left[-\int_0^T (\partial_t v^i + \Delta v^i + \bar{\alpha}^i \cdot Dv^i)(t, X^{N,i}_t)dt + G(X^{N,i}_T)\right].$$

Using the equation of v^i, we obtain

$$\mathbb{E}[v^i(0, Z_i)] = \mathbb{E}\Big[\int_0^T (-H(X_t^{N,i}, Dv^i(t, X_t^{N,i})) - \bar{\alpha}^i(t, X_t^{N,i}) \cdot Dv^i(t, X_t^{N,i}) \\ + F^{N,i}(t, X_t^{N,i}, m_0, \bar{\alpha}))\, dt + G(X_T^{N,i})\Big].$$

As H is locally strictly convex in the second variable (assumption **(H1)**) and $\bar{\alpha}^i$ is bounded, we have therefore

$$\mathbb{E}[v^i(0, Z_i)] \geqslant \mathbb{E}\Big[\int_0^T (L(X_t^{N,i}, \bar{\alpha}^i(t, X_t^{N,i})) + C^{-1}|\bar{\alpha}^i(t, X_t^{N,i}) - D_pH(X_t^{N,i}, Dv^i(t, X_t^{N,i})|^2 \\ + F^{N,i}(t, X_t^{N,i}, m_0, \bar{\alpha}))\, dt + G(X_T^{N,i})\Big] \\ \geqslant \mathbb{E}[v^i(0, Z_i)] + C^{-1}\mathbb{E}\Big[\int_0^T |\bar{\alpha}^i(t, X_t^{N,i}) + D_pH(X_t^{N,i}, Dv^i(t, X_t^{N,i})|^2\Big],$$

where we used equality (17) in the last line. Thus $\bar{\alpha}^i(t, X_t^{N,i}) = -D_pH(X_t^{N,i}, Dv^i(t, X_t^{N,i}))$ a.e. Since the law of $X_t^{N,i}$ has a positive density, we conclude that $\bar{\alpha}^i(t, x) = -D_pH(x, Dv^i(t, x))$ for a.e. (t, x). This shows that the (u^i, m^i) satisfy (1).

The existence of a symmetric Nash equilibrium is obtained by a very similar argument, defining the map $\Phi : \mathcal{A}_b \to \mathcal{A}_b$ by requiring that the strategies of the players are the same.

3 Convergence

Let us fix $m_0 \in \mathcal{P}(\mathbb{T}^d)$ with a C^2 density. For an integer $N \geqslant 2$, we consider a symmetric equilibrium in open-loop form $(v^{N,i}, m^{N,i})_{i\in\{1,\dots,N\}}$ of (1). Note that, by symmetry, there exists actually a single unknown $v^N = v^{N,i}$ for any $i \in \{1, \dots, N\}$.

Our aim is to prove that v^N is close to u, where (u, m) is the solution of the MFG system (12). For this we first compare v^N and u^N, where (u^N, m^N) is the solution of the perturbed MFG system (11).

3.1 Estimates Between $v^{N,i}$ and u^N

Let $(Z_i)_{i\in\{1,\dots,N\}}$ be an i.i.d family of N random variables of law m_0. We set $\boldsymbol{Z} = (Z_i)_{i\in\{1,\dots,N\}}$. Let also $((B_t^i)_{t\in[0,T]})_{i\in\{1,\dots,N\}}$ be a family of N independent d-dimensional Brownian Motions which is also independent of $(Z_i)_{i\in\{1,\dots,N\}}$. We consider the systems of SDEs with variables $(\boldsymbol{X}_t = (X_{i,t})_{i\in\{1,\dots,N\}})_{t\in[0,T]}$ and $(\boldsymbol{Y}_t = (Y_{i,t})_{i\in\{1,\dots,N\}})_{t\in[0,T]}$:

$$\begin{cases} dX_{i,t} = -D_pH\big(X_{i,t}, Du^N(t, X_{i,t})\big)dt + \sqrt{2}dB_t^i & t \in [0, T], \\ X_{i,0} = Z_i, \end{cases} \tag{18}$$

and
$$\begin{cases} dY_{i,t} = -D_pH\big(Y_{i,t}, Dv^N(t, Y_{i,t})\big)dt + \sqrt{2}dB_t^i & t \in [0, T], \\ Y_{i,0} = Z_i. \end{cases} \tag{19}$$

Note that the $(X_{i,t})$ are i.i.d. with law $m^N(t)$. The X_i and the Y_i depend on N, but we do not write this dependence explicitly for the sake of simplicity.

Theorem 3.1 *Assume that K_N is such that*

$$\begin{cases} K_N N^{-\frac{1}{d}} \leq \bar{C}^{-1} & \text{if } d \geqslant 3 \\ K_N N^{-\frac{1}{2}} \log(N) \leq \bar{C}^{-1} & \text{if } d = 2 \end{cases} \tag{20}$$

for some constant $\bar{C}$. Then, for any $i \in \{1, \dots, N\}$,

$$\mathbb{E}\left[\sup_{t\in[0,T]} |X_{i,t} - Y_{i,t}|\right] \leq \begin{cases} CK_N^{\frac{1}{2}}N^{-\frac{1}{2d}} & \text{if } d \geqslant 3 \\ CK_N^{\frac{1}{2}}N^{-\frac{1}{4}} \log^{\frac{1}{2}}(N) & \text{if } d = 2 \end{cases}$$

and

$$\mathbb{E}\Big[|u^N(0, Z_i) - v^N(0, \mathbf{Z})|\Big] \leq \begin{cases} C\left(K_N^{\frac{3}{2}}N^{-\frac{1}{2d}} + K_N N^{-\frac{1}{d}}\right) & \text{if } d \geqslant 3 \\ C\left(K_N^{\frac{3}{2}}N^{-\frac{1}{4}}\log^{\frac{1}{2}}(N) + K_N N^{-\frac{1}{2}}\log(N)\right) & \text{if } d = 2 \end{cases}$$

where the constant C depends on m_0 but not on N.

Proof Following Proposition 1.2, we know that u^N is bounded in $C^{2,\alpha}$ (for some $\alpha \in (0, 1)$): we will use this uniform regularity all along the section.

By Itô's formula, we have

$$\mathbb{E}^{Z_i}\Big[u^N(T, X_{i,T})\Big] = \mathbb{E}^{Z_i}\left[u^N(0, Z_i) + \int_0^T \partial_t u^N + \Delta u^N - Du^N \cdot D_pH(X_{i,t}, Du^N(t, X_{i,t}))\, dt\right],$$

where u^N and its derivatives are evaluated at $(t, X_{i,t})$ and where $\mathbb{E}^{Z_i}[\cdot]$ denotes the conditional expectation with respect to Z_i. As u^N solves (11), we get therefore

$$\begin{aligned} u^N(0, Z_i) &= \mathbb{E}^{Z_i}\Big[\int_0^T (-H(X_{i,t}, Du^N) \\ &\qquad + D_pH(X_{i,t}, Du^N) \cdot Du^N + F^N(X_{i,t}, m^N(t)))\, dt + G(X_{i,T})\Big]. \end{aligned}$$

We now compute the variation of $v^N(t, X_{i,t})$. We have, using the equation satisfied by v^N,

$$
\begin{aligned}
dv^N(t, X_{i,t}) &= (\partial_t v^N + \Delta v^N \\
&\quad - Dv^N \cdot D_p H(X_{i,t}, Du^N(t, X_{i,t}))dt + \sqrt{2} Dv^N \cdot dB_t^i \\
&= (H(X_{i,t}, Dv^N) - Dv^N \cdot D_p H(X_{i,t}, Du^N(t, X_{i,t})))dt \\
&\quad - F^{N,i}(t, X_{i,t}; \mathbf{Z}, \bar{\alpha})dt + \sqrt{2} Dv^N \cdot dB_t^i
\end{aligned}
$$

where v^N and its derivatives are evaluated at $(t, X_{i,t})$ and $F^{N,i}$ is defined by

$$
\begin{aligned}
F^{N,i}(t, x_i; \mathbf{Z}, \bar{\alpha}) &= \int_{(\mathbb{T}^d)^{N-1}} F^N(x_i, m_{\boldsymbol{x}}^{N,i}) \prod_{j \neq i} m^j(t, x_j) dx_j \\
&= \mathbb{E}\left[F^N(x_i, m_{\mathbf{Y}_t}^{N,i}) \right]
\end{aligned}
$$

Note that, since $X_{i,t}$ and $(Y_{j,t})_{j \neq i}$ are independent, we have

$$
\mathbb{E}\left[F^{N,i}(t, X_{i,t}; \mathbf{Z}, \bar{\alpha}) \right] = \mathbb{E}\left[F^N(X_{i,t}, m_{\mathbf{Y}_t}^{N,i}) \right].
$$

Hence

$$
\begin{aligned}
v^N(0, Z_i) &= \mathbb{E}^{Z_i}\left[\int_0^T (-H(X_{i,t}, Dv^N) + Dv^N \cdot D_p H(X_{i,t}, Du^N(t, X_{i,t})) \right. \\
&\quad \left. + F^N(X_{i,t}, m_{\mathbf{Y}_t}^{N,i}) \, dt + G(X_{i,T}) \right].
\end{aligned}
$$

So

$$
\begin{aligned}
& u^N(0, Z_i) - v^N(0, Z_i) \\
&\quad = \mathbb{E}^{Z_i}\left[\int_0^T \left[H(X_{i,t}, Dv^N) - H(X_{i,t}, Du^N) - D_p H(X_{i,t}, Du^N) \cdot (Dv^N - Du^N) \right. \right. \\
&\qquad \left. \left. + (F^N(X_{i,t}, m^N(t)) - F^N(X_{i,t}, m_{\mathbf{Y}_t}^{N,i}) \right] dt \right].
\end{aligned} \tag{21}
$$

Recall that Du^N is bounded by some constant R independently of N. Let us set, for $z \geqslant 0$,

$$
\Psi(z) = \begin{cases} z^2 & \text{if } z \in [0, 1] \\ 2z - 1 & \text{if } z \geqslant 1 \end{cases} \tag{22}
$$

From Lemma 3.3 below, there exists $C_0 > 0$ (which depends on R) such that

$$
H(x, q) - H(x, p) - D_p H(x, p) \cdot (q - p) \geqslant C_0^{-1} \Psi(|q - p|) \qquad \forall p, q \text{ with } |p| \leq R.
$$

Therefore

$$
\begin{aligned}
& u^N(0, Z_i) - v^N(0, Z_i) \\
&\quad \geqslant \mathbb{E}^{Z_i}\left[\int_0^T C_0^{-1} \Psi\left(|Dv^N(t, X_{i,t}, \mathbf{Y}_t^i) - Du^N(t, X_{i,t})| \right) \right. \\
&\qquad \left. + (F^N(X_{i,t}, m^N(t)) - F^N(X_{i,t}, m_{\mathbf{Y}_t}^{N,i}) \right] dt \right].
\end{aligned} \tag{23}
$$

Computing in the same way the variation of the terms $-u^N(t, Y_{i,t}) + v^N(t, Y_{i,t})$, we find

$$
\begin{aligned}
&-u^N(0, Z_i) + v^N(0, Z_i) \\
&= \mathbb{E}^{Z_i}\Big[\int_0^T \big[H(Y_{i,t}, Du^N) - H(Y_{i,t}, Dv^N) - D_pH(Y_{i,t}, Dv^N)\cdot(Du^N - Dv^N) \\
&\qquad +(F^N(Y_{i,t}, m_{\mathbf{Y}_t}^{N,i}) - F^N(Y_{i,t}, m^N(t))\big]dt\Big],
\end{aligned}
$$

where Du^N and Dv^N are computed at $(t, Y_{i,t})$.

In order to estimate the first term in the right-hand side, we use Lemma 3.3 below to infer the existence of a constant $c_0 > 0$ (which depends on the uniform bound R on $\|Du^N\|_\infty$) such that

$$
H(x, q) - H(x, p) - D_pH(x, p)\cdot(q - p) \geqslant c_0 \min\{|p - q|^2, c_0\} \qquad \forall p, q \text{ with } |q| \le R.
$$

Therefore

$$
\begin{aligned}
&\mathbb{E}[-u^N(0, Z_i) + v^N(0, \mathbf{Z})] \\
&\quad\geqslant \mathbb{E}\Big[\int_0^T c_0 \min\{|Du^N(t, Y_{i,t}) - Dv^N(t, Y_{i,t})|^2, c_0\} \\
&\qquad\qquad +(F^N(Y_{i,t}, m_{\mathbf{Y}_t}^{N,i}) - F^N(Y_{i,t}, m^N(t)))\big]dt\Big].
\end{aligned}
$$

Combining this inequality with (23), we obtain

$$
\begin{aligned}
0 \geqslant\ &\mathbb{E}\Big[\int_0^T C_0^{-1}\Psi\left(|Dv^N(t, X_{i,t}) - Du^N(t, X_{i,t})|\right)\Big] \\
&+\mathbb{E}\Big[\int_0^T c_0 \min\{|Du^N(t, Y_{i,t}) - Dv^N(t, Y_{i,t})|^2, c_0\}\Big] \\
&+\mathbb{E}\Big[\int_0^T F^N(X_{i,t}, m^N(t)) - F^N(X_{i,t}, m_{\mathbf{Y}_t}^{N,i}) - F^N(Y_{i,t}, m^N(t)) + F^N(Y_{i,t}, m_{\mathbf{Y}_t}^{N,i})\, dt\Big].
\end{aligned}
$$

Let us set $m_{\mathbf{X}_t}^N = \frac{1}{N}\sum_j \delta_{X_{j,t}}$ and $m_{\mathbf{Y}_t}^N = \frac{1}{N}\sum_j \delta_{Y_{j,t}}$. We note that $\mathbf{d}_1(m_{\mathbf{Y}_t}^{N,i}, m_{\mathbf{Y}_t}^N) \le CN^{-1}$. Moreover, as the $(X_{i,t})$ are i.i.d. with law $m^N(t)$, a result by Dereich, Scheutzow and Schottstedt [5] implies that, for $d \geqslant 3$,

$$
\mathbb{E}\left[\mathbf{d}_1\left(m_{\mathbf{X}_t}^N, m^N(t)\right)\right] \le CN^{-\frac{1}{d}}.
$$

For $d = 2$, the estimate becomes (see Ajtai, Komlos and Tusnády [1]),

$$
\mathbb{E}\left[\mathbf{d}_1\left(m_{\mathbf{X}_t}^N, m^N(t)\right)\right] \le CN^{-\frac{1}{2}}\log(N).
$$

As F^N is K_N-Lipschitz continuous (recall (8)) we obtain (for $d \geqslant 3$)

$$
\begin{aligned}
CK_N N^{-\frac{1}{d}} \geqslant\ & \mathbb{E}\left[\int_0^T C_0^{-1}\Psi\left(|Dv^N(t,X_{i,t}) - Du^N(t,X_{i,t})|\right)\right] \\
&+\mathbb{E}\left[\int_0^T c_0 \min\{|Du^N(t,Y_{i,t}) - Dv^N(t,Y_{i,t})|^2, c_0\}\right] \\
&+\mathbb{E}\left[\int_0^T F^N(X_{i,t}, m^N_{\mathbf{X}_t}) - F^N(X_{i,t}, m^N_{\mathbf{Y}_t}) - F^N(Y_{i,t}, m^N_{\mathbf{X}_t}) + F^N(Y_{i,t}, m^N_{\mathbf{Y}_t}))dt\right].
\end{aligned}
$$

We now sum these expressions over i. Since

$$
\begin{aligned}
&\sum_i F^N(X_{i,t}, m^N_{\mathbf{X}_t}) - F^N(X_{i,t}, m^N_{\mathbf{Y}_t}) - F^N(Y_{i,t}, m^N_{\mathbf{X}_t}) + F^N(Y_{i,t}, m^N_{\mathbf{Y}_t}) \\
&\quad = \int_{\mathbb{T}^d} (F^N(x, m^N_{\mathbf{X}_t}) - F^N(x, m^N_{\mathbf{Y}_t}))d(m^N_{\mathbf{X}_t} - m^N_{\mathbf{Y}_t})(x) \geqslant 0,
\end{aligned}
$$

we obtain:

$$
\begin{aligned}
CK_N N^{1-\frac{1}{d}} \geqslant\ & \sum_i \mathbb{E}\left[\int_0^T C_0^{-1}\Psi\left(|Dv^N(t,X_{i,t}) - Du^N(t,X_{i,t})|\right)\right] \\
&+\sum_i \mathbb{E}\left[\int_0^T c_0 \min\{|Du^N(t,Y_{i,t}) - Dv^N(t,Y_{i,t})|^2, c_0\}\right].
\end{aligned}
$$

The random variables

$$
Dv^N(t, X_{i,t}) - Du^N(t, X_{i,t})
$$

have the same law for any i. In the same way, the random variables

$$
Du^N(t, Y_{i,t}) - Dv^N(t, Y_{i,t})
$$

have the same law for any i. We have therefore, for any $i \in \{1, \ldots, N\}$ and $d \geqslant 3$,

$$
\begin{aligned}
CK_N N^{-\frac{1}{d}} \geqslant\ & \mathbb{E}\left[\int_0^T C_0^{-1}\Psi\left(|Dv^N(t,X_{i,t}) - Du^N(t,X_{i,t})|\right)\right] \\
&+\mathbb{E}\left[\int_0^T c_0 \min\{|Du^N(t,Y_{i,t}) - Dv^N(t,Y_{i,t})|^2, c_0\}\right].
\end{aligned}
\tag{24}
$$

In view of the SDEs satisfied by the $(X_{i,t})$ and by the $(Y_{i,t})$, we have

$$
\begin{aligned}
|X_{i,t} - Y_{i,t}| &\leq \int_0^t |-D_pH(X_{i,s}, Du^N(s,X_{i,s})) + D_pH(Y_{i,s}, Dv^N(s,Y_{i,s}))|\, ds \\
&\leq \int_0^t |-D_pH(X_{i,s}, Du^N(s,X_{i,s})) + D_pH(Y_{i,s}, Du^N(s,Y_{i,s}))|\, ds \\
&\quad + \int_0^t |-D_pH(Y_{i,s}, Du^N(s,Y_{i,s})) + D_pH(Y_{i,s}, Dv^N(s,Y_{i,s}))|\, ds \\
&\leq C\int_0^t |X_{i,s} - Y_{i,s}|\, ds + C\int_0^T \min\{|Du^N(s,Y_{i,s}) - Dv^N(s,Y_{i,s})|, \|D_pH\|_\infty\}\, ds
\end{aligned}
$$

where we have used the bound and Lipschitz regularity of D_pH as well as the uniform Lipschitz bound of Du^N in the space variable x given in Proposition 1.2. So, by Gronwall's inequality and (24), we obtain, for any $i \in \{1, \dots, N\}$ and $d \geqslant 3$,

$$\mathbb{E}\left[\sup_{t\in[0,T]} |X_{i,t} - Y_{i,t}|\right] \le CK_N^{\frac{1}{2}} N^{-\frac{1}{2d}}. \tag{25}$$

When $d = 2$, the right-hand side has to be replaced by $CK_N^{\frac{1}{2}} N^{-\frac{1}{4}} \log^{\frac{1}{2}}(N)$.

In order to estimate $u^N(0, Z_i) - v^N(0, Z_i)$, we come back to (21). By the Lipschitz continuity of H and F^N we have:

$$\begin{aligned} &\mathbb{E}\left[|u^N(0, Z_i) - v^N(0, Z_i)|\right] \\ &\quad \le \mathbb{E}\left[\int_0^T C|Dv^N(t, X_{i,t}) - Du^N(t, X_{i,t})| + K_N \mathbf{d}_1(m^N(t), m^{N,i}_{\mathbf{Y}_t})dt\right]. \end{aligned} \tag{26}$$

Let us first estimate the first term in the right-hand side of (26) (for $d \geqslant 3$): we use inequality (24) and Jensen's inequality (since Ψ is convex and increasing):

$$\begin{aligned} &T^{-1}\mathbb{E}\left[\int_0^T |Dv^N(t, X_{i,t}) - Du^N(t, X_{i,t})|dt\right] \\ &\quad \le \Psi^{-1}\left(T^{-1}\mathbb{E}\left[\int_0^T \Psi\left(|Dv^N(t, X_{i,t}) - Du^N(t, X_{i,t})|\right) dt\right]\right) \\ &\quad \le \Psi^{-1}\left(CK_N N^{-\frac{1}{d}}\right) \end{aligned}$$

So, if we suppose as in assumption (20) that $K_N N^{-\frac{1}{d}} \le C^{-1}$, we obtain

$$\mathbb{E}\left[\int_0^T |Dv^N(t, X_{i,t}) - Du^N(t, X_{i,t})|dt\right] \le CK_N^{\frac{1}{2}} N^{-\frac{1}{2d}}.$$

To estimate the second term in the right-hand side of (26), we note that

$$\begin{aligned} \mathbb{E}\left[\mathbf{d}_1(m^N(t), m^{N,i}_{\mathbf{Y}_t})\right] &\le \mathbb{E}\left[\mathbf{d}_1(m^N(t), m^{N,i}_{\mathbf{X}_t}) + \mathbf{d}_1(m^{N,i}_{\mathbf{X}_t}, m^{N,i}_{\mathbf{Y}_t})\right] \\ &\le \mathbb{E}\left[\mathbf{d}_1(m^N(t), m^{N,i}_{\mathbf{X}_t}) + \frac{1}{N}\sum_{j\neq i} |X_{j,t} - Y_{j,t}|\right]. \end{aligned}$$

So, using, on the one hand, the fact that the $(X_{i,t})$ are i.i.d. with law $m^N(t)$ and the result by Dereich, Scheutzow and Schottstedt [5] and, on the other hand, inequality (25), we have

$$\mathbb{E}\left[\mathbf{d}_1(m^N(t), m^{N,i}_{\mathbf{Y}_t})\right] \le C\left(N^{-\frac{1}{d}} + K_N^{\frac{1}{2}} N^{-\frac{1}{2d}}\right).$$

This proves that, if $d \geqslant 3$,

$$\mathbb{E}\left[|u^N(0,Z_i) - v^N(0,Z_i)|\right] \leq C\left(K_N^{\frac{3}{2}} N^{-\frac{1}{2d}} + K_N N^{-\frac{1}{d}}\right).$$

When $d = 2$, the right-hand side becomes $C\left(K_N^{\frac{3}{2}} N^{-\frac{1}{4}} \log^{\frac{1}{2}}(N) + K_N N^{-\frac{1}{2}} \log(N)\right)$.

□

Remark 3.2 A variant of Theorem 3.1 can be obtained by replacing u^N by u in the definition of the X_i and in the whole proof, thus avoiding the approximation argument of Proposition 1.2. As the assumption $\partial_m F \geqslant \delta$ is only used in Proposition 1.2, this condition can then be removed. The price to pay is a deterioration of the convergence rate because the left-hand side of (24) has to involve a term of the form $\sup_t \|F(\cdot, m(t,\cdot)) - F^N(\cdot, m(t))\|_\infty$.

In the proof we used the

Lemma 3.3 *Assume that* $D^2_{pp}H > 0$ *and let* Ψ *be defined by* (22)*. Then, for any* $R > 0$*, there exists* $C_0, c_0 > 0$ *such that*

$$H(x,q) - H(x,p) - D_pH(x,p)\cdot(q-p) \geqslant C_0^{-1}\Psi(|q-p|) \qquad \forall p, q \text{ with } |p| \leq R$$

and

$$H(x,q) - H(x,p) - D_pH(x,p)\cdot(q-p) \geqslant c_0 \min\{|p-q|^2, c_0\} \qquad \forall p, q \text{ with } |q| \leq R.$$

Proof As $D^2_{pp}H > 0$, there exists $\theta > 0$, depending on R, such that $D^2_{pp}H \geqslant \theta$ in $\mathbb{T}^d \times B_{2R}(0)$. Let $x \in \mathbb{T}^d$, $p, q \in \mathbb{R}^d$ with $|p| \leq R$. If $|q-p| \leq R$, then by the lower bound $D^2_{pp}H \geqslant \theta$ we have

$$H(x,q) - H(x,p) - D_pH(x,p)\cdot(q-p) \geqslant \frac{\theta}{2}|p-q|^2.$$

Now assume that $|q-p| > R$. Let $\hat{q}$ be the projection of q onto the ball $B_R(p)$. Then (omitting the x dependence which plays no role)

$$\begin{aligned}
&H(q) - H(p) - D_pH(p)\cdot(q-p) \\
&\quad = H(q) - H(\hat{q}) - D_pH(\hat{q})\cdot(q-\hat{q}) + H(\hat{q}) - H(p) - D_pH(p)\cdot(\hat{q}-p) \\
&\qquad + (D_pH(\hat{q}) - D_pH(p))\cdot(q-\hat{q}) \\
&\quad \geqslant \frac{\theta}{2}|p-\hat{q}|^2 + R^{-1}(|q-p| - R)(D_pH(\hat{q}) - D_pH(p))\cdot(\hat{q}-p)
\end{aligned}$$

since $q - \hat{q}$ and $\hat{q} - p$ are collinear and $|\hat{q} - p| = R$. Using once more the lower bound on $D^2_{pp}H$ in $B_R(p)$, we get

$$H(q) - H(p) - D_pH(p)\cdot(q-p) \geqslant \frac{\theta}{2}R^2 + (|q-p| - R)R\theta.$$

This gives the first result. The second one is obtained in the same way: the inequality holds if $|q-p| \le R$. Otherwise, let $\hat p$ be the projection of p onto $B_R(q)$. Then

$$\begin{aligned} &H(q) - H(p) - D_pH(p)\cdot(q-p) \\ &\quad = H(q) - H(\hat p) - D_pH(\hat p)\cdot(q-\hat p) + H(\hat p) - H(p) - D_pH(p)\cdot(\hat p - p) \\ &\qquad + (D_pH(\hat p) - D_pH(p))\cdot(q-\hat p) \\ &\quad \geqslant \frac{\theta}{2}|p-\hat p|^2 = \frac{\theta}{2}R^2. \end{aligned}$$

3.2 *Putting the Estimates Together*

Here we fix an initial condition $m_0 \in \mathcal{P}(\mathbb{T}^d)$, where m_0 has a positive density of class C^2. Let $(v^{N,i})$ be a symmetric solution of the open-loop Nash system (1) and u be the solution to the MFG system (12). Recall that the $v^{N,i}$ are equal and we call v^N this map. Combining, Proposition 1.2 and Theorem 3.1 we have:

Theorem 3.4 *If condition* (20) *holds, then*

$$\left\| v^N(0,\cdot) - u(0,\cdot) \right\|_{L^1_{m_0}(\mathbb{T}^d)} \le \begin{cases} C\left(K_N^{\frac{3}{2}} N^{-\frac{1}{2d}} + K_N N^{-\frac{1}{d}} + k_N^{R,\alpha}\right) & \text{if } d \geqslant 3 \\ C\left(K_N^{\frac{3}{2}} N^{-\frac{1}{4}} \log^{\frac{1}{2}}(N) + K_N N^{-\frac{1}{2}} \log(N) + k_N^{R,\alpha}\right) & \text{if } d = 2 \end{cases}, \tag{27}$$

where R and α do not depend on N (but depend on m_0). In particular, $v^N(0,\cdot)$ converges to $u(0,\cdot)$ in $L^1_{m_0}(\mathbb{T}^d)$ as soon as $K_N = o(N^{\frac{1}{3d}})$ if $d \geqslant 3$ and $K_N = o(N^{\frac{1}{6}}/\log^{\frac{1}{3}}(N))$ if $d = 2$.

Next we discuss the convergence of the optimal solutions. Let (Z_i) be an i.i.d family of N random variables of law m_0. We set $\mathbf{Z} = (Z_1, \dots, Z_N)$. Let also $((B^i_t)_{t\in[0,T]})_{i\in\{1,\dots,N\}}$ be a family of N independent Brownian motions which is also independent of (Z_i). We consider the optimal trajectories $(\boldsymbol{Y}_t = (Y_{1,t}, \dots, Y_{N,t}))_{t\in[0,T]}$ for the N-player game:

$$\begin{cases} dY_{i,t} = -D_pH(Y_{i,t}, Dv^N(t, Y_{i,t}))dt + \sqrt{2}dB^i_t, & t \in [0,T] \\ Y_{i,0} = Z_i \end{cases}$$

and the optimal solution $(\tilde{\boldsymbol{X}}_t = (\tilde X_{1,t}, \dots, \tilde X_{N,t}))_{t\in[0,T]}$ to the limit MFG system:

$$\begin{cases} d\tilde X_{i,t} = -D_pH\left(\tilde X_{i,t}, Du\big(t, \tilde X_{i,t}\big)\right)dt + \sqrt{2}dB^i_t, & t \in [0,T] \\ \tilde X_{i,0} = Z_i. \end{cases}$$

The next result provides an estimate of the distance between the solutions:

Theorem 3.5 *Under the assumption of Theorem 3.4, we have*

$$\mathbb{E}\Big[\sup_{t\in[0,T]}\big|Y_{i,t}-\tilde{X}_{i,t}\big|\Big] \leq \begin{cases} C\left[K_N^{\frac{1}{2}}N^{-\frac{1}{2d}}+\left(k_N^{R,\alpha}\right)^{\frac{2}{d+2}}\right] & \text{if } d \geqslant 3 \\ C\left[K_N^{\frac{1}{2}}N^{-\frac{1}{4}}\log^{\frac{1}{2}}(N)+\left(k_N^{R,\alpha}\right)^{\frac{1}{2}}\right] & \text{if } d=2 \end{cases} \tag{28}$$

where the constant $C>0$ is independent of N. In particular, Y_i converges to $\tilde{X}_i$ if $K_N=o(N^{\frac{1}{d}})$ if $d\geqslant 3$ and $K_N=o(N^{\frac{1}{2}}/\log(N))$ if $d=2$.

The proof is an immediate application of Corollary 1.3 and Theorem 3.1. We finally apply the above estimates to our main example:

Corollary 3.6 *Assume that $F^N=F^{\epsilon_N}$ where*

$$F^{\epsilon}(x,m)=F(\cdot,\xi^{\epsilon}\star m(\cdot))\star\xi^{\epsilon}(x)$$

and where ξ^ϵ is as in the example in Proposition 1.1. If one chooses $\epsilon_N=N^{-\beta}$, with $\beta\in(0,(3d(2d+12+3\alpha))^{-1})$, then there exists $\gamma\in(0,1)$ such that

$$\left\|w^{N,i}(0,\cdot,m_0)-u(0,\cdot)\right\|_{L^1_{m_0}(\mathbb{T}^d)}\leq CN^{-\gamma}$$

and

$$\mathbb{E}\Big[\sup_{t\in[0,T]}\big|Y_{i,t}-\tilde{X}_{i,t}\big|\Big]\leq CN^{-\gamma}.$$

Proof From Proposition 1.1, we can choose

$$k_N^{R,\alpha}\leq C(1+R)\epsilon_N^{\alpha}=CN^{-\alpha\beta},\qquad K_N\leq C\epsilon_N^{-2d-12-3\alpha}=CN^{\beta(2d+12+3\alpha)}.$$

Inserting these inequality into (27) gives (for $d\geqslant 3$),

$$\left\|w^{N,i}(0,\cdot,m_0)-u(0,\cdot)\right\|_{L^1(m_0)}\leq C\left(N^{3\beta(2d+12+3\alpha)/2-\frac{1}{2d}}+N^{\beta(2d+12+3\alpha)-\frac{1}{d}}+N^{-\alpha\beta}\right),$$

where the right-hand side is of order $N^{-\gamma}$ for some $\gamma\in(0,1)$ thanks to our choice of β. In the same way, by (28),

$$\mathbb{E}\Big[\sup_{t\in[0,T]}\big|Y_{i,t}-\tilde{X}_{i,t}\big|\Big]\leq C\left[N^{\beta(2d+12+3\alpha)/2-\frac{1}{2d}}+N^{-\alpha\beta\frac{2}{d+2}}\right],$$

which also yield to an algebraic rate of convergence. Computation for the case $d=2$ is similar.

Acknowledgements The author thanks the anonymous referee for his/her comments which helped to improve the paper. The author was partially supported by the ANR (Agence Nationale de la Recherche) project ANR-16-CE40-0015-01. This work was also supported by a public grant as part of the Investissement d'avenir project, reference ANR-11-LABX-0056-LMH, LabEx LMH.

References

1. Ajtai, M., Komlós, J., Tusnády, G.: On optimal matchings. Combinatorica **4**, 259–264 (1984)
2. Cardaliaguet, P.: The convergence problem in mean field games with local coupling. Appl. Math. Optim. **76**, 177–215 (2017)
3. Cardaliaguet, P., Delarue, F., Lasry, J.-M., Lions, P.-L..: The master equation and the convergence problem in mean field games (2015). arXiv preprint arXiv:1509.02505
4. Carmona, R., Delarue, F.: Probabilistic Theory of Mean Field Games with Applications. Springer Nature, Basingstoke (2018)
5. Dereich, S., Scheutzow, M., Schottstedt, R.: Constructive quantization: approximation by empirical measures. In: Annales de l'Institut Henri Poincaré, Probabilités et Statistiques, vol. 49, Institut Henri Poincaré, pp. 1183–1203 (2013)
6. Fischer, M.: On the connection between symmetric n-player games and mean field games. Ann. Appl. Probab. **27**, 757–810 (2017)
7. Gomes, D.A., Pimentel, E., Sánchez-Morgado, H.: Time-dependent mean-field games in the superquadratic case. ESAIM: Control Optim. Calc. Var. **22**, 562–580 (2016)
8. Gomes, D.A., Pimentel, E.A., Sánchez-Morgado, H.: Time-dependent mean-field games in the subquadratic case. Commun. Partial Differ. Equ. **40**, 40–76 (2015)
9. Lacker, D.: A general characterization of the mean field limit for stochastic differential games. Probab. Theory Relat. Fields **165**, 581–648 (2016)
10. Lasry, J.-M., Lions, P.-L.: Jeux à champ moyen. i–le cas stationnaire. Comptes Rendus Mathématique **343**, 619–625 (2006)
11. Lasry, J.-M., Lions, P.-L.: Jeux à champ moyen. ii–horizon fini et contrôle optimal. Comptes Rendus Mathématique **343**, 679–684 (2006)
12. Lasry, J.-M., Lions, P.-L.: Mean field games. Jpn. J. Math. **2**, 229–260 (2007)
13. Lions, P.-L.: Cours au college de france. Available at www.college-de-france.fr (2007)
14. Porretta, A.: Weak solutions to fokker-planck equations and mean field games. Arch. Rat. Mech. Anal. **216**, 1–62 (2015)

Remarks on the Control of Family of b–Equations

Enrique Fernández-Cara and Diego A. Souza

Abstract In this paper, we deal with the control of the viscous b–equation in a one-dimensional bounded domain. For $b = 2$ and $b = 0$, we get in particular the Camassa–Holm and the Burgers-α equations, respectively. We prove that, for any real number b, we can steer the solution to the equation to zero at any given time, using a distributed control, locally supported in space, when the initial data are sufficiently small. Also, for $b = 0$, we prove the global null controllability for large time.

Keywords Null controllability · Carleman inequalities · Camassa–Holm model · Burgers-α equation · b–equations

Mathematics Subject Classification 93B05 · 35Q35 · 35G25 · 93B07

1 Introduction

Let $L > 0$ and $T > 0$ be given. Let $\omega \subset (0, L)$ be a (small) nonempty open interval which will be referred to as the *control domain*. Let us present the notations used along this work. The symbols C, $\hat{C}$ and C_i, $i = 0, 1, \ldots$ stand for positive constants (usually depending on ω, L and T). For any $r \in [1, +\infty]$ and any given Banach space X, $\| \cdot \|_{L^r(X)}$ will denote the usual norm in Lebesgue-Bochner space $L^r(0, T; X)$. In particular, the norm in $L^r(0, L)$ will be denoted by $\| \cdot \|_r$. We will also need the Hilbert spaces $K^s(0, L) := H^s(0, L) \cap H_0^1(0, L)$, with $s \in (1, 3]$.

E. Fernández-Cara (✉)
Departamento Ecuaciones Diferenciales y Análisis Numérico, Instituto de Matemáticas de la Universidad de Sevilla (IMUS), Universidad de Sevilla, C/ Tarfia S/N, 41012 Seville, Spain
e-mail: cara@us.es

D. A. Souza
Department of Mathematics, Federal University of Pernambuco,
Av. Jornalista Aníbal Fernandes, Cidade Universitária, Recife 50740-560, Brazil
e-mail: diego.souza@dmat.ufpe.br

F. Alabau-Boussouira et al. (eds.), *Trends in Control Theory and Partial Differential Equations*, Springer INdAM Series 32,
https://doi.org/10.1007/978-3-030-17949-6_7

In this paper, we are concerned with the null controllability of the b–equations, which are described in [10, 23]:

$$\begin{cases} z_t - \alpha^2 z_{xxt} - \gamma z_{xx} + \gamma \alpha^2 z_{xxxx} + (b+1) z z_x - \alpha^2 z z_{xxx} - b\alpha^2 z_x z_{xx} = v & \text{in } (0, L) \times (0, T), \\ z(0, \cdot) = z_{xx}(0, \cdot) = z(L, \cdot) = z_{xx}(L, \cdot) = 0 & \text{in } (0, T), \\ z(\cdot, 0) = z_0 & \text{in } (0, L). \end{cases} \tag{1.1}$$

The physical motivation of this family is explained in [12, 13], where it is shown that (1.1) can be viewed as an asymptotically equivalent approximation of the shallow water equations. Thus, z can be viewed as the fluid velocity in the x direction (or equivalently the height of the free surface of the fluid above a flat bottom), $\gamma > 0$ is the fluid viscosity, $\alpha > 0$ and $b \in \mathbb{R}$. When $b = 2$, this equation is the so-called *one-dimensional viscous Camassa–Holm equation*; it describes the unidirectional surface waves at a free surface of shallow water under the influence of gravity, see [5]). When $b = 3$, we are dealing with the *viscous Degasperis–Procesi equation* (it plays a similar role in water wave theory, see [9]). Finally, when $b = 0$, this equation is the so called *Burgers–α equation* (that can be regarded as a nonlinear smoothing regularization of the viscous Burgers equation, see [3]).

Here, we rewrite (1.1) as a controlled parabolic-elliptic system:

$$\begin{cases} y_t - \gamma\, y_{xx} + z y_x + b\, z_x y = v 1_\omega & \text{in } (0, L) \times (0, T), \\ z - \alpha^2 z_{xx} = y & \text{in } (0, L) \times (0, T), \\ y(0, \cdot) = z(0, \cdot) = y(L, \cdot) = z(L, \cdot) = 0 & \text{in } (0, T), \\ y(\cdot, 0) = y_0 & \text{in } (0, L). \end{cases} \tag{1.2}$$

The function $v = v(x, t)$ (usually in $L^2(\omega \times (0, T))$) is the control acting on the system and 1_ω denotes the characteristic function of ω. This way, we see that the role of α is to regularize the velocity and it is natural to try to deduce control properties and/or estimates independent of α. For simplicity, throughout this paper we will take $\gamma = 1$ (all the results can be extended without difficulty to the case where γ is an arbitrary positive number).

The null controllability problem for (1.2) at time $T > 0$ is the following:

For any $y_0 \in H_0^1(0, L)$, find $v \in L^2(\omega \times (0, T))$ such that the associated solution to (1.2) satisfies

$$y(\cdot, T) = 0 \quad in \quad (0, L). \tag{1.3}$$

Our first main result deals with the local uniform (with respect to α) null controllability for all $b \in \mathbb{R}$ (a generalization of [1, Theorem 1] for $b \neq 0$). It is the following:

Theorem 1.1 *Let $b \in \mathbb{R}$ be fixed. Then, for each $T > 0$, the system (1.2) is locally uniformly null-controllable at time T. More precisely, there exists $\delta > 0$ (independent of α) such that, for any $y_0 \in H_0^1(0, L)$ with $\|y_0\|_{H_0^1} \le \delta$, there exist controls*

$v_\alpha \in L^\infty(0, T; L^2(\omega))$ and associated states (y_α, z_α) satisfying (1.3). Moreover, one can find the v_α uniformly bounded:

$$\|v_\alpha\|_{L^\infty(L^2)} \leq C \quad \forall \alpha > 0. \tag{1.4}$$

In our second main result, we stablish the controllability in the limit, as $\alpha \to 0^+$. More precisely, one has:

Theorem 1.2 *Let $b \in \mathbb{R}$ be fixed, let $T > 0$ be given and let $\delta > 0$ be the constant furnished by Theorem 1.1. Assume that $y_0 \in H_0^1(0, L)$ and $\|y_0\|_{H_0^1} \leq \delta$, let v_α be a null control for (1.2) satisfying (1.4) and let (y_α, z_α) be an associated state satisfying (1.3). Then, at least for a subsequence, one has*

$$\begin{aligned} v_\alpha &\to v \text{ weakly-* in } L^\infty(0, T; L^2(\omega)), \\ z_\alpha &\to y \text{ and } y_\alpha \to y \text{ weakly-* in } L^\infty(0, T; H_0^1(0, L)) \end{aligned} \tag{1.5}$$

as $\alpha \to 0^+$, where (v, y) is a control-state pair for the viscous Burgers equation

$$\begin{cases} y_t - \gamma\, y_{xx} + (1+b) y y_x = v 1_\omega & \text{in } (0, L) \times (0, T), \\ y(0, \cdot) = y(L, \cdot) = 0 & \text{in } (0, T), \\ y(\cdot, 0) = y_0 & \text{in } (0, L) \end{cases} \tag{1.6}$$

and y satisfies (1.3).

These theorems generalize other previous results on the control of nonlocal nonlinear parabolic equations, see [1]. In particular, we are able here to handle the case $b \neq 0$, where a "new" nonlinear term appears. This makes the difference with respect to many other previous works, such as [11, 14, 16, 18, 33].

For completeness, let us mention some previous works on the control of (1.1) and other similar systems. The controllability properties of the inviscid Camassa–Holm and the inviscid Burgers equations were widely studied in [21, 28] and [6, 24], respectively. The optimal control of the viscous Camassa–Holm equation with homogeneous Dirichlet boundary conditions is studied in [29, 32]. On the other hand, the results on the controllability of the Camassa–Holm equation previous to this paper have been established on the one-dimensional torus: see [19, 27]. On the other hand, the controllability of the viscous Burgers equation has been analyzed in [6, 15, 18, 22, 25, 26].

The rest of this paper is organized as follows. In Sect. 2, we recall some results concerning the controllability of parabolic equations. Sections 3 and 4 deal with the proofs of Theorems 1.1 and 1.2, respectively. Finally, in Sect. 5, we present some additional results and comments.

2 Carleman Inequalities and Null Controllability

In this section, we will recall a null controllability result for a general parabolic linear system of the form

$$\begin{cases} y_t - y_{xx} + (Ay)_x + By = v1_\omega & \text{in } (0, L) \times (0, T), \\ y(0, \cdot) = y(L, \cdot) = 0 & \text{on } (0, T), \\ y(\cdot, 0) = y_0 & \text{in } (0, L). \end{cases} \tag{2.1}$$

where $y_0 \in L^2(0, L)$, $A \in L^\infty((0, L) \times (0, T))$ and $B \in L^\infty(0, T; L^2(0, L))$ and the control v is searched in $L^2(\omega \times (0, T))$.

It is well known that, for each v and y_0, there exists exactly one solution y to (2.1), with

$$y \in C^0([0, T]; L^2(0, L)) \cap L^2(0, T; H_0^1(0, L)).$$

The null controllability problem for (2.1) at time $T > 0$ is the following:

For any $y_0 \in L^2(0, L)$, find $v \in L^2(\omega \times (0, T))$ such that the associated solution to (2.1) satisfies (1.3).

Let us present the main steps to construct a control for (2.1). This will be later useful to control the Camassa–Holm equation. Thus, let us present an appropriate Carleman inequality for the adjoint system of (2.1), which is the following:

$$\begin{cases} -\varphi_t - \varphi_{xx} - A\varphi_x + B\varphi = f & \text{in } (0, L) \times (0, T), \\ \varphi(0, \cdot) = \varphi(L, \cdot) = 0 & \text{on } (0, T), \\ \varphi(\cdot, T) = \varphi_T & \text{in } (0, L). \end{cases} \tag{2.2}$$

Proposition 2.1 *There exist constants $\lambda_0 > 0$, $s_0 > 0$ and $C_0 > 0$ (depending on L and ω) such that, for any $\lambda \geq \lambda_0$, any $s \geq s_0(T + T^2)$, any $\psi_T \in L^2(0, L)$ and any $g \in L^2((0, L) \times (0, T))$, the unique weak solution to*

$$\begin{cases} -\psi_t - \psi_{xx} = g & \text{in } (0, L) \times (0, T), \\ \psi(0, \cdot) = \psi(L, \cdot) = 0 & \text{on } (0, T), \\ \psi(\cdot, T) = \psi_T & \text{in } (0, L) \end{cases}$$

satisfies

$$\begin{aligned} &\int_0^T \int_0^L \left[(s\xi)^{-1}(|\psi_{xx}|^2 + |\psi_t|^2) + (s\xi)\lambda^2|\psi_x|^2 + (s\xi)^3\lambda^4|\psi|^2 \right] e^{-2s\alpha}\, dx\, dt \\ &\quad \leq C_0 \left(\int_0^T \int_0^L |g|^2 e^{-2s\alpha}\, dx dt + \int_0^T \int_\omega (s\xi)^3\lambda^4 |\psi|^2 e^{-2s\alpha}\, dx\, dt \right), \end{aligned} \tag{2.3}$$

where

$$\alpha(x,t) := \frac{e^{2\lambda\|\eta_0\|_\infty} - e^{\lambda\eta_0(x)}}{(T-t)}, \quad \xi(x,t) := \frac{e^{\lambda\eta_0(x)}}{(T-t)} \; \forall (x,t) \in (0,L)\times(0,T) \tag{2.4}$$

and $\eta_0 \in C^2([0,L])$ *is such that* $\eta_0 > 0$ *in (0,L),* $\eta_0(0) = \eta_0(L) = 0$ *and* $|\eta_{0,x}| > 0$ *in* $[0,L]\backslash\omega$.

This result was established in [18, Lemma 1.1].

Now, let us introduce $g := f + A\varphi_x - B\varphi$, $\psi_T := \varphi_T$ and let us apply (2.3) to (2.2). We see that

$$\begin{aligned}&\int_0^T\int_0^L \left[(s\xi)^{-1}(|\varphi_{xx}|^2 + |\varphi_t|^2) + (s\xi)\lambda^2|\varphi_x|^2 + (s\xi)^3\lambda^4|\varphi|^2\right] e^{-2s\alpha}\,dx\,dt \\ &\quad \le C_0\left(\int_0^T\int_0^L (|f|^2 + |A\varphi_x|^2 + |B\varphi|^2)e^{-2s\alpha}\,dxdt + \int_0^T\int_\omega (s\xi)^3\lambda^4|\varphi|^2e^{-2s\alpha}\,dx\,dt\right).\end{aligned}$$

Using the assumptions on A and B, we get

$$\begin{aligned}&\int_0^T\int_0^L \left[(s\xi)^{-1}(|\varphi_{xx}|^2 + |\varphi_t|^2) + (s\xi)\lambda^2|\varphi_x|^2 + (s\xi)^3\lambda^4|\varphi|^2\right] e^{-2s\alpha}\,dx\,dt \\ &\quad \le C_0\left(\int_0^T\int_0^L |f|^2e^{-2s\alpha}\,dxdt + \int_0^T\int_\omega (s\xi)^3\lambda^4|\varphi|^2e^{-2s\alpha}\,dx\,dt + \int_0^T\int_0^L |\varphi_x|^2e^{-2s\alpha}\,dx\,dt\right. \\ &\qquad + (\|A\|^2_{L^\infty(L^\infty)} + \|B\|^2_{L^\infty(L^2)})\int_0^T\int_0^L |\varphi_x|^2e^{-2s\alpha}\,dx\,dt \\ &\qquad \left. + \|B\|^2_{L^\infty(L^2)}\int_0^T\int_0^L (s\xi)^2\lambda^2|\varphi|^2e^{-2s\alpha}\,dx\,dt\right).\end{aligned}$$

But these last three terms in the right hand side can be absorbed by the left-hand side if we take s large enough. Indeed, it suffices to take

$$s_1 = \max(s_0, C(L,\omega)(\|A\|^2_{L^\infty(L^\infty)} + \|B\|^2_{L^\infty(L^2)})$$

to have

$$\begin{aligned}&\int_0^T\int_0^L [(s\xi)^{-1}(|\varphi_{xx}|^2 + |\varphi_t|^2) + (s\xi)\lambda^2|\varphi_x|^2 + (s\xi)^3\lambda^4|\varphi|^2]\,e^{-2s\alpha}\,dx\,dt \\ &\quad \le C_0\left(\int_0^T\int_0^L |f|^2e^{-2s\alpha}\,dxdt + \int_0^T\int_\omega (s\xi)^3\lambda^4|\varphi|^2e^{-2s\alpha}\,dx\,dt\right),\end{aligned} \tag{2.5}$$

for any $\lambda \ge \lambda_0$ and any $s \ge s_1(T+T^2)$.

Now, we are going to construct a null-control for (2.1), like in [18]. First, let us introduce the auxiliary extremal problem

$$\begin{cases} \text{Minimize } \dfrac{1}{2}\left\{\displaystyle\int_0^T\int_0^L e^{2s\alpha}|y|^2\,dx\,dt + \int_0^T\int_\omega (s\xi)^{-3}\lambda^{-4}e^{2s\alpha}|v|^2\,dx\,dt\right\} \\ \text{Subject to } (y,v) \in \mathcal{H}(y_0,T), \end{cases} \tag{2.6}$$

where the linear manifold $\mathcal{H}(y_0, T)$ is given by

$$\mathcal{H}(y_0, T) = \{ (y, v) : v \in L^2(\omega \times (0, T)),\ y \text{ solves } (2.7) \}.$$

It can be proved that (2.6) possesses exactly one solution (y, v) satisfying

$$\|(s\xi)^{-3/2}\lambda^{-2}e^{s\alpha}v\|_{L^2(L^2)} \leq C_1\|y_0\|_2,$$

where

$$C_1 = e^{C_2\left(1+1/T+(1+T)\left(\|A\|^2_{L^\infty(L^\infty)}+\|B\|^2_{L^\infty(L^2)}\right)\right)}$$

and $C_2 > 0$ only depends on L and ω.

Moreover, thanks to the classical Euler–Lagrange characterization, the solution to the extremal problem (2.6) is given by

$$y = e^{-2s\alpha}(-\varphi_t - \varphi_{xx} - A\varphi_x + B\varphi), \quad v := -(s\xi)^3\lambda^4 e^{-2s\alpha}\varphi 1_{\omega\times(0,T)}.$$

From the Carleman inequality (2.5), we can conclude that

$$(s\xi)^{-1/2}e^{-s\alpha}\varphi \in H^1(0, T; L^2(0, L)) \cap L^2(0, T; K^2(0, L))$$

and, in particular,

$$\|(s\xi)^{-1/2}e^{-s\alpha}\varphi\|_{L^\infty(L^2)} \leq C\|(s\xi)^{3/2}\lambda^2 e^{-s\alpha}\varphi\|_{L^2(L^2)}.$$

Hence, we actually have that $v \in L^\infty(0, T; L^2(0, L))$ and

$$\|v\|_{L^\infty(L^2)} \leq C_3\|y_0\|_2, \tag{2.7}$$

where again C_3 takes the form

$$C_3 = e^{C_4\left(1+1/T+(1+T)\left(\|A\|^2_{L^\infty(L^\infty)}+\|B\|^2_{L^\infty(L^2)}\right)\right)},$$

for some C_4 only depending on L and ω.

This way, we get the following result:

Theorem 2.1 *Assume that $A \in L^\infty((0, L) \times (0, T))$ and $B \in L^\infty(0, T; L^2(0, L))$. Then, the linear parabolic equation(2.1) is null-controllable at any time $T > 0$. Specifacally, for each $y_0 \in L^2(0, L)$ there exists $v \in L^\infty(0, T; L^2(\omega))$ such that the corresponding solution to (2.1) satisfies (1.3). Furthermore, the control v can be chosen satisfying the estimate (2.7).*

3 Local Null Controllability of the b-Equations

In this section, we present the proof of Theorem 1.1.

For the moment, let us assume that $y_0 \in K^2(0, L)$. Let us set $Z := L^\infty(0, T; H_0^1(0, L))$. The argument is as follows: first, we fix $\overline{y}$ in Z and we solve the family of elliptic problems:

$$\begin{cases} z - \alpha^2 z_{xx} = \overline{y}, & \text{in } (0, L) \times (0, T), \\ z(0, \cdot) = z(L, \cdot) = 0 \text{ in } (0, T); \end{cases} \tag{3.1}$$

then, we control exactly to zero the linear system (2.1) with $A = z$, $B = (b - 1)z_x$ (by solving the corresponding extremal problem (2.6)); at this point, we set $\Lambda_\alpha(\overline{y}) := y$ and we consider the mapping $\Lambda_\alpha : Z \mapsto Z$; then the task is to solve the fixed-point equation $y = \Lambda_\alpha(y)$.

Several fixed-point theorems can be applied here. In this paper, we have preferred to use Schauder's Theorem, although other results also lead to the good conclusion; for instance, an argument relying on Kakutani's Theorem, like in [11], is possible.

In order to get good properties of Λ_α, it is very appropriate to choose controls belonging to the space $L^\infty(0, T; L^2(0, L))$, as we have done in Sect. 2 (see Theorem 2.1).

Observe that, if $\overline{y} \in Z$, it is clear that $z \in L^\infty(0, T; K^3(0, L))$. Then, thanks to Sobolev embedding, we have $z,\ z_x,\ z_{xx} \in L^\infty((0, L) \times (0, T))$ and the following is satisfied:

$$\begin{aligned} \|z\|^2_{L^\infty(H_0^1)} + 2\alpha^2 \|z\|^2_{L^\infty(K^2)} &\le \|\overline{y}\|^2_{L^\infty(H_0^1)}, \\ 2\alpha^2 \|z_{xx}\|^2_{L^\infty(L^2)} + \alpha^4 \|z_{xxx}\|^2_{L^\infty(L^2)} &\le \|\overline{y}_x\|^2_{L^\infty(L^2)}, \end{aligned} \tag{3.2}$$

Let us consider the system (2.1) with $A = z$, $B = (b - 1)z_x$. As already said, we can associate to z the null control $v \in L^\infty(0, T; L^2(0, L))$ furnished by Theorem 2.1 and, then, the corresponding state y.

Since $y_0 \in K^2(0, L)$ and $z \in Z$, it is clear that

$$\begin{gathered} y \in L^2(0, T; K^2(0, L)) \cap C^0([0, T]; H_0^1(0, L)), \\ y_t \in L^2(0, T; L^2(0, L)) \end{gathered}$$

and, from standard energy estimates and (2.7), we deduce that

$$\begin{aligned} \|y_t\|_{L^2(L^2)} + \|y\|_{L^2(H^2)} + \|y\|_{L^\infty(H_0^1)} &\le (\|y_0\|_{H_0^1} + \|v\|_{L^2(L^2)}) e^{C_5(1 + \|z\|^2_{L^\infty(L^\infty)} + \|z_x\|^2_{L^\infty(L^2)})} \\ &\le (\|y_0\|_{H_0^1} + \|v\|_{L^2(L^2)}) e^{C_6(1 + \|\overline{y}\|^2_Z)} \\ &\le \|y_0\|_{H_0^1} e^{C_7(1 + \|\overline{y}\|^2_Z)}, \end{aligned} \tag{3.3}$$

where C_5, C_6 and C_7 depend on L, ω and T but are independent of α.

We will use the following result, whose proof is given at the end of this section:

Lemma 3.1 *One has $y \in L^\infty(0,T;K^{2\epsilon}(0,L))$, for all $\epsilon \in (0,1)$, with*

$$\|y\|_{L^\infty(K^{2\epsilon})} \le (\|y_0\|_{K^2} + \|v\|_{L^\infty(L^2)})e^{C_\epsilon(1+\|\overline{y}\|^2_{L^\infty(H_0^1)})}. \tag{3.4}$$

Now, let us fix $\epsilon \in (0,1)$, let us introduce the Banach space

$$W = \{w \in L^\infty(0,T;K^{2\epsilon}(0,L)) : w_t \in L^2(0,T;L^2(0,L))\} \tag{3.5}$$

and the closed ball

$$M = \{w \in L^\infty(0,T;H_0^1(0,L)) : \|w\|_{L^\infty(H_0^1)} \le 1\} \tag{3.6}$$

and let us observe that $\tilde{\Lambda}_\alpha$ maps the whole space Z into W.

Notice that the embedding $W \hookrightarrow Z$ is compact, in view of classical results of the Aubin-Lions kind, see [30].

On the other hand, thanks to (3.3), if $\|y_0\|_{H_0^1} \le \delta$ (with $\delta > 0$, small enough, independent of α), Λ_α maps M into itself. Consequently, Schauder's Fixed-Point Theorem can be applied in this context and Λ_α possesses at least one fixed point in M.

This ends the proof of Theorem 1.1 when $y_0 \in K^2(0,L)$.

In the general case, when $y_0 \in H_0^1(0,L)$, we can use a standard approximation argument. Indeed, is $\{y_{0,n}\}$ is a sequence in $K^2(0,L)$ with $\|y_{0,n}\|_{H_0^1} \le \delta$ and $y_{0,n} \to y_0$ in $H_0^1(0,L)$ and the controls v_n and states y_n and z_n are constructed as above, it is clear that the v_n are uniformly bounded in $L^\infty(0,T;L^2(0,L))$ and the y_n and z_n are uniformly bounded in good spaces, with respect to α and n, in such a way that we can pass to the limit in n in the equations. Thus, we get again a control such that (1.3) holds in this case.

Let us now return to Lemma 3.1 and let us establish its proof.

Proof of Lemma 3.1. In view of (2.1), y solves the following abstract initial value problem in $L^2(0,L)$:

$$\begin{cases} y_t = \partial^2_{xx} y - z y_x - b z_x y + v 1_\omega \text{ in } [0,T], \\ y(0) = y_0. \end{cases}$$

This equation can be rewritten as a nonlinear integral equation:

$$y(t) = e^{t\partial^2_{xx}} y_0 + \int_0^t e^{(t-s)\partial^2_{xx}}(-z y_x - b z_x y + v 1_\omega)(s)\,ds.$$

Consequently, applying the operator $(\partial^2_{xx})^\epsilon$ to both sides, we have

$$(\partial^2_{xx})^\epsilon y(t) = (\partial^2_{xx})^\epsilon e^{t\partial^2_{xx}} y_0 + \int_0^t (\partial^2_{xx})^\epsilon e^{(t-s)\partial^2_{xx}}(-z y_x - b z_x y + v 1_\omega)(s)\,ds. \tag{3.7}$$

Recall that, for any $r > 0$, there exists $C(r) > 0$ such that

$$\|(\partial^2_{xx})^r e^{t\partial^2_{xx}}\|_{\mathcal{L}(L^2(0,L);L^2(0,L))} \le C(r)\, t^{-r} \quad \forall t > 0;$$

for the proof, see for instance [4].

Taking L^2–norms in both sides of (3.7), we see that

$$\begin{aligned}\|(\partial^2_{xx})^\epsilon y(t)\|_{L^2} &\le \|y_0\|_{K^{2\epsilon}} + C\int_0^t (t-s)^{-\epsilon}\left[(\|z(s)\|_{L^\infty} + \|z_x\|_{L^2})\|y(s)\|_{H^1_0} + \|v(s)1_\omega\|_{L^2}\right] ds \\ &\le \|y_0\|_{K^2} + C\left((\|z\|_{L^\infty(L^\infty)} + \|z_x\|_{L^2(L^2)})\|y\|_{L^\infty(H^1_0)} + \|v1_\omega\|_{L^\infty(L^2)}\right)\int_0^t (t-s)^{-\epsilon}\, ds\end{aligned}$$

Finally, using (3.2) and (3.3) and taking into account that $\epsilon \in (0, 1)$, we easily obtain that

$$\|(\partial^2_{xx})^\epsilon y(t)\|_{L^2} \le \left(\|y_0\|_{K^2} + \|v1_\omega\|_{L^\infty(L^2)}\right) e^{C_\epsilon(1+\|\overline{y}\|^2_Z)}$$

and the proof. □

4 Controllability in the Limit

In this section, we prove Theorem 1.2.

For the null controls v_α furnished by Theorem 1.1 and the associated solutions (y_α, z_α) to (1.2), we have the uniform estimates (3.2), (3.3) and (3.4). Recall that $y \in M$ and M is given by (3.6). Consequently, there exist $y \in L^2(0, T; K^2(0, L))$ with $y_t \in L^2(0, T; L^2(0, L))$ and $v \in L^\infty(0, T; L^2(\omega))$ such that, at least for a subsequence, one has:

$$\begin{aligned} &y_\alpha \to y \text{ weakly in } L^2(0, T; K^2(0, L)),\\ &(y_\alpha)_t \to y_t \text{ weakly in } L^2(0, T; L^2(0, L))\\ &v_\alpha \to v \text{ weakly} - * \text{ in } L^\infty(0, T; L^2(\omega)).\end{aligned} \tag{4.1}$$

As before, we can apply a compactness argument of the Aubin-Lions kind and deduce that

$$y_\alpha \to y \text{ strongly in } L^2(0, T; H^1_0(0, L)). \tag{4.2}$$

Using the second equation in (1.2), we see that

$$(z_\alpha - y) - \alpha^2(z_\alpha - y)_{xx} = (y_\alpha - y) + \alpha^2 y_{xx}.$$

Multiplying this equation by $-(z_\alpha - y)_{xx}$ and integrating in $(0, L) \times (0, T)$, we deduce

$$\int_0^T\int_0^L |(z_\alpha - y)_x|^2\,dx\,dt + \alpha^2\int_0^T\int_0^L |(z_\alpha - y)_{xx}|^2\,dx\,dt$$
$$= \int_0^T\int_0^L (y_\alpha - y)_x(z_\alpha - y)_x\,dx\,dt$$
$$- \alpha^2\int_0^T\int_0^L y_{xx}(z_\alpha - y)_{xx}\,dx\,dt,$$

whence

$$\int_0^T\int_0^L |(z_\alpha - y)_x|^2\,dx\,dt \le \int_0^T\int_0^L |(y_\alpha - y)_x|^2\,dx\,dt + \alpha^2\|y_{xx}\|_2^2.$$

This shows that

$$z_\alpha \to y \text{ strongly in } L^2(0,T;H_0^1(0,L)) \tag{4.3}$$

and the transport and reaction terms in (1.2) satisfy

$$\begin{aligned} z_\alpha(y_\alpha)_x &\to yy_x \text{ strongly in } L^1((0,L)\times(0,T)) \\ (z_\alpha)_x y_\alpha &\to y_x y \text{ strongly in } L^1((0,L)\times(0,T)). \end{aligned} \tag{4.4}$$

In this way, for each $\psi \in L^\infty(0,T;H_0^1(0,L))$, we obtain

$$\int_0^T\int_0^L ((y_\alpha)_t\psi + (y_\alpha)_x\psi_x + z_\alpha(y_\alpha)_x\psi + b(z_\alpha)_x y_\alpha\psi)\,dx\,dt = \int_0^T\int_0^L v_\alpha 1_{(a,b)}\psi\,dx\,dt. \tag{4.5}$$

Using (4.1) and (4.4), we can pass to the limit as $\alpha \to 0^+$ in all the terms of (4.5) to find

$$\int_0^T\int_0^L (y_t\psi + y_x\psi_x + (1+b)yy_x\psi)\,dx\,dt = \int_0^T\int_0^L v1_{(a,b)}\psi\,dx\,dt. \tag{4.6}$$

That is, y is the unique solution to (1.6) and y satisfies (1.3).

5 Some Additional Results and Comments

5.1 Exponential Decay and Large Time Null Controllability for the Burgers-α Equation

Let us see that, when $b = 0$, the large time null controllability of (1.2) holds. This result provides a positive answer to an open question in [1, Remark 2]. More precisely, we have the following:

Theorem 5.1 *Assume that $b = 0$. For each $y_0 \in H_0^1(0, L)$, there exists $\alpha_0 > 0$ such that (1.2) is null-controllable at large time for any $\alpha \in (0, \alpha_0)$. In other words, there exist $T > 0$ (independent of α), controls $v_\alpha \in L^\infty(\omega \times (0, T))$ and associated states (y_α, z_α) satisfying*

$$\|v_\alpha\|_{L^\infty(L^\infty)} \le C \tag{5.1}$$

and

$$y_\alpha(\cdot, T) = 0 \quad in \quad (0, L). \tag{5.2}$$

Proof The task is to replace the assumption "y_0 is small" by an assumption imposing that T is large enough (and independent of α).

Thus, let us assume that $b = 0$. If $y_0 \in H_0^1(0, L)$, then the associated uncontrolled solution to (1.2) satisfies

$$\begin{gathered} y_\alpha \in L^2(0, T; H^2(0, L)) \cap C^0([0, T]; H_0^1(0, L)), \\ z_\alpha \in L^2\big(0, T; H^4(0, L)\big) \cap L^\infty\big(0, T; H_0^1(0, L) \cap H^3(0, L)\big), \\ (y_\alpha)_t \in L^2((0, L) \times (0, T)), \ (z_\alpha)_t \in L^2\big(0, T; H^2(0, L)\big). \end{gathered}$$

This is a straightforward consequence of elliptic and parabolic regularity theory. Furthermore, the following estimates hold:

$$\begin{aligned} \|y_\alpha\|_{L^\infty(L^\infty)} &\le \|y_0\|_\infty, \\ \|z_\alpha\|_{L^\infty(L^\infty)} &\le \|y_0\|_\infty. \end{aligned} \tag{5.3}$$

These regularity properties allow to rewrite (1.2) with $b = 0$ and $v = 0$ in the form

$$\begin{cases} z_t - \alpha^2 z_{xxt} - z_{xx} + \alpha^2 z_{xxxx} + z z_x - \alpha^2 z z_{xxx} = 0 \text{ in } (0, L) \times (0, T), \\ z(0, \cdot) = z(L, \cdot) = z_{xx}(0, \cdot) = z_{xx}(L, \cdot) = 0 \quad \text{on } (0, T), \\ z(\cdot, 0) = (I - \alpha^2 \partial_{xx})^{-1} y_0 \quad \text{in } (0, L), \end{cases} \tag{5.4}$$

where, by simplicity, we have omitted the index α.

Let us multiply (5.4) by z and let us integrate in $(0, L)$. Then, introducing the instantaneous energy $E_\alpha(t) := \|z(\cdot, t)\|_2^2 + \alpha^2 \|z_x(\cdot, t)\|_2^2$, we see that

$$\frac{1}{2}\frac{d}{dt} E_\alpha(t) + \|z_x(\cdot, t)\|_2^2 + \alpha^2 \|z_{xx}(\cdot, t)\|_2^2 = -2\alpha^2 \int_0^L z z_x z_{xx} \, dx$$

for all $t > 0$ and, therefore,

$$\frac{d}{dt} E_\alpha(t) + 2\left(1 - 2\alpha^2 \|y_0\|_\infty^2\right) \|z_x(\cdot, t)\|_2^2 + \alpha^2 \|z_{xx}(\cdot, t)\|_2^2 \le 0$$

and

$$\frac{d}{dt} E_\alpha(t) + (\pi/L)^2 C(y_0, \alpha) E_\alpha(t) \le 0,$$

where $C(y_0, \alpha) = \min\{1, 2(1 - 2\alpha^2\|y_0\|_\infty^2)\}$. Taking $\alpha_0 = 1/(2\|y_0\|_\infty)$, we find that

$$E_\alpha(t) \le e^{-(\pi/L)^2 t} E_\alpha(0) \qquad \forall t > 0.$$

and, consequently,

$$\|z(\cdot\,,t)\|_2^2 \le e^{-(\pi/L)^2 t}\|y_0\|_2^2 \qquad \forall t > 0 \tag{5.5}$$

whenever $\alpha \in (0, \alpha_0)$.

Let us multiply (1.2) by y and let us integrate in $(0, L)$. Then

$$\frac{d}{dt}\|y(\cdot\,,t)\|_2^2 + \|y_x(\cdot\,,t)\|_2^2 \le \|y_0\|_\infty^2\|z(\cdot\,,t)\|_2^2 \tag{5.6}$$

and, using Poincaré's inequality and (5.5), we get

$$\frac{d}{dt}\|y(\cdot\,,t)\|_2^2 + (\pi/L)^2\|y(\cdot\,,t)\|_2^2 \le e^{-(\pi/L)^2 t}\|y_0\|_\infty^2\|y_0\|_2^2.$$

Hence, the L^2–norm of y decays exponentially:

$$\|y(\cdot\,,t)\|_2^2 \le e^{-(\pi/L)^2 t}\|y_0\|_2^2(1 + \|y_0\|_\infty^2) \qquad \forall t > 0. \tag{5.7}$$

Let us see that the same is true for the L^2–norm of y_x.

Let us introduce $r = \frac{1}{2}(\pi/L)^2$. It follows that

$$\|y(\cdot\,,t)\|_2^2 \le \|y_0\|_2^2(1 + \|y_0\|_\infty^2)e^{-2rt}. \tag{5.8}$$

Hence, combining (5.5), (5.6) and (5.8), it is easy to see that

$$\frac{d}{dt}\left(e^{rt}\|y(\cdot\,,t)\|_2^2\right) + e^{rt}\|y_x(\cdot\,,t)\|_2^2 \le (r(1 + \|y_0\|_\infty^2) + \|y_0\|_\infty^2)\|y_0\|_2^2\, e^{-rt}.$$

Integrating from 0 to t yields

$$\int_0^t e^{r\sigma}\|y_x(\cdot\,,\sigma)\|_2^2\,d\sigma \le \left(2 + \|y_0\|_\infty^2 + \frac{\|y_0\|_\infty^2}{r}\right)\|y_0\|_2^2. \tag{5.9}$$

Now, we take the L^2-inner product of (1.2) and $-(y_\alpha)_{xx}$ and get

$$\frac{d}{dt}\|y_x(\cdot\,,t)\|_2^2 \le \|y_0\|_\infty^2\|y_x(\cdot\,,t)\|^2.$$

Multiplying this inequality by e^{rt}, we deduce that

$$\frac{d}{dt}\left(e^{rt}\|y_x(\cdot\,,t)\|_2^2\right) \le (r + \|y_0\|_\infty^2)\, e^{rt}\,\|y_x(\cdot\,,t)\|_2^2$$

and, consequently, we see from (5.9) that

$$\|y_x(\cdot\,,t)\|_2^2 \le \left[(r+\|y_0\|_\infty^2)\left(2+\|y_0\|_\infty^2+\frac{\|y_0\|_\infty^2}{r}\right)\|y_0\|_2^2+\|y_0\|_{H_0^1}^2\right]e^{-rt}.$$

This ends the proof. □

5.2 A Boundary Controllability Result

Thanks to an extension argument, it can be proved that boundary uniform null controllability results similar to Theorem 1.1 hold for the b–family.

For instance, let us see that the analog of Theorem 1.1 remains true. Thus, let us introduce the controlled system

$$\begin{cases} y_t - y_{xx} + zy_x + bz_x y = 0 & \text{in } (0,L)\times(0,T),\\ z-\alpha^2 z_{xx} = y & \text{in } (0,L)\times(0,T),\\ z(0,\cdot)=y(0,\cdot)=0,\ \ y(L,\cdot)=z(L,\cdot)=u & \text{on } (0,T),\\ y(\cdot,0)=y_0 & \text{in } (0,L),\end{cases} \tag{5.10}$$

where $u=u(t)$ stands for the control function and $y_0\in H_0^1(0,L)$ is given.

Let ω and $\widetilde{L}$ be given, with $\widetilde{L}>L$ and $\overline{\omega}\subset(L,\widetilde{L})$. Let $\tilde{y}_0$ be the extension-by-zero of y_0; note that $\tilde{y}_0\in H_0^1(0,\widetilde{L})$. Arguing as in Theorem 1.1, it can be proved that there exists $(\tilde{y},\tilde{v})$, with $\tilde{v}\in L^\infty(0,T;L^2(0,\widetilde{L}))$,

$$\begin{cases} \tilde{y}_t - \tilde{y}_{xx} + z1_{[0,L]}\,\tilde{y}_x + bz_x 1_{[0,L]}\tilde{y} = \tilde{v}1_\omega & \text{in } (0,\widetilde{L})\times(0,T),\\ z-\alpha^2 z_{xx} = \tilde{y} & \text{in } (0,L)\times(0,T),\\ \tilde{y}(0,\cdot)=\tilde{y}(\widetilde{L},\cdot)=0,\quad z(0,\cdot)=0,\quad z(L,\cdot)=\tilde{y}(L,\cdot) & \text{on } (0,T),\\ \tilde{y}(\cdot,0)=\tilde{y}_0 & \text{in } (0,\widetilde{L}),\end{cases}$$

and $\tilde{y}(x,T)\equiv 0$. Then, if we take $y:=\tilde{y}|_{(0,L)}$, z and $u(t):=\tilde{y}(L,t)$, we see that $u\in L^\infty(0,T)$, y, z and u satisfy (5.10) and $y(x,T)\equiv 0$.

5.3 The Situation in Higher Spatial Dimensions

Let $\Omega\subset\mathbb{R}^N$ be a bounded connected and regular open set ($N=2$ or $N=3$) and let $\omega\subset\Omega$ be a (small) open set. We will use the notation $Q:=\Omega\times(0,T)$ and $\Sigma:=\partial\Omega\times(0,T)$ and we will use bold symbols for vector-valued functions and spaces of vector-valued functions.

For any $\mathbf{v}$ and any $\mathbf{y}_0$ in appropriate spaces, let us consider the following controlled Navier–Stokes system

$$\begin{cases} \mathbf{y}_t - \Delta\mathbf{y} + (\mathbf{y}\cdot\nabla)\mathbf{y} + \nabla p = \mathbf{v}1_\omega & \text{in } Q, \\ \nabla\cdot\mathbf{y} = 0 & \text{in } Q, \\ \mathbf{y} = \mathbf{0} & \text{on } \Sigma, \\ \mathbf{y}(0) = \mathbf{y}_0 & \text{in } \Omega. \end{cases} \tag{5.11}$$

For an explanation of the meaning of the equations and unknowns, see for instance [31]. Here, $\mathbf{v} = \mathbf{v}(x, t)$ stands for the control function.

Introducing a smoothing kernel and a related modification of (5.11), we obtain the so called Navier–Stokes-α model or N–dimensional viscous Camassa–Holm equations:

$$\begin{cases} \mathbf{y}_t - \Delta\mathbf{y} + (\mathbf{z}\cdot\nabla)\mathbf{y} + b\nabla\mathbf{z}^t\cdot\mathbf{y} + \nabla p = \mathbf{v}1_\omega & \text{in } Q, \\ \mathbf{z} - \alpha^2\Delta\mathbf{z} + \nabla\pi = \mathbf{y} & \text{in } Q, \\ \nabla\cdot\mathbf{y} = \nabla\cdot\mathbf{z} = 0 & \text{in } Q, \\ \mathbf{y} = \mathbf{z} = \mathbf{0} & \text{on } \Sigma, \\ \mathbf{y}(0) = \mathbf{y}_0 & \text{in } \Omega, \end{cases} \tag{5.12}$$

for more informations about this equation, see [7, 8, 17, 20].

Let us recall the definitions of some function spaces that are frequently used in the analysis of the systems (5.11) and (5.12):

$$\mathbf{H} = \left\{ \varphi \in \mathbf{L}^2(\Omega) : \nabla\cdot\varphi = 0 \text{ in } \Omega,\ \ \varphi\cdot\mathbf{n} = 0 \text{ on } \partial\Omega \right\},$$
$$\mathbf{V} = \left\{ \varphi \in \mathbf{H}_0^1(\Omega) : \nabla\cdot\varphi = 0 \text{ in } \Omega \right\}.$$

With arguments similar to those in [2] and Theorem 1.1, it can be proved that, for any $T > 0$, there exists $\epsilon > 0$ such that, if $\|\mathbf{y}_0\|_{\mathbf{V}} < \epsilon$ we can find controls $\mathbf{v}_\alpha \in \mathbf{L}^2(\omega\times(0,T))$ (bounded independently of α) and associate states $(\mathbf{y}_\alpha, p_\alpha, \mathbf{z}_\alpha, \pi_\alpha)$ satisfying

$$\mathbf{y}_\alpha(x, T) = \mathbf{0} \quad \text{in} \quad \Omega.$$

Furthermore, it can also be seen that, at least for a subsequence, the $\mathbf{v}_\alpha$ converge, in an appropriate sense, to a null control of (5.11).

Funding

D. A. Souza was partially supported by CNPq (Brazil) by the grants 313148/2017-1 and Propesq (UFPE)- Edital Qualis A. E. Fernández-Cara was partially supported by grant MTM2016-76990-P, Ministry of Economy and Competitiveness (Spain).

References

1. Araruna, F.D., Fernández-Cara, E., Souza, D.A.: On the control of the Burgers-alpha model. Adv. Differ. Equ. **18**(9–10), 935–954 (2013)
2. Araruna, F.D., Fernández-Cara, E., Souza, D.A.: Uniform local null control of the Leray-α model. ESAIM Control Optim. Calc. Var. **20**(4), 1181–1202 (2014)

3. Bhat, H.S., Fetecau, R.C.: A Hamiltonian regularization of the Burgers equation. J. Nonlinear Sci. **16**(6), 615–638 (2006)
4. Brezis, H.: Functional Analysis, Sobolev Spaces and Partial Differential Equations. Springer, Berlin (2011)
5. Camassa, R., Holm, D.D.: An integrable shallow water equation with peaked solitons. Phys. Rev. Lett. **71**(11), 1661–1664 (1993)
6. Chapouly, M.: Global controllability of nonviscous and viscous Burgers-type equations. SIAM J. Control Optim. **48**(3), 1567–1599 (2009)
7. Chen, S.Y., Holm, D., Margolin, L.G., Zhang, R.: Direct numerical simulations of the Navier-Stokes alpha model. Phys. D **133**, 66–83 (1999)
8. Cheskidov, A., Holm, D., Olson, E., Titi, E.: On a Leray-α model of turbulence. Proc. R. Soc. A **461**, 629–649 (2005)
9. Degasperis, A., Procesi, M.: Asymptotic integrability. In: Degasperis, A., Gaeta, G. (eds.) Symmetry and Perturbation Theory, pp. 23–37. World Scientific, River Edge (1999)
10. Degasperis, A., Holm, D.D., Hone, A.N.W.: Integrable and non-integrable equations with peakons. Nonlinear Physics: Theory and Experiment, II (Gallipoli, 2002), pp. 37–43. World Scientific Publishing, River Edge (2003)
11. Doubova, A., Fernández-Cara, E., González-Burgos, M., Zuazua, E.: On the controllability of parabolic systems with a nonlinear term involving the state and the gradient. SIAM J. Control Optim. **41**, 798–819 (2002)
12. Dullin, H.R., Gottwald, G.A., Holm, D.D.: Camassa-Holm, Korteweg-de Vries-5 and other asymptotically equivalent equations for shallow water waves. Fluid Dyn. Res. **33**, 73–95 (2003)
13. Dullin, H.R., Gottwald, G.A., Holm, D.D.: On asymptotically equivalent shallow water wave equations. Phys. D **190**, 1–14 (2004)
14. Fabre, C., Puel, J.-M., Zuazua, E.: Approximate controllability of the semilinear heat equation. Proc. R. Soc. Edinb. **125A**, 31–61 (1995)
15. Fernández-Cara, E., Guerrero, S.: Null controllability of the Burgers system with distributed controls. Syst. Control. Lett. **56**, 366–372 (2007)
16. Fernández-Cara, E., Zuazua, E.: Null and approximate controllability for weakly blowing up semilinear heat equations. Ann. Inst. H. Poincaré Anal. Non Linéaire **17**, 583–616 (2000)
17. Foias, C., Holm, D., Titi, E.: The three dimensional viscous Camassa-Holm equation and their relation to the Navier-Stokes equation and turbulence theory. J. Dyn. Differ. Equ. **14**, 1–36 (2002)
18. Fursikov, A.V., Imanuvilov, O.Y.: Controllability of Evolution Equations. Lecture Notes, vol. 34. Seoul National University, Korea (1996)
19. Gao, P.: Null controllability of the viscous Camassa-Holm equation with moving control. Proc. Indian Acad. Sci. Math. Sci. **126**(1), 99–108 (2016)
20. Geurts, B.J., Holm, D.: Leray and LANS-α modelling of turbulent mixing. J. Turbul. **7**(10), 33 (2006). (electronic)
21. Glass, O.: Controllability and asymptotic stabilization of the Camassa-Holm equation. J. Differ. Equ. **245**(6), 1584–1615 (2008)
22. Guerrero, S., Imanuvilov, O.Y.: Remarks on global controllability for the Burgers equation with two control forces. Ann. Inst. H. Poincaré Anal. Non Linéaire **24**, 897–906 (2007)
23. Holm, D., Staley, M.: Wave structure and nonlinear balances in a family of evolutionary PDEs. SIAM J. Appl. Dyn. Syst. **2**(3), 323–380 (2003)
24. Horsin, T.: On the controllability of the Burgers equation. ESAIM Control Optim. Calc. Var. **3**, 83–95 (1998)
25. Marbach, F.: Small time global null controllability for a viscous Burgers' equation despite the presence of a boundary layer. J. Math. Pures et Appl. **102**(2), 364–384 (2014)
26. Marbach, F.: An obstruction to small time local null controllability for a viscous Burgers' equation, accepted in Annales scientifiques de l'École normale supérieure (2018)
27. Mitra, D.: Local null controllability of viscous Camassa-Holm equation. J. Evol. Equ. **18**, 627–657 (2018)

28. Perrollaz, V.: Initial boundary value problem and asymptotic stabilization for the Camassa-Holm equation on an interval. J. Funct. Anal. **259**(9), 2333–2365 (2010)
29. Shen, C., Gao, A., Tian, L.: Optimal control of the viscous generalized Camassa-Holm equation. Nonlinear Anal. Real World Appl. **11**, 1835–1846 (2010)
30. Simon, J.: Compact sets in the space $L^p(0, T; B)$. Ann. Mat. Pura Appl. **146**, 65–96 (1987)
31. Temam, R.: Navier-Stokes equations. Theory and numerical analysis. AMS Chelsea Publishing, Providence (2001)
32. Tian, L., Shen, C., Ding, D.: Optimal control of the viscous Camassa-Holm equation. Nonlinear Anal. Real World Appl. **10**, 519–530 (2009)
33. Zuazua, E.: Controllability and observability of partial differential equations: some results and open problems. Handbook of Differential Equations: Evolutionary Equations, vol. III, pp. 527–621. Elsevier/North-Holland, Amsterdam (2007)

1-d Wave Equations Coupled via Viscoelastic Springs and Masses: Boundary Controllability of a Quasilinear and Exponential Stabilizability of a Linear Model

Günter Leugering, Tatsien Li and Yue Wang

Abstract We consider the out-of-the-plane displacements of nonlinear elastic strings which are coupled through point masses attached to the ends and viscoelastic springs. We provide the modeling, the well-posedness in the sense of classical semi-global C^2-solutions together with some extra regularity at the masses and then prove exact boundary controllability and velocity-feedback stabilizability, where controls act on both sides of the mass-spring-coupling.

Keywords Coupled system of quasilinear wave equations · Dynamical boundary condition · Visoelastic springs · Exact boundary controllability

AMS subject classifications 93B05 · 35L05 · 35L72

Yue Wang—Project supported by the DFG EXC315 Engineering of Adcanced Materials, National Basic Research Program of China (No 2013CB834100), and the National Natural Science Foundation of China (11121101).

G. Leugering (✉)
Department of Mathematics, Friedrich-Alexander University, Erlangen-Nuremberg,
Cauerstrasse 11, 91058 Erlangen, Germany
e-mail: leugering@am.uni-erlangen.de

T. Li · Y. Wang
School of Mathematical Sciences, Fudan University, Shanghai 200433,
People's Republic of China
e-mail: dqli@fudan.edu.cn

Y. Wang
e-mail: yuewang15@fudan.edu.cn

T. Li
Shanghai Key Laboratory for Contemporary Applied Mathematics, Fudan University,
Shanghai 200433, People's Republic of China

Nonlinear Mathematical Modeling and Methods Laboratory, Fudan University,
Shanghai 200433, People's Republic of China

F. Alabau-Boussouira et al. (eds.), *Trends in Control Theory and Partial Differential Equations*, Springer INdAM Series 32,
https://doi.org/10.1007/978-3-030-17949-6_8

1 Introduction

Controllability properties for elastic strings with attached tip-masses have been under consideration for quite some time. In [8] an in-span mass has been considered and controllability results in asymmetric spaces have been estbalished that reflect a smoothing property according the presence of the point-mass. See also [9]. In [2] the authors consider inverse problems for networks of strings, where the transmission conditions at multiple joints involve point-masses. The combination of elastic strings coupled via elastic springs and tip-masses has been considered by the authors of this article in [18], where exact boundary controllability was shown. In this article we extend the results of [18] to a coupling via viscoelastic springs. The method is based on the fundamental concept described in [12–14]. See also the recent work [11].

For a single 1-D quasilinear wave equation, based on a result concerning semi-global C^2 solutions, Li and Yu [15] used a direct constructive method with modular structure [13, 14] to establish local exact boundary controllability with Dirichlet, Neumann, Robin and dissipative boundary controls, respectively. For elastic strings, where a tip mass is attached to one of the ends, dynamical boundary condition appear according to Newton's law, see [3]. Exact boundary controllability for 1-D quasilinear single wave equations with dynamical boundary conditions has been obtained in [18]. We begin with two nonlinear elastic strings of common length L coupled at $x = 0$ via an elastic linear spring with stiffness κ. If we restrict ourselves to out-of-the-plane displacements the equations governing the motion of the strings become scalar. At the end points, i.e. at $x = 0$, $x = L$, we attach masses, which for the sake of simplicity we take as being equal to 1. At the free ends, i.e. at $x = L$, we apply boundary controls acting as forces. See Fig. 1.

Fig. 1 Two strings coupled via an elastic spring and masses

We introduce the stiffness of the strings as $K_i(u^i_x)$, $i = 1, 2$ and, correspondingly, $V_i(r) := \int_0^r K_i(s)ds$. We introduce the Lagrange function

$$\mathcal{L}(u) := \int_0^T \left\{ \sum_{i=1}^2 \left[\int_0^L \frac{1}{2}(u^i_t)^2(x,t) - V_i(u^i_x)(x,t)dx \right] + \frac{1}{2}\sum_{i=1}^2 (u^i_t)^2(0,t) - \frac{1}{2}\kappa\left(u^1(0,t) - u^2(0,t)\right)^2 \right\} dt.$$

Then, upon standard variational calculations, we obtain the following coupled system of two linear wave equations:

$$\begin{cases} u^i_{tt} - K_i(u^i_x)_x = 0, \quad x \in (0,L),\ t \in (0,T),\ i = 1,2, \\ x = 0 : u^1_{tt}(0,t) = K_1(u^1_x)(0,t) - \kappa(u^1(0,t) - u^2(0,t)), \\ \qquad u^2_{tt}(0,t) = K_2(u^2_x)(0,t) + \kappa(u^1(0,t) - u^2(0,t)), \\ x = L : u^1_{tt} = -K_1(u^1_x)(L,t) + h^1(t), \\ \qquad u^2_{tt} = -K_2(u^2_x)(L,t) + h^2(t),\ t \in (0,T) \\ t = 0 : u^i(x,0) = \phi^i_0(x), u^i_t(x,0) = \psi^i_1(x),\ x \in [0,L], i = 1,2, \end{cases} \tag{1.1}$$

where κ stands for the stiffness (Hooke's constant) of the spring. We are to find two boundary controls $(h^1(t), h^2(t))$ on $x = L$ in order to achieve exact boundary controllability for the coupled system (1.1). We assume that zero is at equilibrium such that

$$K_i(0) = 0,\ K_i'(0) > 0.$$

For non-constant equilibria, we need to work around such equilibria. We refer to a forthcoming publication for more complicated networks and non-constant equilibria. We now extend the model problem in that we introduce a linear viscoelastic behavior of Kelvin–Voigt type to the coupling spring. To this end we note that a Maxwell element, as shown in Fig. 2, satisfies the constitutive equation in continuum mechanics

$$\sigma = E\epsilon + \tau\dot{\epsilon},$$

where σ, ϵ signify the stress and the strain, respectively. See e.g. [17]. In this context the strain in the spring is given by the difference between the mass points. We, therefore, introduce besides $\kappa = E$ the parameter τ, representing the dash-pot. See Fig. 2 for a cartoon of this situation.

The corresponding system of quasilinear wave equations with this coupling is given by.

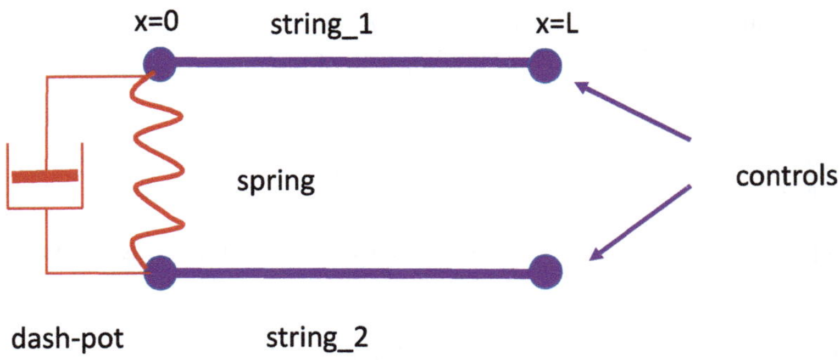

Fig. 2 Two strings coupled via a viscoelastic element and masses

$$\begin{cases} u^i_{tt} - K_i(u^i_x)_x = 0, \quad x \in (0, L),\ t \in (0, T),\ i = 1, 2, \\ x = 0 : u^1_{tt}(0, t) = K_1(u^1_x)(0, t) - \kappa(u^1(0, t) - u^2(0, t)) - \tau(u^1_t(0, t) - u^2_t(0, t)), \\ \qquad u^2_{tt}(0, t) = K_2(u^2_x)(0, t) + \kappa(u^1(0, t) - u^2(0, t)) + \tau(u^1_t(0, t) - u^2_t(0, t)), \\ x = L : u^1_{tt} = -K_1(u^1_x)(L, t) + h^1(t), \\ \qquad u^2_{tt} = -K_2(u^2_x)(L, t) + h^2(t),\ t \in (0, T), \\ t = 0 : u^i(x, 0) = \phi^i_0(x), u^i_t(x, 0) = \psi^i_1(x), x \in [0, L], i = 1, 2. \end{cases} \tag{1.2}$$

Kelvin-type visocelasticity as seen above is described by an ordinary differential equation between the stress and the strains. The corresponding boundary condition is still local, but second order in time. General viscoelastic springs would involve a convolution with a relaxation kernel $a(\cdot)$. The corresponding model then takes the following format.

$$\begin{cases} u^i_{tt} - K_i(u^i_x)_x = 0, \quad x \in (0, L),\ t \in (0, T),\ \ i = 1, 2, \\ x = 0 : u^1_{tt}(0, t) = K_1(u^1_x)(0, t) - \kappa(u^1(0, t) - u^2(0, t)) - \frac{\partial}{\partial t}\int\limits_0^t a(t - s)(u^1(0, s) - u^2(0, s))ds, \\ \qquad u^2_{tt}(0, t) = K_2(u^2_x)(0, t) + \kappa(u^1(0, t) - u^2(0, t)) + \frac{\partial}{\partial t}\int\limits_0^t a(t - s)(u^1(0, s) - u^2(0, s))ds, \\ x = L : u^1_{tt} = -K_1(u^1_x)(L, t) + h^1(t), \\ \qquad u^2_{tt} = -K_2(u^2_x)(L, t) + h^2(t),\ t \in (0, T), \\ t = 0 : u^i(x, 0) = \phi^i(x), u^i_t(x, 0) = \psi^i(x), u^i(x, s) = 0, s < 0, x \in [0, L], i = 1, 2. \end{cases} \tag{1.3}$$

In the case of (1.3), we need to add a zero displacement history. The alternative is to assume that $a(s) = 0, t < 0$. In this case the boundary condition is non-local in time already. The classical boundary controllability problem then consists in finding suitably smooth controls $h^i(\cdot), i = 1, 2$ such that in a given time $T > 0$ the controls drive the system (1.1), (1.2) or (1.3) to a given displacement and velocity profile at the final time T:

$$
\begin{aligned}
&\exists? \quad (h^1, h^2) \text{ such that } u^i \text{ satisfy (1.1), (1.2) or (1.3) and} \\
&t = 0: \quad (u^i(x,0), u^i_t(x,0))^{\mathrm{T}} = (\phi^i(x), \psi^i(x))^{\mathrm{T}}, \quad i = 1,2, \; 0 \le x \le L \\
&t = T: \quad (u^i(x,T), u^i_t(x,T))^{\mathrm{T}} = (\Phi^i(x), \Psi^i(x))^{\mathrm{T}}, \quad i = 1,2, \; 0 \le x \le L.
\end{aligned} \tag{1.4}
$$

While the question of exact controllability is natural for (1.1) and (1.2), it is more complicated in the case (1.3), as the final targets Φ, Ψ need to be holdable states. This means that after hitting the targets, the solution should stay there, possibly under applying constant controls. This is true for (1.1) and (1.2), but may fail to hold in the case (1.3), as the convolution dives the system beyond the final time T if the controls are switched off.

We integrate the second-order in time boundary conditions appearing in (1.1), (1.2) or (1.3) with respect to time. We obtain at $x = 0$

$$
u^1_t(0,t) = u^1_t(0,0) + \int_0^t \big(K_1(u^1_x)(0,s) - \kappa(u^1(0,s) - u^2(0,s))\big)\,ds \tag{1.5}
$$

$$
\begin{aligned}
u^1_t(0,t) = u^1_t(0,0) &+ \int_0^t \big(K_1(u^1_x)(0,s) - \kappa(u^1(0,s) - u^2(0,s))\big)\,ds \\
&-\tau(u^1(0,t) - u^2(0,t)) + \tau(u^1(0,0) - u^2(0,0))
\end{aligned} \tag{1.6}
$$

$$
\begin{aligned}
u^1_t(0,t) = u^1_t(0,0) &+ \int_0^t \big(K_1(u^1_x)(0,s) - \kappa(u^1(0,s) - u^2(0,s))\big)\,ds \\
&+ \int_0^t a(t-s)(u^1(0,s) - u^2(0,s))ds.
\end{aligned} \tag{1.7}
$$

In case of (1.5), (1.6), the boundary conditions can be put into the format

$$
u^1_t(0,t) = G_{11}(\psi^1(0), \phi^1(0), \phi^2(0)) + G_{21}(u^1(0,t), u^2(0,t)) + \int_0^t G_{31}(s, u^1(0,s), u^2(0,s), u^1_x(0,s))ds, \tag{1.8}
$$

whereas in case (1.7), the corresponding boundary condition is given by:

$$
u^1_t(0,t) = G_{11}(\psi^1(0), \phi^1(0), \phi^2(0)) + \int_0^t G_{31}(t, s, u^1(0,s), u^2(0,s), u^1_x(0,s))ds, \tag{1.9}
$$

where now the kernel G_{31} explicitly depends on the actual time t. The situation for u^2 is analogous. We may summarize as follows.

$$
\begin{cases}
u^i_{tt} - K_i(u^i_x)_x = 0, \quad x \in (0, L),\ t \in (0, T) \quad i = 1, 2, \\
x = 0: \ u^1_t(0, t) = G_{11}(\psi^1(0), \phi^1(0), \phi^2(0)) + G_{21}(u^1(0, t), u^2(0, t)) \\
\qquad + \int_0^t G_{31}(t, s, u^1(0, s), u^2(0, s), u^1_x(0, s))ds, \\
\quad u^2_t(0, t) = G_{12}(\psi^2(0), \phi^1(0), \phi^2(0)) + G_{22}(u^1(0, t), u^2(0, t)) \\
\qquad + \int_0^t G_{32}(t, s, u^1(0, s), u^2(0, s), u^2_x(0, s))ds, \\
x = L: \ u^1_t(L, t) = \psi^1(L) + \int_0^t \bar{G}_{21}(u^1_x)(L, s)ds + \int_0^t h^1(s)ds, \\
\quad u^2_t(L, t) = \psi^2(L) + \int_0^t \bar{G}_{22}(u^2_x)(L, s)ds + \int_0^t h^2(s)ds, \ t \in (0, T), \\
t = 0: \ u^i(x, 0) = \phi^i_0(x), u^i_t(x, 0) = \psi^i_1(x), \ x \in [0, L], i = 1, 2.
\end{cases}
\tag{1.10}
$$

Thus, the basic model to be discussed consists of coupled quasilinear wave equations, where the coupling is given by a non-local in time boundary condition of first order. In the case of general viscoelasticity, the kernels depend on the actual time, whereas in the elastic case and the Maxwell-type viscoelastic case the kernel does not depend on the actual time.

2 Well-Posedness and Dissipativity of the Viscoeleastic Model

2.1 Well-Posedness

In order to prove existence and uniqueness of semi-global classical solutions, we introduce the new variables

$$v^i := u^i_x, \ w^i := u^i_t, \ i = 1, 2.$$

We have

$$v^i_t = w^i_x = u^i_{xt}, \ w^i_t = K_i(v^i)_x = K_i'(v^i)v^i_x = K_i'(u^i_x)u^i_{xx}, \ i = 1, 2.$$

We further introduce $h_i(z) := \int_0^z \sqrt{K_i'(s)}ds$ and define the following Riemann invariants

$$r_-^i(x,t) := w^i(x,t) + h_i(v^i(x,t)),\ r_+(x,t) := w^i(x,t) - h_i(v^i(x,t)),\ r_0^i(x,t) := u^i(x,t). \tag{2.1}$$

We deduce the following equations for these Riemann invariants

$$\partial_t r_-^i(x,t) - \sqrt{K_i'(v^i(x,t)}\partial_x r_-^i(x,t) = 0 \tag{2.2}$$

$$\partial_t r_+^i(x,t) + \sqrt{K_i'(v^i(x,t)}\partial_x r_+^i(x,t) = 0$$

$$\partial_t r_0^i(x,t) = w^i(x,t). \tag{2.3}$$

We have the relations

$$w^i = \frac{1}{2}(r_-^i + r_+^i),\ h_i(v^i) = \frac{1}{2}(r_-^i - r_+^i),\ i = 1, 2. \tag{2.4}$$

As we assume $K_i'(s) > 0$, we have $D_{v^i} h_i(v^i) = \sqrt{K_i'(v^i)} > 0$ and, thus, h_i is strictly monotone. Therefore, there is an inverse mapping such that $v^i = p^i(r_-^i - r_+^i)$. The Riemann invariants obviously diagonalize our system of equations transformed into a first order system. We are going to write the coupling and boundary conditions in terms of the Riemann invariants. To this end, we insert the definitions (2.1) and the relations (2.2), (2.4) and the expression for v^i into (1.2). We assume $\phi = (\phi_1, \ldots, \phi_n)^{\mathrm{T}}$ is C^2 a vector-valued function of x with small $C^2[0, L]$ norm, $\psi = (\psi_1, \ldots, \psi_n)^{\mathrm{T}}$ is C^1 a vector-valued function of x with small C^1 norm, such that the conditions of C^2 compatibility at the points $(t, x) = (0, 0)$ and $(0, L)$ are satisfied, respectively.

$$\begin{aligned}\frac{1}{2}(r_+^1 + r_-^1)_t(0,t) &= K_1(p^1(r_-^1 - r_+^1)(0,t)) - \kappa\left(r_0^1(0,t) - r_0^2(0,t)\right)\\ &\quad - \tau\left(\frac{1}{2}(r_-^1 + r_+^1)_t(0,t) - \frac{1}{2}(r_-^2 + r_+^2)(0,t))\right)\end{aligned} \tag{2.5}$$

$$\begin{aligned}\frac{1}{2}(r_+^2 + r_-^2)(0,t) &= K_2(p^2(r_-^2 - r_+^2)(0,t)) + \kappa\left(r_0^1(0,t) - r_0^2(0,t)\right)\\ &\quad + \tau\left(\frac{1}{2}(r_-^1 + r_+^1)(0,t) - \frac{1}{2}(r_-^2 + r_+^2)(0,t))\right).\end{aligned} \tag{2.6}$$

We integrate (2.5) with respect to time and leave the Riemann variable $r_+^i(0, t)$ on the left-hand side, as this is the variable that determines the outgoing waves at $x = 0$. We obtain

$$r_+^1(0,t) = (r_-^1(0,0) + r_+^1(0,0)) - r_-^1(0,t) \tag{2.7}$$
$$+2\int_0^t \left\{K_1(p^1(r_-^1 - r_+^1)(0,s)) - \tau\left(r_-^1(0,s) - r_-^2(0,s) + r_+^1(0,s) - r_+^2(0,s)\right)\right\} ds$$
$$-2\kappa\int_0^t \left(r_0^1(0,s) - r_0^2(0,s)\right) ds$$

$$r_+^2(0,t) = (r_-^2(0,0) + r_+^2(0,0)) - r_-^2(0,t) \tag{2.8}$$
$$+2\int_0^t \left\{K_2(p^1(r_-^2 - r_+^2)(0,s)) + \tau\left(r_-^1(0,s) - r_-^2(0,s) + r_+^1(0,s) - r_+^2(0,s)\right)\right\} ds$$
$$+2\kappa\int_0^t \left(r_0^1(0,s) - r_0^2(0,s)\right) ds$$

Similarly, we obtain the boundary conditions at $x = L$ as follows

$$r_-^1(L,t) = (r_-^1(L,0) + r_+^1(L,0)) - r_+^1(L,t)$$
$$-2\int_0^t \left\{K_1(p^1(r_-^1 - r_+^1)(L,s)) - \left(r_-^1(L,s) + r_+^1(L,s)\right)\right\} ds + 2\int_0^t h_1(s)ds$$
$$r_-^2(L,t) = (r_-^2(L,0) + r_+^2(L,0)) - r_+^2(L,t) \tag{2.9}$$
$$-2\int_0^t \left\{K_2(p^2(r_-^2 - r_+^2)(L,s)) - \left(r_-^2(L,s) + r_+^2(L,s)\right)\right\} ds + 2\int_0^t h_1(s)ds.$$

We may now introduce the kernels

$$g_1(s, r_-^i, r_0^i, r_+^i, i = 1,2) := K_1(p^1(r_-^1 - r_+^1)(0,s)) - \tau\left(r_-^1(0,s) - r_-^2(0,s) + r_+^1(0,s) - r_+^2(0,s)\right)$$
$$-2\kappa\left(r_0^1(0,s) - r_0^2(0,s)\right)$$
$$g_2(s, r_-^i, r_0^i, r_+^i, i = 1,2) := K_2(p^2(r_-^1 - r_+^1)(0,s)) + \tau\left(r_-^1(0,s) - r_-^2(0,s) + r_+^1(0,s) - r_+^2(0,s)\right)$$
$$+2\kappa\left(r_0^1(0,s) - r_0^2(0,s)\right) \tag{2.10}$$
$$\bar g_1(s, r_-^i, r_0^i, r_+^i, i = 1,2) := K_1(p^1(r_-^1 - r_+^1)(L,s)) - \left(r_-^1(L,s) + r_+^1(L,s)\right)$$
$$\bar g_2(s, r_-^i, r_0^i, r_+^i, i = 1,2) := K_2(p^2(r_-^2 - r_+^2)(L,s)) - \left(r_-^2(L,s) + r_+^2(L,s)\right).$$

We also introduce the initial value functions:

$$f^i(\psi^i(0), \phi_x(0)) := (r_-^1(0,0) + r_+^1(0,0)) - r_-^1(0,t). \tag{2.11}$$

With this notation we are in the position to rewrite the system (1.2) as follows.

$$\begin{cases} \partial_t \begin{pmatrix} r_0^i \\ r_- \\ r_+ \end{pmatrix} + \begin{pmatrix} 0 & 0 & 0 \\ 0 & -\sqrt{K_i(p^i(r_-, r_+)} & 0 \\ 0 & 0 & \sqrt{K_i(p^i(r_-, r_+)} \end{pmatrix} \partial_x \begin{pmatrix} r_0^i \\ r_- \\ r_+ \end{pmatrix} = \begin{pmatrix} \frac{1}{2}(r_- + r_+) \\ 0 \\ 0 \end{pmatrix} \\ r_+^i(0,t) = f^i(\phi^i(0), \phi_x(0)) + \int_0^t g_i(s, r_-, r_0, r_+) ds \\ r_-^i(L,t) = \bar{f}^i(\phi^i(L), \phi_x(L)) + \int_0^t \bar{g}(s, r_-, r_0, r_+) ds + \int_0^t h_i(s) ds \\ r_0(x,0) = \psi^i(x),\ r_-(x,0) = \psi^i(x) + h_i(\phi'(x,0)),\ r_+(x,0) = \psi^i(x) - h_i(\phi'(x,0)). \end{cases} \tag{2.12}$$

This is precisely the format requested in [18] in order to show well-posedness of (2.12) and, hence, of (1.2). In order to apply the results of [18], we need to assume C^2-compatibility of the initial data. That is

$$\begin{cases} K_1'(\phi_x^1(0))\phi_{xx}^1(0) = K_1(\phi_x^1(0)) - \kappa(\phi^1(0) - \phi^2(0)) - \tau(\psi^1(0) - \psi^2(0)) \\ K_2'(\phi_x^1(0))\phi_{xx}^2(0) = K_2(\phi_x^1(0)) + \kappa(\phi^1(0) + \phi^2(0)) - \tau(\psi^1(0) - \psi^2(0)) \\ K_1'(\phi_x^1(L))\phi_{xx}^1(L) = K_1(\phi_x^1(L)) - k_1\psi^1(L) \\ K_1'(\phi_x^2(L))\phi_{xx}^2(L) = K_2(\phi_x^2(L)) - k_1\psi^1(L). \end{cases} \tag{2.13}$$

Theorem 2.1 *For any given $T > 0$, suppose that $\|(\phi, \psi)\|_{(C^2[0,L])^2 \times (C^1[0,L])^2}$, $\|h\|_{(C^0[0,T])^2}$ and $\|\bar{h}[0,T]\|_{(C^0[0,T])^2}$ are small enough (depending on T), and the conditions of C^2 compatibility* (2.13) *are satisfied at the points $(t, x) = (0, 0)$ and $(0, L)$, respectively. Then, the forward mixed initial-boundary value problems* (1.2) *admit a unique semi-global C^2 solution $\mathbf{u} = \mathbf{u}(t, x)$ with small C^2 norm on the domain $\mathcal{R}(T) = \{(t, x) | 0 \le t \le T, 0 \le x \le L\}$.*

We also obtain an additional regularity with respect to the time at $x = 0$, due to the masses there.

Remark 2.1 For the semi-global C^2 solution $\mathbf{u} = \mathbf{u}(t, x)$ given in Theorem 2.1, if $h^i(t) \equiv 0 (i = 1, 2)$, or more generally, $h^i(t) \in C^1[0, T]$ with small $C^1[0, T]$ norm, there is a hidden regularity on $x = 0$ that $u^i(t, 0) \in C^3[0, T] (i = 1, 2)$ with small C^3 norm.

2.2 Dissipativity of the Nonlinear Model

We now consider the following total energy related with the original system: (1.2).

$$E(t) = \sum_{i=1,2} \int_0^L \left(\frac{1}{2}(u_t^i)^2 + V^i(u_x^i) \right) dx + \frac{1}{2}\left(u_t^i(0,t)^2 + u_t^i(L,t)^2 \right) + \frac{1}{2}\left(u^1(0,t) - u^2(0,t) \right)^2, \tag{2.14}$$

where the potential $V^i(r)$ satisfies $V^i(r) = \int_0^r K_i(s)ds$. However, for $h_1(t), h_2(t)$ in (1.2), we choose velocity feedack controls

$$\begin{aligned} u_{tt}^1(L,t) &= -K_1(u_x^1)(L,t) - k_1 u_t^1(L,t), \\ u_{tt}^2(L,t) &= -K_2(u_x^2)(L,t) - k_2 u_t^2(L,t), \quad t \in (0,T) \end{aligned} \tag{2.15}$$

Assuming second order regularity, we obtain.

$$\begin{aligned} \frac{d}{dt}E(t) &= \sum_{i=1,2}\left\{\int_0^L \left(u_t^i u_{tt}^i + K_i(u_x^i)u_{xt}^i\right)dx + u_t^i(0,t)u_{tt}^i(0,t)\right\} \\ &\quad + \kappa\left(u^1(0,t) - u^2(0,t)\right)\left(u_t^1(0,t) - u_t^2(0,t)\right) \\ &= \sum_{i=1,2}\int_0^L u_t^i\left(u_{tt}^i - (K_i u_x^i)_x\right)dx + \sum_{i=1,2} K_i(u_x^i)u_t^i(x,t)|_0^L + u_t^1(L,t)u_{tt}^1(L,t) + u_t^2(L,t)u_{tt}^2(L,t) \\ &\quad + u_t^1(0,t)\left(u_{tt}^1 + \kappa(u^1(0,t) - u^2)0,t))\right) + u_t^1(0,t)\left(u_{tt}^2 - \kappa(u^1(0,t) - u^2(0,t))\right) \\ &= u_t^1(L,t)(u_{tt}^1(L,t) + K_1(u_x^1)(L,t)) + u_t^2(L,t)(u_{tt}^2(L,t) + K_2(u_x^2)(L,t)) \\ &\quad + u_t^1(0,t)\left(u_{tt}^1 - K_1(u_x^1)(0,t) + \kappa(u^1(0,t) - u^2(0,t))\right) \\ &\qquad + u_t^1(0,t)\left(u_{tt}^2 - K_2(u_x^2)(0,t) - \kappa(u^1(0,t) - u^2(0,t))\right) \\ &= -k_1(u_t^1(L,t))^2 - k_2(u_t^2(L,t)^2 + u_t^1(0,t)(-\tau(u_t^1(0,t) - u_t^2(0,t)) + u_t^2(0,t)(\tau(u_t^1(0,t) - u_t^2(0,t)) \\ &= -k_1(u_t^1(L,t))^2 - k_2(u_t^2(L,t))^2 - \tau\left(u_t^1(0,t) - u_t^2(0,t)\right)^2 \leq 0. \end{aligned} \tag{2.16}$$

This shows dissipativity. It is clear from (2.16) that the uncontrolled and purely elastic case leads to energy conservation. This suggests that boundary exponential stabilizability should hold. In the case of the linear model, we provide a proof of this fact. The investigation of the nonlinear case will be the subject of a forthcoming publication.

3 Exact Boundary Controllability for the Kelvin-Type Viscoelastic Coupling

In this section, we examine the problem of exact boundary controllability for a coupled system of two 1-D quasilinear wave equations, where the coupling is given by a Maxwell-type visocelastic spring-dash-pot system.

To this end, we provide final data Φ, Ψ, where $\Phi = (\Phi_1, \Phi_2)^{\mathrm{T}}$ is a C^2 vector-valued function of x with small $C^2[0,L]$ norm, $\Psi = (\Psi_1, \Psi_2)^{\mathrm{T}}$ is a $C^1[0,L]$ vector-valued function of x with small $C^1[0,L]$ norm, such that the conditions of C^2 compatibility (2.13) at the points $(t,x) = (T,0)$ and (T,L) are satisfied, respectively. Obviously, $\mathbf{u} = \mathbf{0}$ is an equilibrium state of (1.2), and we will establish local one-sided exact boundary controllability around $\mathbf{u} = \mathbf{0}$. By the results in [18], we obtain:

Theorem 3.1 *Let*

$$T > 2L \max_{i=1,2} \left(\frac{1}{\sqrt{K_i'(0)}} \right). \tag{3.1}$$

For any given initial data (ϕ, ψ) *and final data* (Φ, Ψ) *with small norms* $\|(\phi, \psi)\|_{(C^2[0,L])^2 \times (C^1[0,L])^2}$ *and* $\|(\Phi, \Psi)\|_{(C^2[0,L])^2 \times (C^1[0,L])^2}$ *and boundary controls* $h^i \equiv 0 (i = 1, 2)$*, such that the conditions of* C^2 *compatibility are satisfied at the points* $(t, x) = (0, 0)$ *and* $(T, 0)$*, respectively. Then, there exist boundary controls* $\overline{H} = (\bar{h}^1, \bar{h}^2)$ *with small norm* $\|\overline{H}\|_{(C^0[0,T])^2}$ *on* $x = L$*, such that the mixed initial-boundary value problem for* (1.2) *admits a unique* C^2 *solution* $\mathbf{u} = \mathbf{u}(t, x)$ *with small* C^2 *norm on the domain* $\mathcal{R}(T) = \{(t, x) | 0 \leq t \leq T, 0 \leq x \leq L\}$*, which exactly satisfies* (1.4).

Remark 3.1 More generally, if $h^i(t) \in C^1[0, T] (i = 1, 2)$ with small C^1 norm, Theorem 3.1 still holds.

4 Exponential Boundary Stabilization of a Linear Kelvin–Voigt-Model

To fix ideas, let us consider the following linear model of two strings coupled via a Kelvin–Voigt-type viscoelastic spring without tip-masses and velocity boundary feedbacks at $x = L$:

$$\begin{cases} u_{tt}^i - u_{xx}^i = 0, \quad x \in (0, L), \ t \in (0, T), \ i = 1, 2, \\ x = 0 : u_x^1(0, t) = \kappa(u^1(0, t) - u^2(0, t)) + \tau(u_t^1(0, t) - u_t^2(0, t)), \\ \qquad u_x^2(0, t) = -\kappa(u^1(0, t) - u^2(0, t)) - \tau(u_t^1(0, t) - u_t^2(0, t)), \\ x = L : u_x^1(L, t) = -k_1 u_t^1(L, t), \\ \qquad u_x^2(L, t) = -k_2 u_t^2(L, t), \ t \in (0, T), \\ t = 0 : u^i(x, 0) = \phi^i(x), u_t^i(x, 0) = \psi^i(x), x \in [0, L], i = 1, 2. \end{cases} \tag{4.1}$$

Here, the feedback parameters k_1, k_2 are positive numbers. As for existence and uniqueness of solutions, in case that $\phi^i(x), i = 1, 2$ are not constant, we refer to the previous section, where the result trivially follows from the nonlinear case. It is, however, also possible to achieve the wellposedness results via semi-group theory. We wish to prove exponential decay via an appropriate Liapunov function. As we ultimately intend to prove such property for the nonlinear model (1.2), we do not rely on results about linear equations, where uniform exponential stabilizability can be determined form exact boundary controllability. According to Theorem 3.1, exact controllability can be inferred in principle also for the linear problem considered here. However, for nonlinear equations no such implication is known. For that matter it is important to retrieve exponential stabilizability by Liapunov-techniques. Moreover, such techniques are much more precise about the decay rates. We refer to [4, 5, 7] for the techniques and their applications. We introduce the new variables

$$v^i := u^i_x,\ w^i := u^i_t, i = 1, 2$$

and the Riemann invariants

$$r^i_- := v^i + w^i,\ r^i_+ := v^i - w^i,\ i = 1, 2.$$

With this, we obtain

$$\partial_t r^i_- - \partial_x r^i_- = 0,\ \partial_t r^i_+ + \partial_x r^i_+ - = 0.$$

We consider a candidate Liapunov function:

$$\begin{aligned} E(t) :=& \frac{1}{2} \sum_{i=1,2} \int_0^L \left\{A^i_+ \exp(-\mu x)(r^i_+)^2 + A^i_- \exp(\mu x)(r^i_-)^2\right\} dx \\ &+ \frac{1}{2}\kappa(u^1(0,t) - u^2(0,t))^2 =: E_0(t) + E_1(t), \end{aligned} \tag{4.2}$$

where μ, A^i_+, $A^i_- > 0$ are still to be determined. We obtain

$$\begin{aligned} \frac{d}{dt}E_0 =& \sum_{i=1,2} \int_0^L \left\{A^i_+ \exp(-\mu x) r^i_+ \partial_t r^i_+ + A^i_- \exp(\mu x) r^i_- \partial_t r^i_-\right\} dx \qquad (4.3) \\ =& \sum_{i=1,2} \int_0^L \left\{A^i_+ \exp(-\mu x)\left(-\frac{1}{2}\partial_x (r^i_+)^2\right) + A^i_- \exp(\mu x)\left(-\frac{1}{2}\partial_x (r^i_-)^2\right)\right\} dx \\ &- \sum_{i=1,2} \frac{1}{2} A^i_+ \exp(-\mu x)(r^i_+)^2|_0^L + \sum_{i=1,2} \frac{1}{2} A^i_- \exp(\mu x)(r^i_-)^2|_0^L - \mu E_0(t). \end{aligned}$$

Moreover

$$\begin{aligned} \frac{d}{dt}E_1 &= \kappa\left(u^1(0,t) - u^2(0,t)\right)\left(u^1_t(0,t) - u^2_t(0,t)\right) \qquad (4.4) \\ &= \kappa\left(u^1(0,t) - u^2(0,t)\right)\left(\frac{1}{\tau}\left(u^1_x(0,t)\right) - \frac{\kappa}{\tau}\left(u^1(0,t) - u^2(0,t)\right)\right) \\ &= -\frac{\kappa^2}{\tau}\left(u^1(0,t) - u^2(0,t)\right)^2 + \frac{\kappa}{\tau}\left(u^1(0,t) - u^2(0,t)\right) u^1_x(0,t) \\ &\le -\frac{\kappa^2}{\tau}\left(u^1(0,t) - u^2(0,t)\right)^2 + \frac{\kappa}{\tau}\rho\left(u^1(0,t) - u^2(0,t)\right)^2 + \frac{\kappa}{\tau}\frac{1}{4\rho}\left(r^1_-(0,t) + r^1_+(0,t)\right)^2 \end{aligned}$$

$$\leq -\frac{\kappa}{\tau}(\kappa-\rho)\left(u^1(0,t)-u^2(0,t)\right)^2+\frac{\kappa}{\tau}\frac{1}{4\rho}\left(\frac{1}{\delta}r_-^1(0,t)^2+\delta r_+^1(0,t)^2\right)$$
$$=\frac{\kappa}{\tau}\frac{1}{4\rho}\left(\frac{1}{\delta}r_-^1(0,t)^2+\delta r_+^1(0,t)^2\right)-2\frac{\kappa-\rho}{\tau}E_1(t)$$

We are now concerned with the boundary values. We have

$$u_x^1(0,t)=\kappa(u^1(0,t)-u^2(0,t))+\tau(u_t^1(0,t)-u_t^2(0,t)) \quad \Leftrightarrow$$
$$(r_-^1(0,t)+r_+^1(0,t))=\kappa(u^1(0,t)-u^2(0,t))+\tau\left(r_-^1(0,t)-r_+^1(0,t)-r_-^2(0,t)+r_+^2(0,t)\right) \quad \Leftrightarrow$$
$$(1+\tau)r_+^1(0,t)-\tau r_+^2(0,t)=(\tau-1)r_-^1(0,t)-\tau r_-^2(0,t)+\kappa(u^1(0,t)-u^2(0,t)).$$

The analogous boundary representation holds for the second string. Together we have

$$(1+\tau)r_+^1(0,t)-\tau r_+^2(0,t)=(\tau-1)r_-^1(0,t)-\tau r_-^2(0,t)+\kappa(u^1(0,t)-u^2(0,t))$$
$$-\tau r_+^1(0,t)+(1+\tau)r_+^2(0,t)=(\tau-1)r_-^2(0,t)-\tau r_-^1(0,t)-\kappa(u^1(0,t)-u^2(0,t)),$$

which reads as follows:

$$\begin{pmatrix}1+\tau & -\tau\\ -\tau & 1+\tau\end{pmatrix}\begin{pmatrix}r_+^1(0,t)\\ r_+^2(0,t)\end{pmatrix}=\begin{pmatrix}\tau-1 & -\tau\\ -\tau & \tau-1\end{pmatrix}\begin{pmatrix}r_-^1(0,t)\\ r_-^2(0,t)\end{pmatrix}+\kappa\begin{pmatrix}(u^1(0,t)-u^2(0,t))\\ -(u^1(0,t)-u^2(0,t))\end{pmatrix} \tag{4.5}$$

solving for the Riemann invariants with sign '+' we obtain

$$\begin{pmatrix}r_+^1(0,t)\\ r_+^2(0,t)\end{pmatrix}=-\frac{1}{1+2\tau}\begin{pmatrix}1 & 2\tau\\ 2\tau & 1\end{pmatrix}\begin{pmatrix}r_-^1(0,t)\\ r_-^2(0,t)\end{pmatrix}+\frac{\kappa}{1+2\tau}\begin{pmatrix}(u^1(0,t)-u^2(0,t))\\ -(u^1(0,t)-u^2(0,t))\end{pmatrix}. \tag{4.6}$$

or

$$\begin{pmatrix}r_+^1(0,t)\\ r_+^2(0,t)\end{pmatrix}=-\frac{1}{1+2\tau}\begin{pmatrix}r_-^1(0,t)+2\tau r_-^2(0,t)\\ 2\tau r_-^1(0,t)+r_-^2(0,t)\end{pmatrix}+\frac{\kappa}{1+2\tau}\begin{pmatrix}(u^1(0,t)-u^2(0,t))\\ -(u^1(0,t)-u^2(0,t))\end{pmatrix}. \tag{4.7}$$

We take the 2−norm of both sides and obtain after some calculus.

$$r_+^1(0,t)^2+r_+^2(0,t)^2\leq\frac{3\kappa^2}{(1+2\tau)^2}(u^1(0,t)-u^2(0,t))^2+\left(1+\frac{2(1-2\tau)^2}{(1+2\tau)^2}\right)(r_-^1(0,t)^2+r_-^2(0,t)^2). \tag{4.8}$$

At $x=L$ we have

$$\begin{aligned}u_x^i(L,t)&=-k_iu_t^i(L,t),\ i=1,2 \Leftrightarrow \\ r_-^i(L,t)+r_+^i(L,t)&=-k_i(r_-^i(L,t)-r_+^i(L,t),\ i=1,2 \Leftrightarrow\\ r_-^i(L,t)&=\frac{k_i-1}{k_i+1}r_+^i(L,t),\ i=1,2..\end{aligned} \tag{4.9}$$

Notice that for $k_i = 1$ (4.9) provides transparent boundary feedback conditions such that no energy enters the strings at $x = L$, i.e. waves approaching $x = L$ from inside the strings leave without any reflection. We now go back to (4.2).

$$\begin{aligned} &-\sum_{i=1,2} \frac{1}{2} A_+^i \exp(-\mu x)(r_+^i)^2|_0^L + \sum_{i=1,2} \frac{1}{2} A_-^i \exp(\mu x)(r_-^i)^2|_0^L \\ &= \frac{1}{2}\Bigg\{\sum_{i=1,2} \left(A_+^i (r_+^i(0,t))^2 - A_-^i (r_-^i(0,t))^2\right) \\ &\quad + \sum_{i=1,2} \left(A_-^i \exp(\mu L)(r_-^1(0,t))^2 - A_+^i \exp(-\mu L)(r_+^i(L,t))^2\right)\Bigg\}. \end{aligned} \tag{4.10}$$

With this, we can now estimate

$$\begin{aligned} \frac{d}{dt}E(t) \leq &\frac{1}{2}\Bigg\{\sum_{i=1,2} \left(A_+^i (r_+^i(0,t))^2 - A_-^i (r_-^i(0,t))^2\right) \\ &+ \sum_{i=1,2} \left(A_-^i \exp(\mu L)(r_-^1(0,t))^2 - A_+^i \exp(-\mu L)(r_+^i(L,t))^2\right)\Bigg\} \\ &+ \frac{\kappa}{4}\frac{1}{\rho\delta} r_-^1(0,t)^2 + \frac{\kappa}{4}\frac{\delta}{\rho} r_+^1(0,t)^2 - \mu E_0(t) - 2\frac{\kappa-\rho}{\tau}E_1(t). \\ = &\frac{1}{2}(A_+^1 + \kappa\frac{2\delta}{\rho}) r_+^1(0,t)^2 + \frac{1}{2}A_+^2 r_+^2(0,t)^2 - \frac{1}{2}(A_-^1 - \kappa\frac{1}{2\rho\delta}) r_-^1(0,t)^2 - \frac{1}{2}A_-^2 r_-^2(0,t)^2 \\ &+ \frac{1}{2}A_-^1 \exp(\mu L) r_-^1(L,t)^2 + \frac{1}{2}A_-^2 \exp(\mu L) r_-^2(L,t)^2 \\ &- \frac{1}{2}A_+^1 \exp(-\mu L) r_+^1(L,t)^2 - \frac{1}{2}A_+^2 \exp(-\mu L) r_+^2(L,t)^2 - \mu E_0(t) - 2\frac{\kappa-\rho}{\tau}E_1(t). \end{aligned} \tag{4.11}$$

We now use (4.8), (4.9) in (4.11) and obtain

$$\begin{aligned} &\frac{d}{dt}E(t) \leq -\mu E_0(t) - 2\frac{\kappa-\rho}{\tau}E_1(t) \\ &+ \max(A_+^1 + \kappa\frac{\delta}{\rho}, A_+^2)\left\{\frac{3\kappa^2}{2(1+2\tau)^2}(u^1(0,t) - u^2(0,t))^2 + \frac{1}{2}\left(1 + 2\frac{(1-2\tau)^2}{(1+2\tau)^2}\right)(r_-^1(0,t)^2 + r_-^2(0,t)^2)\right\} \\ &- \frac{1}{2}(A_-^1 - \kappa\frac{1}{2\rho\delta}) r_-^1(0,t)^2 - \frac{1}{2}A_-^2 r_-^2(0,t)^2 + \frac{1}{2}\left(\frac{k-1}{k+1}\right)^2 \exp(\mu L)\left(A_-^1 r_+^1(L,t)^2 + A_-^2 r_+^2(L,t)^2\right) \\ &- \frac{1}{2}\exp(-\mu L)\left(A_+^1 r_+^1(L,t) + A_+^2 r_+^2(L,t)^2\right) \\ &= \left\{\max(A_+^1 + \kappa\frac{\delta}{\rho}, A_+^2)\frac{1}{2}\left(1 + 2\frac{(1-2\tau)^2}{(1+2\tau)^2}\right) + \frac{\kappa}{2}\frac{1}{\delta\rho} - \frac{1}{2}A_-^1\right\} r_-^1(0,t)^2 \\ &\quad + \left\{\max(A_+^1 + \kappa\frac{\delta}{\rho}, A_+^2)\frac{1}{2}\left(1 + 2\frac{(1-2\tau)^2}{(1+2\tau)^2}\right) - \frac{1}{2}A_-^2\right\} r_-^2(0,t)^2 \\ &\quad + \frac{1}{2}\left\{\left(\frac{k-1}{k+1}\right)^2 A_-^1 \exp(\mu L) - A_+^1 \exp(-\mu L)\right\} r_+^1(L,t)^2 \end{aligned} \tag{4.12}$$

$$+\frac{1}{2}\left\{\left(\frac{k-1}{k+1}\right)^2 A_-^2 \exp(\mu L) - A_+^2 \exp(-\mu L)\right\} r_+^2(L,t)^2$$
$$-\mu E_0(t) - \left(2\frac{\kappa-\rho}{\tau} - \frac{3\kappa^2}{2(1+2\tau)^2}\max(A_+^1 + \kappa\frac{\delta}{\rho}, A_+^2)\right) E_1(t).$$

Recall that $\kappa, \tau \geq 0$ are fixed physical parameters, while $A_\pm, \delta, \tau > 0$ can be chosen under given constraints in order to achieve the desired energy estimate. It is clear from (4.12) that if we choose the feedback-gains $k_i = k = 1, i = 1, 2$, the third and the fourth term are automatically negative, regardless how small $A_+^i, i = 1, 2$ are, and the first and the second term become negative for large $A_-^i, i = 1, 2$ and small $A_+^i, i = 1, 2$, small δ, ρ with $\delta \approx \rho$. In this case also the factor of $E_1(t)$ becomes negative, say $-\mu$ for suitably small $\mu > 0$. One can also choose the viscosity parameter τ to improve the estimates. Thus, for the case of *optimal feedback gains*, i.e. $k_i = 1, i = 1, 2$, we obtain the estimate

$$\frac{d}{dt}E(t) \leq -\mu E(t), \forall t > 0, \tag{4.13}$$

which provides us with exponential decay, for suitable choices of the parameters above. In the general case, we have to fulfil the following inequalities, where for the sake of simplicity, we choose $A_-^i = A_-, A_+^i = A_+, i = 1, 2$.

$$\begin{aligned}
&(i) \quad (A_+ + \kappa\frac{\delta}{\rho})\frac{1}{2}\left(1 + 2\frac{(1-2\tau)^2}{(1+2\tau)^2}\right) + \frac{\kappa}{2}\frac{1}{\delta\rho} - \frac{1}{2}A_- \leq 0\\
&(ii) \quad \frac{1}{2}\left(\frac{k-1}{k+1}\right)^2 A_- \exp(\mu L) - A_+ \exp(-\mu L) \leq 0\\
&(iii) \quad \left(2\frac{\kappa-\rho}{\tau} - \frac{3\kappa^2}{2(1+2\tau)^2}(A_+ + \kappa\frac{\delta}{\rho})\right) \geq \mu > 0.
\end{aligned} \tag{4.14}$$

Under the conditions (4.14), we obtain again (4.13). Clearly, small spring stiffness κ and small viscosity τ will improve the exponential decay rate μ which also depends on the relation between A_+ and A_-:

$$\frac{A_+}{A_-} \geq \frac{1}{2}\left(\frac{k-1}{k+1}\right)^2 \exp(2\mu L).$$

We assume the following compatibility conditions.

$$\begin{cases}
\phi_x^1(0) = \kappa(\phi^1(0) - \phi^2(0)) + \tau(\psi^1(0) - \psi^2(0))\\
\phi_x^2(0) = -\kappa(\phi^1(0) - \phi^2(0)) - \tau((\psi^1(0) - \psi^2(0))\\
\phi_x^1(L) = -k_1\psi^1(L)\\
\phi_x^2(L) = -k_2\psi^2(L).
\end{cases} \tag{4.15}$$

Theorem 4.1 *Let $\phi \in C^1(0, L), \psi \in C^0(0, L)$ satisfy the compatibility conditions* (4.15) *and let the assumptions* (4.14) *be fulfilled. Then the unique solution of* (4.1) *decays exponentially.*

Remark 4.1 The result concerns an L^2-type Liapunov function for the linear system (4.1). We conjecture that a similar result, also for H^2-type Liapunov functions hold true. This will be the subject of a forthcoming publication.

5 Conclusion and Outlook

We have analyzed linear and quasilinear strings coupled via visco-elastic springs of standard type. We have provided a framework that allows for generalizations in various directions. First of all, general visco-elastic spring coupling of fading memory type can be considered in the quasilinear context. See [19] for general non-local boundary conditions in the context of exact controllability from both sides of the spring coupling. The situation is more complex for controls appearing only at the end of one string. If the spring stiffness is infinite, in other words, if the strings are directly coupled via a mass, we have to consider asymmetric spaces, due to the smoothing effect of the coupling mass. See e.g. [8]. Such phenomena have not been discussed for the quasilinear wave equation so far. Therefore, this contribution gives a first result concerning controllability of nonlinear strings with point-mass and visco-elastic spring couplings.

We also embarked on stability and stabilization properties of such systems. However, due to space limitations, we just looked at linear strings, no masses and low regularity of solutions. The full system with masses and quasilinear strings is currently open, but subject to a forthcoming publication. Moreover, all that has been said in this contribution concerns out-of-plane-displacement models. There is currently no corresponding result for planar of spatial quasilinear strings and springs. Again, this is subject to current research of the authors. For a general model and corresponding controllability results for 3-d quasilinear string networks see [10]. In [6] quasilinear networks of Timoshenko beams have been considered. Again, these models may be extended to spring-couplings as in this article.

We end with the proposition of a damage model, where we assume that the coupling spring undergoes a damage process which, in turn, is driven by excessive strains at the coupling point. To this end, we consider a time dependent stiffness $\kappa(t)$ of the coupling spring and propose an evolution of damage in due course as follows

$$\begin{cases} u^i_{tt} - K_i(u^i_x)_x = 0, \quad x \in (0,, T)L), \ t \in (0, \quad i = 1, 2, \\ x = 0 : u^1_{tt}(0,t) = K_1(u^1_x)(0,t) - \kappa(t)(u^1(0,t) - u^2(0,t)), \\ \qquad u^2_{tt}(0,t) = K_2(u^2_x)(0,t) + \kappa(t)(u^1(0,t) - u^2(0,t)), \\ \qquad \kappa_t = -\{\frac{1}{2}(u^1(0,t) - u^2(0,t))^2 - \eta\}_+ \kappa(t), \quad \kappa(0) = \kappa_0, \\ x = L : u^1_{tt} = -K_1(u^1_x)(L,t) + h^1(t), \\ \qquad u^2_{tt} = -K_2(u^2_x)(L,t) + h^2(t), \\ t = 0 : u^i(x,0) = \phi^i(x), u^i_t(x,0) = \psi^i(x), u^i(x,s) = 0, s < 0, x \in [0,L], i = 1, 2. \end{cases} \tag{5.1}$$

Here $\{a\}_+ = \max(a, 0)$. The nonlinear ordinary differential equation for the evolution of the damage describes an exponential decay of $\kappa(t)$ for time periods, where the displacement of the spring is excessively large (larger than $\eta >> 0$). Problems of this type are open. They are connected to the general problem of degeneration in the coefficients of wave equations in the sense of [1]. Clearly, if only one control is considered, the problem looses the property of controllability as the spring damage finally leads to break of the spring. The controllability or observability time will tend to infinity as $\kappa(t)$ tends to zero.

References

1. Alabau-Boussouira, F., Cannarsa, P., Leugering, G.: Control and stabilization of degenerate wave equations. SIAM J. Control Optim. **55**(3), 2052–2087 (2017)
2. Avdonin, S., Avdonina, N., Edward, J.: Boundary inverse problems for networks of vibrating strings with attached masses, IMA Preprint #2457 November 2015, Institute for Mathematics and its Applications, University of Minnesota
3. Cindea, N., Micu, S., Pazoto, A.: Periodic solutions for a weakly dissipated hybrid system. J. Math. Anal. Appl. **385**, 399–413 (2012)
4. Dick, M., Gugat, M., Herty, M., Leugering, G., Steffensen, S., Wang, K.: Stabilization of networked hyperbolic systems with boundary feedback. Trends in PDE Constrained Optimization. International Series of Numerical Mathematics, vol. 165, pp. 487–504. Birkhäuser/Springer, Cham (2014)
5. Gugat, M., Leugering, G., Wang, K.: Neumann boundary feedback stabilization for a nonlinear wave equation: a strict H^2-Lyapunov function. Math. Control Relat. Fields **7**(3), 419–448 (2017)
6. Gu, Q., Leugering, G., Li, T.: Exact boundary controllability on a tree-like network of nonlinear planar Timoshenko beams. Chin. Ann. Math. Ser. B **38**(3), 711–740 (2017)
7. Gugat, M., Leugering, G., Tamasoiu, S., Wang, K.: H^2-stabilization of the isothermal Euler equations: a Lyapunov function approach. Chin. Ann. Math. Ser. B **33**(4), 479–500 (2012)
8. Hansen, S., Zuazua, E.: Exact controllability and stabilization of a vibrating string with an interior point mass. SIAM J Control Optim. **33**(5), 1357–1391 (1995)
9. Leugering, G.: On dynamic domain decomposition of controlled networks of elastic strings and joint masses. In: Kappel, F. (ed.) Control of Distributed Parameter Systems. ISNM, vol. 126, pp. 199–205. Birkhäuser, Basel (1998)
10. Leugering, G., Schmidt, E.J.P.G.: On exact controllability of networks of nonlinear elastic strings in 3-dimensional space. Chin. Ann. Math. Ser. B **33**(1), 33–60 (2012)
11. Long, H., Ji, F., Wang, K.: Exact boundary controllability and exact boundary observability for a coupled system of quasilinear wave equations. Chin. Ann. Math. **34B**(4), 479–490 (2013)

12. Li, T.: Controllability and Observability for Quasilinear Hyperbolic Systems. AIMS Series on Applied Mathematics, vol. 3. AIMS & Higher Education Press, Beijing (2010)
13. Li, T., Rao, B.: Local exact boundary controllability for a class of quasilinear hyperbolic systems. Chin. Ann. Math. **23B**, 209–218 (2002)
14. Li, T., Rao, B.: Exact boundary controllability for quasilinear hyperbolic systems. SIAM J. Control Optim. **41**, 1748–1755 (2003)
15. Li, T., Yu, L.: Exact boundary controllability for 1-D quasilinear wave equations. SIAM J. Control Optim. **45**, 1074–1083 (2006)
16. Li, T., Yu, W.: Boundary Value Problems for Quasilinear Hyperbolic systems. Duke University Mathematics Series V (1985)
17. Renardy, M., Hrusa, W.J., Nohel, J.A.: Mathematical problems in viscoelasticity. In: Pitman Monographs and Surveys in Pure and Applied Mathematics, vol. 35, x+273 pp. Longman Scientific & Technical, Harlow; Wiley, New York (1987). ISBN 0-582-00320-2
18. Wang, Y., Leugering, G., Li, T.: Exact boundary controllability for 1-D quasilinear wave equations with dynamical boundary conditions. Math. Methods Appl. Sci. **40**(10), 3808–3820 (2017)
19. Wang, Y., Leugering, G., Li, T.: Exact boundary controllability for a coupled system of quasilinear wave equations with dynamical boundary conditions (2017, to appear)

A Semilinear Integro-Differential Equation: Global Existence and Hidden Regularity

Paola Loreti and Daniela Sforza

Abstract Here we show a hidden regularity result for nonlinear wave equations with an integral term of convolution type and Dirichlet boundary conditions. Under general assumptions on the nonlinear term and on the integral kernel we are able to state results about global existence of strong and mild solutions without any further smallness on the initial data. Then we define the trace of the normal derivative of the solution showing a regularity result. In such a way we extend to integrodifferential equations with nonlinear term well-known results available in the literature for linear wave equations with memory.

Keywords Hidden regularity · Positive definite kernels · Partial differential equations

1 Introduction

Let $\Omega \subset \mathbb{R}^N$ $(N \geq 1)$ be a bounded open domain of class C^2. Let us denote by ν the outward unit normal vector to the boundary Γ. In this paper we will consider the Cauchy problem for nonlinear wave equations with a general integral term and Dirichlet boundary conditions:

$$\begin{cases} u_{tt}(t,x) = \Delta u(t,x) + \displaystyle\int_0^t \dot{a}(t-s)\Delta u(s,x)\,ds + g(u(t,x)), & t \geq 0,\ x \in \Omega, \\ u(t,x) = 0 \quad t \geq 0,\ x \in \Gamma, \\ u(0,x) = u_0(x), \quad u_t(0,x) = u_1(x), & x \in \Omega. \end{cases} \tag{1}$$

P. Loreti · D. Sforza (✉)
Dipartimento di Scienze di Base e Applicate per l'Ingegneria,
Sapienza Università di Roma, Via Antonio Scarpa 16, 00161 Rome, Italy
e-mail: daniela.sforza@sbai.uniroma1.it

P. Loreti
e-mail: paola.loreti@sbai.uniroma1.it

F. Alabau-Boussouira et al. (eds.), *Trends in Control Theory and Partial Differential Equations*, Springer INdAM Series 32,
https://doi.org/10.1007/978-3-030-17949-6_9

According to the physical model as proposed in [23], we will assume that the integral kernel satisfies

$$\begin{aligned} & a : (0, \infty) \to \mathbb{R} \ \text{ is a positive definite function with } a(0) < 1, \\ & a\,, \dot{a} \in L^1(0, +\infty), \end{aligned} \tag{2}$$

and the nonlinear term fulfils the following conditions:

- $g \in C(\mathbb{R})$ such that there exist $\alpha \geq 0$, with $\alpha(N-2) \leq 2$, and $C > 0$ so that

$$\begin{aligned} & g(0) = 0, \\ & |g(x) - g(y)| \leq C(1 + |x|^\alpha + |y|^\alpha)|x - y| \qquad \forall x, y \in \mathbb{R}, \end{aligned} \tag{3}$$

- set $G(t) = \displaystyle\int_0^t g(s)\,ds$, there exists $C_0 > 0$ such that

$$G(t) \leq C_0 |t|^2 \qquad \forall t \in \mathbb{R}. \tag{4}$$

We will establish the following global existence result without any smallness assumption on initial data.

Theorem 1.1 *Under the assumptions* (2)–(4), *for any* $(u_0, u_1) \in H_0^1(\Omega) \times L^2(\Omega)$ *problem* (1) *admits a unique mild solution* u *on* $[0, \infty)$.

In our previous work [17] we study the linear case of (1) where the integral kernel $\dot{a} : [0, \infty) \to (-\infty, 0]$ is a locally absolutely continuous function, $\dot{a}(0) < 0$, $\ddot{a}(t) \geq 0$ for a.e. $t \geq 0$ and $a(0) < 1$.

In this paper the existence result may be stated for more general kernels, as

- $a(t) = a_0 \int_t^\infty \frac{e^{-\alpha s}}{s^\beta}\,ds$, with $\alpha > 0$, $0 \leq \beta < 1$ and $0 \leq a_0 < \frac{\Gamma(1-\beta)}{\alpha^{1-\beta}}$,
- $a(t) = \int_t^\infty (a_0 s + a_1) e^{-\alpha s}\,ds = (\frac{a_0}{\alpha} t + \frac{a_0 + \alpha a_1}{\alpha^2}) e^{-\alpha t}$, with $\alpha > 0$, $a_0, a_1 \geq 0$, $\frac{a_0 + \alpha a_1}{\alpha^2} < 1$, $\alpha a_1 - a_0 \geq 0$,
- $a(t) = k \int_t^\infty \frac{1}{(1+s)^\alpha} ds$, with $k > 0$ such that $a(0) < 1$, $\alpha > 2$.

Examples of fading memory kernels can be found in [8, 10].

Some examples of g satisfying assumptions (3)–(4) are

$$g(x) = c|x|^p x, \quad c < 0, \quad p(N-2) \leq 2, \qquad\qquad g(x) = c \sin x, \quad c \in \mathbb{R}.$$

In addition, for more regular kernels we will prove a so-called hidden regularity result.

Theorem 1.2 *Assume* (3)–(4),

$$C_0 < \lambda(1 - a(0))/2, \qquad \lambda := \inf\{\|\nabla v\|_{L^2}^2, \ v \in H_0^1(\Omega), \ \|v\|_{L^2} = 1\},$$

and

$$\begin{aligned}
&a \in C^1([0,\infty)),\ \dot a(0) < 0,\ a(t) \ge 0,\ \dot a(t) \le 0\ \forall t \ge 0,\\
&\ddot a(t) \in L^1_{loc}(0,+\infty), \ddot a(t) \ge 0,\ a.e.\ t \ge 0, \\
&a\,,\dot a \in L^1(0,+\infty),\ a(0) < 1\,.
\end{aligned} \tag{5}$$

Let $T > 0$, there exists a constant $c = c(T) > 0$ such that for any $u_0 \in H^1_0(\Omega)$ and $u_1 \in L^2(\Omega)$ if u is the mild solution of (1), then, denoting by $\partial_\nu u$ the normal derivative, we have

$$\int_0^T \int_\Gamma |\partial_\nu u|^2\, d\Gamma\, dt \le c(\|\nabla u_0\|^2_{L^2} + \|u_1\|^2_{L^2})\,.$$

Moreover, if the energy $E(t)$ of the solution u satisfies

$$\int_0^t E(s)\, ds \le c_0 E(0) \qquad \forall t \ge 0 \qquad (c_0 > 0\ \textit{independent of}\ t),$$

then we have

$$\partial_\nu u \in L^2(0,\infty; L^2(\Gamma)).$$

The proof of the existence of the solution u relies on energy estimates. Although we use some results obtained in [1, 5, 6], here we are interested to treat initial data without any smallness, so the previous results have to be adapted to consider our setting. We also mention the papers [3, 11].

To understand how to frame our paper in the literature, we recall briefly some known results. Lasiecka and Triggiani [14] established the hidden regularity property for the weak solution u of the wave equation with Dirichlet boundary conditions that is

$$\partial_\nu u \in L^2_{loc}(\mathbb{R}; L^2(\Gamma)).$$

The term hidden was proposed by J.L. Lions [15] for the wave equation in the context of the exact controllability problems. Later in [16] J.L. Lions proved that the weak solution of the nonlinear wave equation

$$u_{tt}(t,x) = \Delta u(t,x) - |u|^p u, \quad t \ge 0,\ x \in \Omega,$$

satisfies a trace regularity result. Milla Miranda and Medeiros [19] enlarged the class of nonlinear terms by means of approximation arguments. However they do not consider memory terms in the equation, that is $\dot a \equiv 0$. To our knowledge it seems that there are not previous papers studying the hidden regularity for solutions of nonlinear integro-differential problems when the integral kernels satisfy the assumptions (5).

The plan of our paper is the following. In Sect. 2 we list some notations and preliminary results. In Sect. 3 we establish existence and uniqueness results of mild and strong solutions. Finally, in Sect. 4 we give hidden regularity results for a nonlinear equation with memory.

2 Preliminaries

Let $L^2(\Omega)$ be endowed with the usual inner product and norm

$$\|u\|_{L^2} = \left(\int_\Omega |u(x)|^2\,dx\right)^{1/2} \qquad u \in L^2(\Omega)\,.$$

Throughout the paper we will use a standard notation for the integral convolution between two functions, that is

$$h * u(t) := \int_0^t h(t-s)u(s)\,ds\,. \tag{6}$$

A well-known result concerning integral equations (see e.g. [9, Theorem 2.3.5]), that we will use later is the following.

Lemma 2.1 *Let $h \in L^1(0,T)$, $T>0$. If the function $\varphi(t) + h*\varphi(t)$ belongs to $L^2(0,T;L^2(\Omega))$ then $\varphi \in L^2(0,T;L^2(\Omega))$ and there exist a positive constant $c_1 = c_1(\|h\|_{L^1(0,T)})$, depending on the norm $\|h\|_{L^1(0,T)}$, such that*

$$\int_0^T \|\varphi(t)\|^2_{L^2}\,dt \le c_1 \int_0^T \left\|\varphi(t) + h*\varphi(t)\right\|^2_{L^2}\,dt\,. \tag{7}$$

Recall that h is a *positive definite kernel* if for any $y \in L^2_{loc}(0,\infty;L^2(\Omega))$ we have

$$\int_0^t\int_\Omega y(\tau,x)\int_0^\tau h(\tau-s)y(s,x)\,ds\,dx\,d\tau \ge 0\,, \qquad t\ge 0\,. \tag{8}$$

Also, h is said to be a *strongly positive definite kernel* if there exists a constant $\delta>0$ such that $h(t)-\delta e^{-t}$ is positive definite. This stronger notion for the integral kernel allows to obtain uniform estimates for solutions of integral equations, see [6, Corollary 2.12]. For completeness we recall here that result, because we will use it later.

Lemma 2.2 *Let $a \in L^1(0,\infty)$ be a strongly positive definite kernel such that $\dot a \in L^1(0,\infty)$ and $a(0) < 1$. If the function $\varphi(t) + \dot a * \varphi(t)$ belongs to $L^2(0,\infty;L^2(\Omega))$ then $\varphi \in L^2(0,\infty;L^2(\Omega))$ and there exist a positive constant c_1, such that*

$$\int_0^\infty \|\varphi(t)\|^2_{L^2}\,dt \le c_1 \int_0^\infty \left\|\varphi(t) + \dot a*\varphi(t)\right\|^2_{L^2}\,dt\,. \tag{9}$$

Regarding the nonlinear term, we will follow the approach pursued in [7] for the nonintegral case when $\dot a \equiv 0$. Precisely, we will consider a function $g \in C(\mathbb{R})$ such that there exist $\alpha \ge 0$, with $(N-2)\alpha \le 2$, and $C>0$ so that

$$
\begin{aligned}
&g(0) = 0, \\
&|g(x) - g(y)| \le C(1 + |x|^\alpha + |y|^\alpha)|x - y| \qquad \forall x, y \in \mathbb{R}.
\end{aligned}
\tag{10}
$$

In [7, Proposition 6.1.5] the following result has been proved.

Proposition 2.3 *If g satisfies the hypotheses* (10), *then g is Lipschitz continuous from bounded subsets of $H_0^1(\Omega)$ to $L^2(\Omega)$. In particular, there exists a positive constant C such that*

$$
\int_\Omega |g(u(x))|^2\, dx \le C \int_\Omega |\nabla u(x)|^2\, dx \qquad \forall u \in H_0^1(\Omega). \tag{11}
$$

We will assume that the integral kernel satisfies the following conditions:

$$
\begin{aligned}
&a : (0, \infty) \to \mathbb{R} \quad \text{is a positive definite function,} \\
&a\,, \dot{a} \in L^1(0, +\infty), \\
&a(0) < 1\,.
\end{aligned}
\tag{12}
$$

For reader's convenience we begin with recalling some known notions and results. First, we write the Laplacian as an abstract operator. Indeed, we define the operator $A : D(A) \subset L^2(\Omega) \to L^2(\Omega)$ as

$$
D(A) = H^2(\Omega) \cap H_0^1(\Omega)
$$

$$
Au(x) = -\Delta u(x) \qquad u \in D(A)\,,\ x \in \Omega \text{ a.e.}
$$

We recall, see e.g. [5, Definition 3.1], that there exists a unique family $\{\mathcal{R}(t)\}_{t\ge 0}$ of bounded linear operators in $L^2(\Omega)$ the so-called resolvent for the linear equation

$$
u''(t) + Au(t) + \int_0^t \dot{a}(t-s)Au(s)\, ds = 0\,, \tag{13}
$$

that satisfy the following conditions:

(i) $\mathcal{R}(0)$ is the identity operator and $\mathcal{R}(t)$ is strongly continuous on $[0, \infty)$, that is, for all $u \in L^2(\Omega)$, $\mathcal{R}(\cdot)u$ is continuous;

(ii) $\mathcal{R}(t)$ commutes with A, which means that $\mathcal{R}(t)D(A) \subset D(A)$ and

$$
A\mathcal{R}(t)u = \mathcal{R}(t)Au\,, \qquad u \in D(A)\,,\ t \ge 0\,;
$$

(iii) for any $u \in D(A)$, $\mathcal{R}(\cdot)u$ is twice continuously differentiable in $L^2(\Omega)$ on $[0, \infty)$ and $\mathcal{R}'(0)u = 0$;

(iv) for any $u \in D(A)$ and any $t \ge 0$,

$$
\mathcal{R}''(t)u + A\mathcal{R}(t)u + \int_0^t \dot{a}(t-\tau)A\mathcal{R}(\tau)u\, d\tau = 0\,.
$$

In the sequel we will use the following uniform estimates for the resolvent, see e.g. [5, Proposition 3.4-(i)], taking into account that $D(A^{1/2}) = H_0^1(\Omega)$.

Proposition 2.4 *For any* $u \in L^2(\Omega)$ *and any* $t > 0$, *we have* $1 * \mathcal{R}(t)u \in H_0^1(\Omega)$ *and*

$$\|\mathcal{R}(t)u\|_{L^2}^2 + \big(1 - a(0)\big) \|\nabla(1 * \mathcal{R})(t)u\|_{L^2}^2 \le \|u\|_{L^2}^2 . \tag{14}$$

In particular, $\nabla(1 * \mathcal{R})(\cdot)$ *is strongly continuous in* $L^2(\Omega)$.

Let $0 < T \le \infty$ be given. We recall some notions of solution for the semilinear equation

$$u_{tt}(t,x) = \Delta u(t,x) + \int_0^t \dot{a}(t-s)\Delta u(s,x)\,ds + g(u(t,x)), \quad t \in [0,T]\,,\ x \in \Omega\,. \tag{15}$$

Definition 2.5 We say that u is a **strong solution** of (15) on $[0, T]$ if

$$u \in C^2([0,T]; L^2(\Omega)) \cap C([0,T]; H^2(\Omega) \cap H_0^1(\Omega))$$

and u satisfies (15) for every $t \in [0, T]$.

Let $u_0, u_1 \in L^2(\Omega)$. A function $u \in C^1([0,T]; L^2(\Omega)) \cap C([0,T]; H_0^1(\Omega))$ is a **mild solution** of (15) on $[0, T]$ with initial conditions

$$u(0) = u_0, \quad u_t(0) = u_1, \tag{16}$$

if

$$u(t) = \mathcal{R}(t)u_0 + \int_0^t \mathcal{R}(\tau)u_1 d\tau + \int_0^t 1 * \mathcal{R}(t-\tau)g(u(\tau))d\tau, \tag{17}$$

where $\{\mathcal{R}(t)\}$ is the resolvent for the linear equation (13).

Notice that the convolution term in (17) is well defined, thanks to Proposition 2.3. A strong solution is also a mild one.

Another useful notion of generalized solution of (15) is the so-called *weak solution*, that is a function $u \in C^1([0,T]; L^2(\Omega)) \cap C([0,T]; H_0^1(\Omega))$ such that for any $v \in H_0^1(\Omega)$, $t \to \int_\Omega u_t v\,dx \in C^1([0,T])$ and

$$\frac{d}{dt}\int_\Omega u_t v\,dx = -\int_\Omega \nabla u \cdot \nabla v\,dx - \int_\Omega \int_0^t \dot{a}(t-s)\nabla u(s)\,ds \cdot \nabla v\,dx + \int_\Omega g(u(t))v\,dx\,, \qquad \forall t \in [0,T]\,. \tag{18}$$

Adapting a classical argument due to Ball [2], one can show that any mild solution of (15) is also a weak solution, and the two notions of solution are equivalent in the linear case when $g \equiv 0$ (see also [22]).

Throughout the paper we denote with the symbol $\cdot$ the Euclidean scalar product in $\mathbb{R}^N$.

3 Existence and Uniqueness of Mild and Strong Solutions

The next proposition ensures the local existence and uniqueness of the mild solution for the Cauchy problem

$$\begin{cases} u_{tt}(t,x) = \Delta u(t,x) + \int_0^t \dot{a}(t-s)\Delta u(s,x)\,ds + g(u(t,x)), & t \geq 0,\ x \in \Omega, \\ u(t,x) = 0 \quad t \geq 0,\ x \in \Gamma, \\ u(0,x) = u_0(x), \quad u_t(0,x) = u_1(x), \quad x \in \Omega. \end{cases} \tag{19}$$

The proof relies on suitable regularity estimates for the resolvent $\{\mathcal{R}(t)\}$ such as (14) (for more details see e.g. [5, Sect. 3]) and a standard fixed point argument (see [4] for an analogous proof).

Proposition 3.1 *If $u_0 \in H_0^1(\Omega)$ and $u_1 \in L^2(\Omega)$, there exists a positive number T such that the Cauchy problem* (19) *admits a unique mild solution on* $[0,T]$.

Assuming more regular data and using standard argumentations, one can show that the mild solution is a strong one.

Proposition 3.2 *Let $u_0 \in H^2(\Omega) \cap H_0^1(\Omega)$ and $u_1 \in H_0^1(\Omega)$. Then, the mild solution of the Cauchy problem* (19) *in* $[0,T]$ *is a strong solution. In addition, u belongs to $C^1([0,T]; H_0^1(\Omega))$.*

To investigate the existence for all $t \geq 0$ of the solutions, for g satisfying (10) we introduce $G \in C(\mathbb{R})$ by means of

$$G(t) = \int_0^t g(s)\,ds\,. \tag{20}$$

We define the energy of a mild solution u of (19) on a given interval $[0,T]$, as

$$E(t) = \frac{1}{2}\int_\Omega |u_t|^2\,dx + \frac{1-a(0)}{2}\int_\Omega |\nabla u|^2\,dx - \int_\Omega G(u)\,dx\,. \tag{21}$$

In view of (10) we have

$$\int_\Omega |G(u_0(x))|\,dx \leq C\int_\Omega |\nabla u_0(x)|^2\,dx \qquad \forall u_0 \in H_0^1(\Omega)\,, \tag{22}$$

and hence

$$E(0) \leq C(\|\nabla u_0\|_{L^2}^2 + \|u_1\|_{L^2}^2) \qquad u_0 \in H_0^1(\Omega),\ u_1 \in L^2(\Omega)\,. \tag{23}$$

About the energy of the solutions, we recall some known results, see [6, Lemma 3.5].

Lemma 3.3 *(i) If $u_0 \in H^2(\Omega) \cap H_0^1(\Omega)$ and $u_1 \in H_0^1(\Omega)$, then the strong solution u of problem* (19) *on* $[0, T]$ *satisfies the identity*

$$E(t)+\int_0^t \int_\Omega a * \nabla u_t(s) \cdot \nabla u_t(s)\, dx\, ds$$
$$= E(0) + a(0) \int_\Omega |\nabla u_0|^2\, dx - a(t) \int_\Omega \nabla u_0 \cdot \nabla u(t)\, dx - \int_0^t \dot{a}(s) \int_\Omega \nabla u_0 \cdot \nabla u(s)\, dx\, ds\,, \tag{24}$$

for any $t \in [0, T]$.

(ii) If $u_0 \in H_0^1(\Omega)$ and $u_1 \in L^2(\Omega)$, then the mild solution u of problem (19) *on* $[0, T]$ *verifies*

$$E(t) \le E(0) + a(0) \int_\Omega |\nabla u_0|^2\, dx - a(t) \int_\Omega \nabla u_0 \cdot \nabla u(t)\, dx - \int_0^t \dot{a}(s) \int_\Omega \nabla u_0 \cdot \nabla u(s)\, dx\, ds\,, \tag{25}$$

for any $t \in [0, T]$.

Assuming an extra condition on G, global existence will follow for all data. For further convenience we introduce the notation

$$\lambda = \inf\{\|\nabla v\|_{L^2}^2,\ v \in H_0^1(\Omega),\ \|v\|_{L^2} = 1\}\,. \tag{26}$$

Theorem 3.4 *Suppose that there exists $C_0 > 0$ such that*

$$G(t) \le C_0 |t|^2 \qquad \forall t \in \mathbb{R}\,. \tag{27}$$

Then for any $(u_0, u_1) \in H_0^1(\Omega) \times L^2(\Omega)$ problem (19) *admits a unique mild solution u on* $[0, \infty)$.

Moreover, if we suppose that the constant $C_0 > 0$ in (27) *satisfies*

$$C_0 < \lambda(1 - a(0))/2, \tag{28}$$

where λ is defined in (26)*, then $E(t)$ is positive and we have for any $t \ge 0$*

$$E(t) \ge \frac{1}{2}\|u_t(t)\|_{L^2}^2 + \frac{C}{2}\|\nabla u(t)\|_{L^2}^2\,, \tag{29}$$

$$E(t) \le C\big(\|u_1\|_{L^2}^2 + \|\nabla u_0\|_{L^2}^2\big)\,, \tag{30}$$

$$\|u_t(t)\|_{L^2}^2 + \|\nabla u(t)\|_{L^2}^2 \le C\big(\|u_1\|_{L^2}^2 + \|\nabla u_0\|_{L^2}^2\big)\,, \tag{31}$$

where the symbol C denotes positive constants, that can be different.

Furthermore, if $u_0 \in H^2(\Omega) \cap H_0^1(\Omega)$ and $u_1 \in H_0^1(\Omega)$, then u is a strong solution of (19) *on* $[0, \infty)$*, $u \in C^1([0, \infty); H_0^1(\Omega))$ and for any $t \ge 0$*

$$E(t) + \int_0^t \int_\Omega a * \nabla u_t(s) \cdot \nabla u_t(s)\, dx\, ds \le C\big(\|u_1\|_{L^2}^2 + \|\nabla u_0\|_{L^2}^2\big)\,. \tag{32}$$

Proof Let $[0, T)$ be the maximal domain of the mild solution u of (19). To prove $T = \infty$, we will argue by contradiction and assume that T is a positive real number. We will show that there exists a constant $C = C(T) > 0$ such that

$$\int_\Omega |u_t|^2\,dx + \int_\Omega |\nabla u|^2\,dx \le C \qquad \forall t \in [0, T)\,. \tag{33}$$

First, thanks to (25) we have

$$\begin{aligned}
&\int_\Omega |u_t|^2\,dx + \big(1 - a(0)\big)\int_\Omega |\nabla u|^2\,dx \\
&\qquad \le \|u_1\|_{L^2}^2 + \big(1 - a(0)\big)\|\nabla u_0\|_{L^2}^2 - 2\int_\Omega G(u_0)\,dx + 2a(0)\|\nabla u_0\|_{L^2}^2 \\
&\qquad - 2a(t)\int_\Omega \nabla u_0 \cdot \nabla u(t)\,dx - 2\int_0^t \dot a(s)\int_\Omega \nabla u_0 \cdot \nabla u(s)\,dx\,ds + 2\int_\Omega G(u(t))\,dx \\
&\qquad \le \|u_1\|_{L^2}^2 + \big(1 + a(0)\big)\|\nabla u_0\|_{L^2}^2 + 2\int_\Omega |G(u_0)|\,dx \\
&\qquad - 2a(t)\int_\Omega \nabla u_0 \cdot \nabla u(t)\,dx - 2\int_0^t \dot a(s)\int_\Omega \nabla u_0 \cdot \nabla u(s)\,dx\,ds + 2\int_\Omega G(u(t))\,dx\,.
\end{aligned} \tag{34}$$

We note that

$$\begin{aligned}
&-2a(t)\int_\Omega \nabla u_0 \cdot \nabla u(t)\,dx \\
&\qquad \le 2\|a\|_\infty \int_\Omega |\nabla u_0|\,|\nabla u(t)|\,dx \le \frac{1 - a(0)}{2}\int_\Omega |\nabla u|^2\,dx + \frac{2\|a\|_\infty^2}{1 - a(0)}\|\nabla u_0\|_{L^2}^2\,.
\end{aligned}$$

Putting the above estimate into (34), we obtain

$$\begin{aligned}
&\int_\Omega |u_t|^2\,dx + \frac{1 - a(0)}{2}\int_\Omega |\nabla u|^2\,dx \\
&\qquad \le \|u_1\|_{L^2}^2 + \left(1 + a(0) + \frac{2\|a\|_\infty^2}{1 - a(0)}\right)\|\nabla u_0\|_{L^2}^2 + 2\int_\Omega |G(u_0)|\,dx \\
&\qquad\qquad - 2\int_0^t \dot a(s)\int_\Omega \nabla u_0 \cdot \nabla u(s)\,dx\,ds + 2\int_\Omega G(u(t))\,dx\,.
\end{aligned} \tag{35}$$

Now, we have to estimate the last two terms on the right-hand side of the previous inequality. As regards the first one, we note that

$$- 2\int_0^t \dot a(s)\int_\Omega \nabla u_0 \cdot \nabla u(s)\,dx\,ds \le \frac{2\|\dot a\|_1}{1 - a(0)}\|\nabla u_0\|_{L^2}^2 + \frac{1 - a(0)}{2}\int_0^t |\dot a(s)|\int_\Omega |\nabla u(s)|^2\,dx\,ds\,. \tag{36}$$

Concerning the other integral, assumption (27) yields for any $t \in [0, T)$

$$\int_\Omega G(u(t))\,dx \le C_0\int_\Omega |u(t)|^2\,dx\,. \tag{37}$$

In addition, we observe that

$$\|u(t)\|_{L^2}^2 = \|u_0\|_{L^2}^2 + \int_0^t \frac{d}{ds} \int_\Omega |u(s)|^2 \, dx \, ds = \|u_0\|_{L^2}^2 + 2\int_0^t \int_\Omega u(s)u_t(s) \, dx \, ds\,.$$

Since, by the definition (26) of λ we have

$$\int_\Omega |\nabla u|^2 \, dx \geq \lambda \int_\Omega |u|^2 \, dx\,, \tag{38}$$

we deduce

$$\|u(t)\|_{L^2}^2 \leq \|u_0\|_{L^2}^2 + \int_0^t \left(\int_\Omega |u_t(s)|^2 \, dx + \frac{1}{\lambda} \int_\Omega |\nabla u(s)|^2 \, dx \right) ds\,.$$

Therefore, by (37)

$$\begin{aligned}\int_\Omega G(u(t)) \, dx &\leq C_0 \|u_0\|_{L^2}^2 + C_0 \int_0^t \left(\int_\Omega |u_t(s)|^2 \, dx + \frac{1}{\lambda} \int_\Omega |\nabla u(s)|^2 \, dx \right) ds \\ &\leq C_0 \|u_0\|_{L^2}^2 + M \int_0^t \left(\int_\Omega |u_t(s)|^2 \, dx + \frac{1-a(0)}{2} \int_\Omega |\nabla u(s)|^2 \, dx \right) ds\,,\end{aligned} \tag{39}$$

where $M = C_0 \max\{1, \frac{2\lambda^{-1}}{1-a(0)}\}$. Plugging (36) and (39) into (35), thanks also to (22) we get

$$\begin{aligned}&\int_\Omega |u_t|^2 \, dx + \frac{1-a(0)}{2} \int_\Omega |\nabla u|^2 \, dx \\ &\quad \leq \|u_1\|_{L^2}^2 + \left(1 + a(0) + 2\frac{\|a\|_\infty^2 + \|\dot a\|_1}{1-a(0)} + \frac{C_0}{\lambda} + C \right) \|\nabla u_0\|_{L^2}^2 \\ &\qquad + \int_0^t (|\dot a(s)| + M) \left(\int_\Omega |u_t(s)|^2 \, dx + \frac{1-a(0)}{2} \int_\Omega |\nabla u(s)|^2 \, dx \right) ds\,.\end{aligned} \tag{40}$$

Applying Gronwall lemma, we obtain for any $t \in [0, T)$

$$\begin{aligned}&\int_\Omega |u_t|^2 \, dx + \frac{1-a(0)}{2} \int_\Omega |\nabla u|^2 \, dx \\ &\quad \leq \left(\|u_1\|_{L^2}^2 + \left(1 + a(0) + 2\frac{\|a\|_\infty^2 + \|\dot a\|_1}{1-a(0)} + \frac{C_0}{\lambda} + C \right) \|\nabla u_0\|_{L^2}^2 \right) e^{\int_0^t (|\dot a(s)| + M) \, ds} \\ &\quad \leq \left(\|u_1\|_{L^2}^2 + \left(1 + a(0) + 2\frac{\|a\|_\infty^2 + \|\dot a\|_1}{1-a(0)} + \frac{C_0}{\lambda} + C \right) \|\nabla u_0\|_{L^2}^2 \right) e^{\|\dot a\|_1 + MT}\,,\end{aligned}$$

and hence, set

$$C(T) = \frac{e^{\|\dot a\|_1 + MT}}{\min\{1, \frac{1-a(0)}{2}\}} \left(\|u_1\|_{L^2}^2 + \left(1 + a(0) + 2\frac{\|a\|_\infty^2 + \|\dot a\|_1}{1-a(0)} + \frac{C_0}{\lambda} + C \right) \|\nabla u_0\|_{L^2}^2 \right)$$

we have that (33) holds true.

To have a contradiction, we will prove that $u \in C([0,T]; H_0^1(\Omega)) \cap C^1([0,T]; L^2(\Omega))$. First, set

$$v(t) = \mathcal{R}(t)u_0 + \int_0^t \mathcal{R}(\tau)u_1 d\tau \qquad t \geq 0\,, \tag{41}$$

we note that, thanks to the properties of the resolvent we have

$$v(t) \in C([0,+\infty); H_0^1(\Omega)) \cap C^1([0,+\infty); L^2(\Omega))\,. \tag{42}$$

Since by (17) and (41) we can write

$$u(t) = v(t) + \int_0^t 1 * \mathcal{R}(t-\tau)g(u(\tau))d\tau, \tag{43}$$

for $h > 0$ and $0 \leq t < t+h < T$ we have

$$\begin{aligned} u(t+h) - u(t) &= v(t+h) - v(t) + \int_0^{t+h} 1 * \mathcal{R}(\tau)g(u(t+h-\tau))d\tau - \int_0^t 1 * \mathcal{R}(\tau)g(u(t-\tau))d\tau \\ &= v(t+h) - v(t) + \int_0^t 1 * \mathcal{R}(\tau)\big[g(u(t+h-\tau)) - g(u(t-\tau))\big]d\tau \\ &\quad + \int_t^{t+h} 1 * \mathcal{R}(\tau)g(u(t+h-\tau))d\tau. \end{aligned}$$

As a consequence, by (14) we have

$$\begin{aligned} \|\nabla u(t+h) - \nabla u(t)\|_{L^2} \leq& \|\nabla v(t+h) - \nabla v(t)\|_{L^2} + \frac{1}{1-a(0)}\int_0^t \|g(u(s+h)) - g(u(s))\|_{L^2} ds \\ &+ \frac{1}{1-a(0)}\int_0^h \|g(u(s))\|_{L^2} ds. \end{aligned}$$

Thanks to Proposition 2.3 and (33) we deduce that

$$\begin{aligned} \|\nabla u(t+h) - \nabla u(t)\|_{L^2} &\leq \|\nabla v(t+h) - \nabla v(t)\|_{L^2} + C\int_0^t \|\nabla u(s+h)) - \nabla u(s)\|_{L^2} ds \\ &\quad + C\int_0^h \|\nabla u(s)\|_{L^2} ds \\ &\leq \|\nabla v(t+h) - \nabla v(t)\|_{L^2} + Ch + C\int_0^t \|\nabla u(s+h)) - \nabla u(s)\|_{L^2} ds, \end{aligned}$$

where $C = C(T) > 0$ is a positive constant. Applying Gronwall lemma, we get

$$\|\nabla u(t+h) - \nabla u(t)\|_{L^2} \leq \big(\|\nabla v(t+h) - \nabla v(t)\|_{L^2} + Ch\big)e^{CT},$$

and hence the function $\nabla u(t)$ is uniformly continuous in $[0,T[$ with values in $L^2(\Omega)$. Therefore $u(t)$ can be also defined in T in a way that $u \in C([0,T]; H_0^1(\Omega))$. Moreover, again by (43) we have

$$u_t(t) = v_t(t) + \int_0^t \mathcal{R}(t-\tau)g(u(\tau))d\tau,$$

and hence, thanks to the regularity of $v(t)$, see (42), and $u \in C([0,T]; H_0^1(\Omega))$ we get $u \in C^1([0,T]; L^2(\Omega))$. Therefore, one can restart by the data $(u(T), u_t(T) \in H_0^1(\Omega) \times L^2(\Omega)$ but this is in contrast with the fact that T is maximal. The contradiction follows by assuming that T is a positive real number and hence $T = \infty$.

Now we suppose that the constant C_0 in (27) satisfies the extra condition (28). By (37) and (38) we get

$$\int_\Omega G(u)\,dx \le \frac{C_0}{\lambda}\int_\Omega |\nabla u|^2\,dx\,. \tag{44}$$

Therefore, putting the previous estimate into the expression (21) of the energy, we obtain

$$E(t) \ge \frac{1}{2}\int_\Omega |u_t|^2\,dx + \left(\frac{1-a(0)}{2} - \frac{C_0}{\lambda}\right)\int_\Omega |\nabla u|^2\,dx\,,$$

that is (29) where $C = \frac{\lambda(1-a(0))-2C_0}{\lambda} > 0$ thanks to the assumption $C_0 < \lambda(1-a(0))/2$. In particular $E(0) \ge 0$.

Again by (25), we get

$$E(t) \le \frac{1}{2}\|u_1\|_{L^2}^2 + \frac{1+a(0)}{2}\|\nabla u_0\|_{L^2}^2 + \int_\Omega |G(u_0)|\,dx \\ - a(t)\int_\Omega \nabla u_0 \cdot \nabla u(t)\,dx - \int_0^t \dot{a}(s)\int_\Omega \nabla u_0 \cdot \nabla u(s)\,dx\,ds\,. \tag{45}$$

If $C > 0$ is the constant in (29), taking into account that

$$-a(t)\int_\Omega \nabla u_0 \cdot \nabla u(t)\,dx \le \|a\|_\infty \int_\Omega |\nabla u_0|\,|\nabla u(t)|\,dx \le \frac{C}{4}\int_\Omega |\nabla u|^2\,dx + \frac{\|a\|_\infty^2}{C}\|\nabla u_0\|_{L^2}^2\,,$$

$$-\int_0^t \dot{a}(s)\int_\Omega \nabla u_0 \cdot \nabla u(s)\,dx\,ds \le \frac{C}{4}\int_0^t |\dot{a}(s)|\int_\Omega |\nabla u(s)|^2\,dx\,ds + \frac{\|\dot{a}\|_1}{C}\|\nabla u_0\|_{L^2}^2\,,$$

from (45) we get

$$E(t) \le \frac{1}{2}\|u_1\|_{L^2}^2 + \left(\frac{1+a(0)}{2} + \frac{\|a\|_\infty^2 + \|\dot{a}\|_1}{C}\right)\|\nabla u_0\|_{L^2}^2 + \int_\Omega |G(u_0)|\,dx \\ + \frac{C}{4}\int_\Omega |\nabla u|^2\,dx + \frac{C}{4}\int_0^t |\dot{a}(s)|\int_\Omega |\nabla u(s)|^2\,dx\,ds\,. \tag{46}$$

Putting together (29) and (46), we get

$$\frac{1}{2}\|u_t(t)\|_{L^2}^2 + \frac{C}{4}\|\nabla u(t)\|_{L^2}^2 \le \frac{1}{2}\|u_1\|_{L^2}^2 + \left(\frac{1+a(0)}{2} + \frac{\|a\|_\infty^2 + \|\dot a\|_1}{C}\right)\|\nabla u_0\|_{L^2}^2 + \int_\Omega |G(u_0)|\,dx$$
$$+\frac{C}{4}\int_0^t |\dot a(s)|\|\nabla u(s)\|_{L^2}^2\,ds\,.$$

Applying Gronwall lemma, we have for any $t \ge 0$

$$\frac{1}{2}\|u_t(t)\|_{L^2}^2 + \frac{C}{4}\|\nabla u(t)\|_{L^2}^2$$
$$\le e^{\|\dot a\|_1}\left(\frac{1}{2}\|u_1\|_{L^2}^2 + \left(\frac{1+a(0)}{2} + \frac{\|a\|_\infty^2 + \|\dot a\|_1}{C}\right)\|\nabla u_0\|_{L^2}^2 + \int_\Omega |G(u_0)|\,dx\right).$$

Moreover, putting the above estimate into (46) and taking into account (22) we obtain that (30) holds true. Finally, (31) follows from (29) and (30), while (32) holds for strong solutions in view of (24). □

Under more regular assumptions on the integral kernel, we can establish a different result concerning the global existence of solutions and the dissipation of energy. Indeed, we will assume that the integral kernel satisfies the following conditions

$$\begin{aligned}
&a \in C^1([0,\infty)),\ \dot a(0) < 0,\ a(t) \ge 0,\ \dot a(t) \le 0\ \forall t \ge 0,\\
&\ddot a(t) \in L^1_{loc}(0,+\infty),\ \ddot a(t) \ge 0,\ \text{a.e. } t \ge 0,\\
&a\,,\dot a \in L^1(0,+\infty),\ a(0) < 1\,.
\end{aligned} \tag{47}$$

It is well known that these conditions imply that a is a strongly positive definite kernel, see [21, Corollary 2.2], and hence (12) holds true. Then we can consider a different expression for the energy of the solutions with respect to (21). More precisely, we will define the energy as follows

$$\begin{aligned}
E(t) =& \frac{1}{2}\int_\Omega |u_t(t,x)|^2\,dx + \frac{1-a(0)+a(t)}{2}\int_\Omega |\nabla u(t,x)|^2\,dx\\
&-\frac{1}{2}\int_\Omega\int_0^t \dot a(t-s)|\nabla u(s,x) - \nabla u(t,x)|^2\,ds\,dx - \int_\Omega G(u)\,dx \qquad t \ge 0\,.
\end{aligned} \tag{48}$$

Thanks to the assumptions (47) $E(t)$ is a decreasing function, see e.g. [1, 20]. In particular, we have

$$E'(t) = \frac{1}{2}\dot a(t)\int_\Omega |\nabla u(t,x)|^2\,dx - \frac{1}{2}\int_\Omega\int_0^t \ddot a(t-s)|\nabla u(s,x) - \nabla u(t,x)|^2\,ds\,dx \qquad \text{a.e. } t \ge 0\,. \tag{49}$$

Theorem 3.5 *Let us assume* (47), (10), (27) *and* (28).

For any $u_0 \in H_0^1(\Omega)$ and $u_1 \in L^2(\Omega)$ there exists a unique mild solution u on $[0,\infty)$ of the Cauchy problem

$$\begin{cases} u_{tt}(t,x) = \Delta u(t,x) + \int_0^t \dot{a}(t-s)\Delta u(s,x)\,ds + g(u(t,x))\,, & t \ge 0,\ x \in \Omega, \\ u(t,x) = 0 \quad t \ge 0,\ x \in \Gamma, \\ u(0,x) = u_0(x), \quad u_t(0,x) = u_1(x), & x \in \Omega. \end{cases} \tag{50}$$

In addition, if the initial data are more regular, that is $u_0 \in H^2(\Omega) \cap H_0^1(\Omega)$ and $u_1 \in H_0^1(\Omega)$ the mild solution of (50) *is a strong one.*

Moreover, the energy of the mild solution u, defined by (48) *is positive and we have for any $t \ge 0$*

$$\frac{1-a(0)}{2}\|\nabla u(t)\|_{L^2}^2 - \int_\Omega G(u)\,dx > C\|\nabla u(t)\|_{L^2}^2 \tag{51}$$

$$E(t) \ge \frac{1}{2}\|u_t(t)\|_{L^2}^2 + C\|\nabla u(t)\|_{L^2}^2\,, \tag{52}$$

$$E(t) \le C\left(\|u_1\|_{L^2}^2 + \|\nabla u_0\|_{L^2}^2\right)\,, \tag{53}$$

$$\|u_t(t)\|_{L^2}^2 + \|\nabla u(t)\|_{L^2}^2 \le C\left(\|u_1\|_{L^2}^2 + \|\nabla u_0\|_{L^2}^2\right)\,, \tag{54}$$

where the symbol C denotes positive constants, maybe different.

4 Hidden Regularity Results

Throughout this section we will assume on the integral kernel and on the nonlinearity the conditions (47), (10), (27) and (28).

We will follow the approach pursued in [12, 13] for linear wave equations without memory and in [17] for the linear case with memory. First, we need to introduce a technical lemma, that we will use later. For the sake of completeness we prefer to give all details of the proof, nevertheless some steps are similar to those of the linear case.

Lemma 4.1 *Let $u \in H^2_{loc}((0,\infty); H^2(\Omega))$ be a function satisfying the following equation*

$$u_{tt}(t,x) = \Delta u(t,x) + \int_0^t \dot{a}(t-s)\Delta u(s,x)\,ds + g(u(t,x))\,, \quad in\ \ (0,\infty)\times\Omega. \tag{55}$$

If $h : \overline{\Omega} \to \mathbb{R}^N$ is a vector field of class C^1, then for any fixed $S, T \in \mathbb{R}$, $0 \le S < T$, the following identity holds true

$$
\begin{aligned}
&\int_S^T \int_\Gamma \left[2\partial_\nu(u+\dot a * u)\, h\cdot\nabla(u+\dot a*u) - h\cdot\nu|\nabla(u+\dot a*u)|^2 + h\cdot\nu\,(u_t)^2\right] d\Gamma\, dt \\
&= 2\Big[\int_\Omega u_t\, h\cdot\nabla(u+\dot a*u)\, dx\Big]_S^T + \int_S^T\int_\Omega \sum_{j=1}^N \partial_j h_j\,(u_t)^2\,dx\,dt \\
&\quad -2\int_S^T\int_\Omega u_t\, h\cdot \int_0^t \ddot a(t-s)(\nabla u(s)-\nabla u(t))ds\,dx\,dt - 2\int_S^T \dot a(t)\int_\Omega u_t\, h\cdot\nabla u\,dx\,dt \\
&\quad +2\int_S^T \sum_{i,j=1}^N \int_\Omega \partial_i h_j \partial_i(u+\dot a*u)\partial_j(u+\dot a*u)\,dx\,dt - \int_S^T\int_\Omega \sum_{j=1}^N \partial_j h_j\,|\nabla(u+\dot a*u)|^2\,dx\,dt \\
&\quad +2\int_S^T\int_\Omega g(u(t))h\cdot\nabla(u+\dot a*u)\,dx\,dt\,.
\end{aligned} \tag{56}
$$

Proof To begin with, we multiply the Eq. (55) by

$$
2h\cdot\nabla\left(u(t)+\int_0^t \dot a(t-s)u(s)\,ds\right)
$$

and integrate over $[S,T]\times\Omega$. For simplicity, here and in the following we often drop the dependence on the variables.

First, we will handle the term with u_{tt}. Indeed, integrating by parts in the variable t gives

$$
\begin{aligned}
&2\int_S^T\int_\Omega u_{tt}\, h\cdot\nabla\left(u(t)+\int_0^t \dot a(t-s)u(s)\,ds\right)dx\,dt \\
&\qquad = 2\Big[\int_\Omega u_t\, h\cdot\nabla\Big(u(t)+\int_0^t \dot a(t-s)u(s)\,ds)\,dx\Big]_S^T \\
&-2\int_S^T\int_\Omega u_t\, h\cdot\nabla u_t\,dx\,dt - 2\int_S^T\int_\Omega u_t\, h\cdot\nabla\left(\int_0^t \ddot a(t-s)u(s)\,ds + \dot a(0)u(t)\right)dx\,dt\,.
\end{aligned} \tag{57}
$$

Now, we note that, if we integrate by parts in the variable x then we obtain

$$
2\int_\Omega u_t\, h\cdot\nabla u_t\,dx = \int_\Omega h\cdot\nabla(u_t)^2\,dx = \int_\Gamma h\cdot\nu\,(u_t)^2\,d\Gamma - \int_\Omega \sum_{j=1}^N \partial_j h_j\,(u_t)^2\,dx\,. \tag{58}
$$

In addition, we can write

$$
\begin{aligned}
\int_0^t \ddot a(t-s)u(s)\,ds &= \int_0^t \ddot a(t-s)(u(s)-u(t))\,ds + \int_0^t \ddot a(s)u(t)\,ds \\
&= \int_0^t \ddot a(t-s)(u(s)-u(t))\,ds + \dot a(t)u(t) - \dot a(0)u(t)\,.
\end{aligned} \tag{59}
$$

Therefore, plugging (58) and (59) into (57) yields

$$
\begin{aligned}
&2\int_S^T\int_\Omega u_{tt}\, h\cdot\nabla\Big(u(t)+\int_0^t \dot a(t-s)u(s)\,ds\Big)\,dx\,dt\\
=&2\Big[\int_\Omega u_t\, h\cdot\nabla\Big(u(t)+\int_0^t \dot a(t-s)u(s)\,ds\Big)\,dx\Big]_S^T-\int_S^T\int_\Gamma h\cdot\nu\,(u_t)^2\,d\Gamma\,dt\\
&+\int_S^T\int_\Omega\sum_{j=1}^N\partial_j h_j\,(u_t)^2\,dx\,dt-2\int_S^T\int_\Omega u_t\,h\cdot\int_0^t \ddot a(t-s)\big(\nabla u(s)-\nabla u(t)\big)\,ds\,dx\,dt\\
&-2\int_S^T \dot a(t)\int_\Omega u_t\,h\cdot\nabla u\,dx\,dt\,.
\end{aligned}
\tag{60}
$$

Now, to manage the terms with $\triangle u$, we set

$$
w(t)=u(t)+\int_0^t \dot a(t-s)u(s)\,ds\,, \tag{61}
$$

so, we have to evaluate the term

$$
2\int_S^T\int_\Omega \triangle w\; h\cdot\nabla w\,dx\,dt\,.
$$

Integrating by parts in the variable x we get

$$
\begin{aligned}
&2\int_S^T\int_\Omega \triangle w\; h\cdot\nabla w\,dx\,dt\\
=&2\int_S^T\int_\Gamma \partial_\nu w\; h\cdot\nabla w\,d\Gamma\,dt-2\int_S^T\int_\Omega\nabla w\cdot\nabla(h\cdot\nabla w)\,dx\,dt\,.
\end{aligned}
\tag{62}
$$

We observe that

$$
\begin{aligned}
2\int_\Omega\nabla w\cdot\nabla(h\cdot\nabla w)\,dx=&2\sum_{i,j=1}^N\int_\Omega\partial_i w\,\partial_i(h_j\partial_j w)\,dx\\
=&2\sum_{i,j=1}^N\int_\Omega\partial_i h_j\partial_i w\partial_j w\,dx+2\sum_{i,j=1}^N\int_\Omega h_j\partial_i w\,\partial_j(\partial_i w)\,dx\,,
\end{aligned}
\tag{63}
$$

and

$$
\begin{aligned}
2\sum_{i,j=1}^N\int_\Omega h_j\partial_i w\,\partial_j(\partial_i w)\,dx=&\sum_{j=1}^N\int_\Omega h_j\,\partial_j\big(\sum_{i=1}^N(\partial_i w)^2\big)\,dx\\
=&\int_\Gamma h\cdot\nu|\nabla w|^2\,d\Gamma-\int_\Omega\sum_{j=1}^N\partial_j h_j\,|\nabla w|^2\,dx\,.
\end{aligned}
\tag{64}
$$

Therefore, by putting (63) and (64) into (62) we obtain

$$
\begin{aligned}
&2\int_S^T\int_\Omega \Delta w\, h\cdot\nabla w\,dx\,dt\\
=&2\int_S^T\int_\Gamma \partial_\nu w\, h\cdot\nabla w\,d\Gamma\,dt-\int_S^T\int_\Gamma h\cdot\nu|\nabla w|^2\,d\Gamma\,dt\\
&-2\int_S^T\sum_{i,j=1}^N\int_\Omega \partial_i h_j\partial_i w\partial_j w\,dx\,dt+\int_S^T\int_\Omega\sum_{j=1}^N\partial_j h_j\,|\nabla w|^2\,dx\,dt\,.
\end{aligned}
\tag{65}
$$

Finally, by (60) and (65), taking into account (61) we have the identity (56). □

Theorem 4.2 *Let $u_0\in H^2(\Omega)\cap H_0^1(\Omega)$, $u_1\in H_0^1(\Omega)$ and u the strong solution of*

$$
\begin{cases}
u_{tt}(t,x)=\Delta u(t,x)+\displaystyle\int_0^t \dot a(t-s)\Delta u(s,x)\,ds+g(u(t,x)), & t\ge 0,\ x\in\Omega,\\
u(t,x)=0 \quad t\ge 0,\ x\in\Gamma, &\\
u(0,x)=u_0(x),\quad u_t(0,x)=u_1(x), \qquad x\in\Omega. &
\end{cases}
\tag{66}
$$

If $T>0$, there is a constant $c_0>0$ independent of T such that u satisfies the inequality

$$
\int_0^T\int_\Gamma \Big|\partial_\nu u+\dot a*\partial_\nu u\Big|^2 d\Gamma dt\le c_0\int_0^T E(t)\,dt+c_0E(0)\,, \tag{67}
$$

where $E(t)$ is the energy of the solution given by (48).

Moreover, for a positive constant $c_0=c_0(T)$ we have

$$
\int_0^T\int_\Gamma \Big|\partial_\nu u+\dot a*\partial_\nu u\Big|^2 d\Gamma dt\le c_0(\|\nabla u_0\|_{L^2}^2+\|u_1\|_{L^2}^2)\,. \tag{68}
$$

Proof To begin with we consider a vector field $h\in C^1(\overline{\Omega};\mathbb{R}^N)$ such that

$$
h=\nu \qquad \text{on}\quad \Gamma\,, \tag{69}
$$

see e.g. [12] for the construction of such vector field. From now on, we will denote with c positive constants, maybe different. In particular, we have

$$
|h(x)|\le c\quad\text{and}\quad \sum_{i,j=1}^N|\partial_i h_j(x)|\,dx\le c\,,\quad \forall x\in\overline{\Omega}. \tag{70}
$$

We will apply the identity (56) with the vector field h satisfying (69) and with $S=0$. First, we observe that

$$
u_t=0\,,\qquad \nabla u=(\partial_\nu u)\nu\quad\text{on}\quad (0,T)\times\Gamma\,. \tag{71}
$$

For a detailed proof of the second identity see e.g. [19, Lemma 2.1]. Therefore, thanks to (71) the left-hand side of (56) with $S = 0$ becomes

$$\int_0^T \int_\Gamma \left|\partial_\nu u + \dot{a} * \partial_\nu u\right|^2 d\Gamma dt \,,$$

and hence (56) can be written as

$$\begin{aligned}
&\int_0^T \int_\Gamma \left|\partial_\nu u + \dot{a} * \partial_\nu u\right|^2 d\Gamma dt \\
&= 2\Big[\int_\Omega u_t\, h\cdot\nabla\big(u + \dot{a} * u\big)\, dx\Big]_0^T + \int_0^T \int_\Omega \sum_{j=1}^N \partial_j h_j\, (u_t)^2\, dx\, dt \\
&\quad - 2\int_0^T \int_\Omega u_t\, h\cdot \int_0^t \ddot{a}(t-s)(\nabla u(s) - \nabla u(t))ds\, dx\, dt - 2\int_0^T \dot{a}(t)\int_\Omega u_t\, h\cdot\nabla u\, dx\, dt \\
&\quad + 2\int_0^T \sum_{i,j=1}^N \int_\Omega \partial_i h_j \partial_i\big(u + \dot{a} * u\big)\partial_j\big(u + \dot{a} * u\big)\, dx\, dt - \int_0^T \int_\Omega \sum_{j=1}^N \partial_j h_j\, |\nabla\big(u + \dot{a} * u\big)|^2\, dx\, dt \\
&\quad + 2\int_0^T \int_\Omega g(u(t))h\cdot\nabla\big(u + \dot{a} * u\big)\, dx\, dt\,.
\end{aligned} \tag{72}$$

To prove (67) we have to estimate every term on the right-hand side of (72). Indeed,

$$\begin{aligned}
&2\Big[\int_\Omega u_t\, h\cdot\nabla\big(u + \dot{a} * u\big)\, dx\Big]_0^T \\
&= 2\int_\Omega u_t(T)\, h\cdot\nabla\big(u + \dot{a} * u\big)(T)\, dx - 2\int_\Omega u_1\, h\cdot\nabla u_0\, dx \\
&\le c\int_\Omega |u_t(T)|^2\, dx + c\int_\Omega |\nabla\big(u + \dot{a} * u\big)(T)|^2\, dx + c\int_\Omega |u_1|^2\, dx + c\int_\Omega |\nabla u_0|^2\, dx\,.
\end{aligned} \tag{73}$$

We proceed to evaluate for all $t \in [0, T]$ the term $\int_\Omega |\nabla\big(u + \dot{a} * u\big)(t)|^2\, dx$, because that evaluation will be also useful later. Since for all $t \in [0, T]$

$$\nabla u(t) + \dot{a} * \nabla u(t) = \Big(1 - a(0) + a(t)\Big)\nabla u(t) + \int_0^t \dot{a}(t-s)\big(\nabla u(s) - \nabla u(t)\big)\, ds\,,$$

we have

$$|\nabla\big(u + \dot{a} * u\big)(t)|^2 \le 2\big(1 - a(0) + a(t)\big)^2|\nabla u(t)|^2 + 2\Big(\int_0^t |\dot{a}(t-s)|\big|\nabla u(s) - \nabla u(t)\big|\, ds\Big)^2\,.$$

In view of $\dot{a}(t) \le 0$, $a(t) \ge 0$ and $a(0) < 1$ we get

$$\begin{aligned}
\Big(\int_0^t |\dot{a}(t-s)|\big|\nabla u(s) - \nabla u(t)\big|\, ds\Big)^2 &\le \int_0^t |\dot{a}(s)|\, ds \int_0^t |\dot{a}(t-s)|\big|\nabla u(s) - \nabla u(t)\big|^2\, ds \\
&\le -\int_0^t \dot{a}(t-s)\big|\nabla u(s) - \nabla u(t)\big|^2\, ds\,,
\end{aligned}$$

and hence

$$|\nabla(u + \dot{a} * u)(t)|^2 \le 2(1 - a(0) + a(t))|\nabla u(t)|^2 - 2\int_0^t \dot{a}(t-s)\big|\nabla u(s) - \nabla u(t)\big|^2\, ds\,.$$

Therefore, taking into account the formula (48) for the energy, by (52) and (51), we get

$$2(1 - a(0))\int_\Omega |\nabla u(t)|^2 \le cE(t)\,,$$

$$2\int_\Omega \left(a(t)|\nabla u(t)|^2 - \int_0^t \dot{a}(t-s)\big|\nabla u(s) - \nabla u(t)\big|^2\, ds\right)\, dx \le 4E(t)$$

and hence

$$\int_\Omega |\nabla(u + \dot{a} * u)(t)|^2\, dx \le cE(t)\,. \tag{74}$$

By putting (74) with $t = T$ into (73) and using again (48), we obtain

$$2\left[\int_\Omega u_t\, h \cdot \nabla(u + \dot{a} * u)\, dx\right]_0^T \le cE(T) + cE(0)\,,$$

and hence, since the energy $E(t)$ is decreasing, see (49), we have

$$2\left[\int_\Omega u_t\, h \cdot \nabla(u + \dot{a} * u)\, dx\right]_0^T \le cE(0)\,.$$

Now, we estimate the second term on the right-hand side of (72) by using (70), the expression of energy (48) and (52), that is

$$\int_0^T \int_\Omega \sum_{j=1}^N |\partial_j h_j|\, |u_t|^2\, dx\, dt \le c\int_0^T E(t)\, dt\,.$$

In order to bound the term

$$2\int_0^T \int_\Omega \left|u_t\, h \cdot \int_0^t \ddot{a}(t-s)\big(\nabla u(s) - \nabla u(t)\big)\, ds\right| dx\, dt$$

we note that, thanks also to (70), we have

$$\begin{aligned} 2c\int_0^T \int_\Omega |u_t| \left|\int_0^t \ddot{a}(t-s)\,(\nabla u(s) - \nabla u(t))ds\right|\, dx\, dt \\ \le c\int_0^T \int_\Omega |u_t|^2\, dx\, dt + c\int_0^T \int_\Omega \left|\int_0^t \ddot{a}(t-s)(\nabla u(s) - \nabla u(t))ds\right|^2\, dx\, dt\,. \end{aligned} \tag{75}$$

To evaluate the second term on the right-hand side of the previous formula, we observe

$$\begin{aligned}\Big|\int_0^t \ddot{a}(t-s)\big(\nabla u(s)-\nabla u(t)\big)ds\Big|^2 &\le \Big(\int_0^t |\ddot{a}(t-s)|^{1/2}|\ddot{a}(t-s)|^{1/2}\big|\nabla u(s)-\nabla u(t)\big|ds\Big)^2\\ &\le \int_0^t \ddot{a}(s)\,ds\int_0^t \ddot{a}(t-s)\big|\nabla u(s)-\nabla u(t)\big|^2 ds\\ &=(\dot{a}(t)-\dot{a}(0))\int_0^t \ddot{a}(t-s)\big|\nabla u(s)-\nabla u(t)\big|^2 ds\,.\end{aligned}$$

Therefore, in view of $\dot{a}\le 0$ and formula (49), giving the derivative of the energy, from the above inequality we obtain

$$\begin{aligned}&\int_0^T\int_\Omega\Big|\int_0^t \ddot{a}(t-s)\big(\nabla u(s)-\nabla u(t)\big)ds\Big|^2 dx\,dt\\ &\le -\dot{a}(0)\int_0^T\int_\Omega\int_0^t \ddot{a}(t-s)\big|\nabla u(s)-\nabla u(t)\big|^2 ds\,dx\,dt \le 2\dot{a}(0)\int_0^T E'(t)dt \le -2\dot{a}(0)E(0)\,.\end{aligned} \tag{76}$$

Plugging (76) into (75) and using (52) yield

$$\begin{aligned}&2\int_0^T\int_\Omega\Big|u_t\,h\cdot\int_0^t \ddot{a}(t-s)\big(\nabla u(s)-\nabla u(t)\big)\,ds\Big|dx\,dt\\ &\le c\int_0^T\int_\Omega |u_t|^2\,dx\,dt + cE(0) \le c\int_0^T E(t)\,dt + cE(0)\,.\end{aligned}$$

Keeping in mind that $\dot{a}(t)\ge\dot{a}(0)$ and by using again (70) and (52), we get

$$\begin{aligned}&-2\int_0^T \dot{a}(t)\int_\Omega |u_t\,h\cdot\nabla u|\,dx\,dt\\ &\le -2\dot{a}(0)\,c\int_0^T\int_\Omega |u_t||\nabla u|\,dx\,dt \le -\dot{a}(0)\,c\int_0^T\int_\Omega |u_t|^2+|\nabla u|^2\,dx\,dt \le c\int_0^T E(t)\,dt\,.\end{aligned}$$

To evaluate the next two terms on the right-hand side of (72) we will use the estimate (74). Indeed, as regards the first one, by means of (70) we have that

$$\begin{aligned}&\int_0^T\sum_{i,j=1}^N\int_\Omega |\partial_i h_j\partial_i(u+\dot{a}*u)\partial_j(u+\dot{a}*u)|\,dx\,dt\\ &\le c\int_0^T\int_\Omega\Big(\sum_{i=1}^N|\partial_i(u+\dot{a}*u)|\Big)^2 dx\,dt \le 2^{N-1}c\int_0^T\int_\Omega |\nabla(u+\dot{a}*u)|^2\,dx\,dt\,.\end{aligned}$$

Since, from (74), we obtain

$$\int_0^T\int_\Omega |\nabla(u+\dot{a}*u)|^2\,dx\,dt \le c\int_0^T E(t)\,dt, \tag{77}$$

thus it follows

$$\int_0^T \sum_{i,j=1}^N \int_\Omega |\partial_i h_j \partial_i\big(u + \dot a * u\big)\partial_j\big(u + \dot a * u\big)|\,dx\,dt \le c\int_0^T E(t)\,dt\,.$$

In a similar way, thanks again to (70) and (77) we have

$$\int_0^T \int_\Omega \sum_{j=1}^N |\partial_j h_j|\,|\nabla\big(u + \dot a * u\big)|^2\,dx\,dt \le c\int_0^T \int_\Omega |\nabla\big(u + \dot a * u\big)|^2\,dx\,dt \le c\int_0^T E(t)\,dt\,.$$

Finally, to estimate the last term on the right-hand side of (72) first we use (70)

$$2\int_0^T \int_\Omega g(u(t))h\cdot\nabla\big(u + \dot a * u\big)\,dx\,dt \le c\int_0^T \int_\Omega |g(u(t))|^2\,dx\,dt + c\int_0^T \int_\Omega |\nabla\big(u + \dot a * u\big)|^2\,dx\,dt\,.$$

Since by (11) and (52) we have

$$\int_0^T \int_\Omega |g(u(t))|^2\,dx\,dt \le c\int_0^T \int_\Omega |\nabla u(t)|^2\,dx\,dt \le c\int_0^T E(t)\,dt\,,$$

thanks also to (77), we obtain

$$2\int_0^T \int_\Omega g(u(t))h\cdot\nabla\big(u + \dot a * u\big)\,dx\,dt \le c\int_0^T E(t)\,dt\,.$$

In conclusion, the previous argumentations show that the sum of all terms on the right-hand side of (72) can be majorized by $c_0\int_0^T E(t)\,dt + c_0E(0)$, with $c_0 > 0$ independent of T, and hence (67) holds true. In addition, since $E(t)$ is a decreasing function and

$$E(0) = \frac{1}{2}\|\nabla u_0\|_{L^2}^2 + \frac{1}{2}\|u_1\|_{L^2}^2 - \int_\Omega G(u_0)\,dx\,,$$

thanks also to (22), (68) follows from (67). □

Corollary 4.3 *For any $T > 0$ there exists a unique continuous linear map*

$$\mathcal{L} : H_0^1(\Omega) \times L^2(\Omega) \to L^2((0,T); L^2(\Gamma))$$

such that for any $u_0 \in H^2(\Omega) \cap H_0^1(\Omega)$ and $u_1 \in H_0^1(\Omega)$, called u the strong solution of (66)*, we have*

$$\mathcal{L}(u_0, u_1) = \partial_\nu u\,.$$

Proof For $u_0 \in H^2(\Omega) \cap H_0^1(\Omega)$ and $u_1 \in H_0^1(\Omega)$, if we denote by u the strong solution of problem (66) and apply Lemma 2.1 with $X = L^2(\Gamma)$, then for any $T > 0$, thanks to (68) and (7) there exists a constant $c_0 = c_0(T, \|\dot a\|_{L^1}) > 0$ such that

$$\int_0^T \int_\Gamma |\partial_\nu u|^2 d\Gamma dt \le c_0(\|\nabla u_0\|_{L^2}^2 + \|u_1\|_{L^2}^2) .$$

By density our claim follows. □

Remark 4.4 For the mild solution u of (66) we can introduce the notation $\partial_\nu u$ instead of $\mathcal{L}(u_0, u_1)$, thanks to Corollary 4.3. So, for any $T > 0$ we have the following trace theorem:

$$(u_0, u_1) \in H_0^1(\Omega) \times L^2(\Omega) \Rightarrow \partial_\nu u \in L^2((0, T); L^2(\Gamma)) ,$$

and there is a positive constant c_0 depending on T and $\|\dot{a}\|_{L^1}$ such that

$$\int_0^T \int_\Gamma |\partial_\nu u|^2 d\Gamma dt \le c_0(\|\nabla u_0\|_{L^2}^2 + \|u_1\|_{L^2}^2) \qquad \forall (u_0, u_1) \in H_0^1(\Omega) \times L^2(\Omega) . \tag{78}$$

This result does not follow from the usual trace theorems of the Sobolev spaces. For this reason it is called a hidden regularity result. The corresponding inequality (78) is often called a direct inequality.

Theorem 4.5 *Assume there exists $c_0 > 0$ independent of t such that*

$$\int_0^t E(s)\, ds \le c_0 E(0) \qquad \forall t \ge 0. \tag{79}$$

Then, a constant $C > 0$ exists such that for any $u_0 \in H_0^1(\Omega)$ and $u_1 \in L^2(\Omega)$ the mild solution u of (66) *satisfies*

$$\int_0^\infty \int_\Gamma |\partial_\nu u|^2 d\Gamma dt \le C(\|\nabla u_0\|_{L^2}^2 + \|u_1\|_{L^2}^2) , \tag{80}$$

that is

$$\partial_\nu u \in L^2(0, \infty; L^2(\Gamma)) . \tag{81}$$

Proof In view of (67) and (79) we have

$$\int_0^T \int_\Gamma |\partial_\nu u + \dot{a} * \partial_\nu u|^2 \, d\Gamma dt \le C E(0) \qquad \forall T > 0 ,$$

where the constant C is independent of T, and hence

$$\int_0^\infty \int_\Gamma |\partial_\nu u + \dot{a} * \partial_\nu u|^2 \, d\Gamma dt \le C E(0) .$$

Finally, thanks to (9) and (22) the estimate (80) follows. □

Remark 4.6 For example, the assumption (79) holds if the energy decays exponentially. Indeed, if there exists $m > 0$ such that

$$-\ddot{a}(t) \leq m\,\dot{a}(t) \qquad \text{for any } t \geq 0,$$

that is the kernel $-\dot{a}$ decays exponentially, we can apply [1, Theorem 3.5] to have that the energy of the mild solution also decays exponentially. Therefore, there exist $\alpha > 0$ such that

$$E(t) \leq e^{1-\alpha t} E(0) \qquad \forall t \geq 0\,.$$

Also in the case the integral kernel decays polynomially then (79) holds (see [1]).

Remark 4.7 If one assumes more regularity on the integral kernel $k = -\dot{a}$, then it is possible to approach the study of the equation

$$u_{tt} = \Delta u - \int_0^t k(t-s)\Delta u(s,x)\,ds + g(u)\,,$$

by using the so-called MacCamy's trick, see [18]. Adapted to our case, the trick consists in setting

$$v = u - k * u\,,$$

to obtain

$$u = v + \rho_k * v,$$

(where ρ_k is the resolvent kernel of k and has the same regularity of k), so v is the solution of the equation

$$v_{tt} + \rho_k(0)v_t + \ddot{\rho}_k * v + \dot{\rho}_k(0)v = \Delta v + g(v + \rho_k * v)\,. \tag{82}$$

However, in (82) the terms $\ddot{\rho}_k * v$ and $\dot{\rho}_k(0)v$ have a meaning only if k, and hence ρ_k, is more regular than in our case. For example, a class of kernels fitting our assumptions (see [21, Corollary 2.2]), but not suitable for applying the MacCamy's trick is given by

$$k(t) = k_0 e^{-\sqrt{t}}$$

for a suitable $k_0 > 0$.

Acknowledgements The authors would like to thank the anonymous referee for helpful comments improving the presentation of the paper.

References

1. Alabau-Boussouira, F., Cannarsa, P., Sforza, D.: Decay estimates for second order evolution equations with memory. J. Funct. Anal. **254**, 1342–1372 (2008)
2. Ball, J.M.: Strongly continuous semigroups, weak solutions, and the variation of constants formula. Proc. Am. Math. Soc. **63**, 370–373 (1977)
3. Berrimi, S., Messaoudi, S.A.: Existence and decay of solutions of a viscoelastic equation with a nonlinear source. Nonlinear Anal. **64**, 2314–2331 (2006)
4. Cannarsa, P., Sforza, D.: An existence result for semilinear equations in viscoelasticity: the case of regular kernels. In: Fabrizio, M., Lazzari, B., Morro, A. (eds.) Mathematical Models and Methods for Smart Materials. Series on Advances in Mathematics for Applied Sciences, vol. 62, pp. 343–354. World Scientific, Singapore (2002)
5. Cannarsa, P., Sforza, D.: A stability result for a class of nonlinear integrodifferential equations with L^1 kernels. Appl. Math. (Warsaw) **35**, 395–430 (2008)
6. Cannarsa, P., Sforza, D.: Integro-differential equations of hyperbolic type with positive definite kernels. J. Differ. Equ. **250**, 4289–4335 (2011)
7. Cazenave, T., Haraux, A.: An Introduction to Semilinear Evolution Equations. Oxford Lecture Series in Mathematics and its Applications, vol. 13. The Clarendon Press, Oxford University Press, New York (1998). Translated from the 1990 French original by Yvan Martel and revised by the authors
8. Doi, M., Edwards, S.F.: Dynamics of concentrated polymer systems, Parts 1, 2 and 3. J. Chem. Soc. Faraday II **74**, 1789–1832 (1978); Parts 4, J. Chem. Soc. Faraday II **75**, 38–54 (1979)
9. Gripenberg, G., Londen, S.O., Staffans, O.J.: Volterra Integral and Functional Equations. Encyclopedia of Mathematics and its Applications, vol. 34. Cambridge University Press, Cambridge (1990)
10. Kawashima, S.: Global solutions to the equation of viscoelasticity with fading memory. J. Differ. Equ. **101**, 388–420 (1993)
11. Kim, J.U.: On the local regularity of solutions in linear viscoelasticity of several space dimensions. Trans. Am. Math. Soc. **346**, 359–398 (1994)
12. Komornik, V.: Exact Controllability and Stabilization: The Multiplier Method. RAM: Research in Applied Mathematics. Masson, Paris; John Wiley and Sons, Ltd., Chichester (1994)
13. Komornik, V., Loreti, P.: Fourier Series in Control Theory. Springer Monographs in Mathematics. Springer, New York (2005)
14. Lasiecka, I., Triggiani, R.: Regularity of hyperbolic equations under $L_2(0, T; L_2(\Gamma))$ boundary terms. Appl. Math. Optim. **10**, 275–286 (1983)
15. Lions, J.-L.: Contrôle des systèmes distribués singuliers. Gauthiers-Villars, Paris (1983)
16. Lions, J.-L.: Hidden regularity in some nonlinear hyperbolic equations. Mat. Apl. Comput. **6**, 7–15 (1987)
17. Loreti, P., Sforza, D.: Hidden regularity for wave equations with memory. Riv. Math. Univ. Parma (N.S.) **7**, 391–405 (2016)
18. MacCamy, R.C.: A model Riemann problem for Volterra equations. Arch. Rat. Mech. Anal. **82**, 71–86 (1983)
19. Milla Miranda, M., Medeiros, L.A.: Hidden regularity for semilinear hyperbolic partial differential equations. Ann. Fac. Sci. Toulouse Math. **9**(5), 103–120 (1988)
20. Muñoz Rivera, J.E., Peres Salvatierra, A.: Asymptotic behaviour of the energy in partially viscoelastic materials. Q. Appl. Math. **59**, 557–578 (2001)
21. Nohel, J.A., Shea, D.F.: Frequency domain methods for Volterra equations. Adv. Math. **22**, 278–304 (1976)
22. Prüss, J.: Evolutionary Integral Equations and Applications. Monographs in Mathematics, vol. 87. Birkhäuser Verlag, Basel (1993)
23. Renardy, M., Hrusa, W.J., Nohel, J.A.: Mathematical Problems in Viscoelasticity, Pitman Monographs Pure and Applied Mathematics, vol. 35. Longman Sci. Tech, Harlow, Essex (1988)

Lyapunov's Theorem via Baire Category

Marco Mazzola and Khai T. Nguyen

Abstract Lyapunov's theorem is a classical result in convex analysis, concerning the convexity of the range of nonatomic measures. Given a family of integrable vector functions on a compact set, this theorem allows to prove the equivalence between the range of integral values obtained considering all possible set decompositions and all possible convex combinations of the elements of the family. Lyapunov type results have several applications in optimal control theory: they are used to prove bang-bang properties and existence results without convexity assumptions. Here, we use the dual approach to the Baire category method in order to provide a "quantitative" version of such kind of results applied to a countable family of integrable functions.

Keywords Lyapunov's convexity theorem · Extremal solutions · Baire category · Nonconvex optimal control problems

1 Introduction

The use of Baire categories in the analysis of nonconvex differential inclusions started with the seminal paper by A. Cellina [4]. These methods were later developed and adapted to various problems involving nonconvex ordinary and partial differential inclusions, notably in a series of articles by F. S. De Blasi and G. Pianigiani (see e.g. [6] and the bibliography therein). It is now known, for example, that the set S^{ext} of extremal solutions of a differential inclusion, associated to a Lipschitz continuous multifunction with nonempty, compact and convex images, is residual in the set of

M. Mazzola
CNRS, Institut de Mathématiques de Jussieu - Paris Rive Gauche,
Sorbonne Université, Case 247, 4 Place Jussieu, 75252 Paris, France
e-mail: marco.mazzola@imj-prg.fr

K. T. Nguyen (✉)
Department of Mathematics, North Carolina State University, Raleigh, NC, USA
e-mail: khai@math.ncsu.edu

F. Alabau-Boussouira et al. (eds.), *Trends in Control Theory and Partial Differential Equations*, Springer INdAM Series 32,
https://doi.org/10.1007/978-3-030-17949-6_10

all solutions S, i.e. it contains the intersection of countably many open dense subsets of S.

The same problem has been more recently approached by A. Bressan [2] from a "dual" point of view. The procedure is the following: introduce auxiliary functions v belonging to some complete space V; associate to each $v \in V$ a nonempty subset $S^v \subseteq S$; finally, show that the set of functions $v \in V$ satisfying $S^v \subseteq S^{ext}$ is residual in V. An advantage of this approach with respect to the "direct" one is that it could work even in the case when S^{ext} is not dense in S. For the differential inclusion problem mentioned above, this situation can appear when no Lipschitzianity assumptions are imposed on the multifunction.

The dual approach was employed in [3] in order to derive an extension of the classical bang-bang theorem in linear control theory. In very broad terms, it was proved that for almost every v in a space of auxiliary functionals, there is a unique control minimizing v and steering the system between two given points; furthermore, this control arc takes values almost everywhere within the extremal points of the set of admissible controls. The classical proof of the bang-bang principle is actually based on a Lyapunov type theorem (see [5]). This result can be stated as follows. Consider a finite family of Lebesgue integrable functions $f_1, \dots, f_m$ from a compact subset $K \subset I\!R^d$ to $I\!R^n$ and the simplex of $I\!R^m$

$$\Delta_m \doteq \left\{ \zeta = (\zeta_1, \dots, \zeta_m) \in I\!R^m \;\middle|\; \zeta_i \geq 0 \;\; \forall i = 1, \dots, m, \; \sum_{i=1}^m \zeta_i = 1 \right\}.$$

Denote by $\mathscr{M}(K, \Delta_m)$ the set of Lebesgue measurable functions from K to Δ_m. Then, for any $\theta = (\theta_1, \dots, \theta_m) \in \mathscr{M}(K, \Delta_m)$ there exists a measurable partition $\{E_1, \dots, E_m\}$ of K such that

$$\int_{E_1} f_1(x)\, dx + \dots + \int_{E_m} f_m(x)\, dx \;=\; \sum_{i=1}^m \int_K \theta_i(x) f_i(x)\, dx\,.$$

An alternative "extremal" formulation of this theorem is the following. Given $\bar{\theta} = (\bar{\theta}_1, \dots, \bar{\theta}_m) \in \mathscr{M}(K, \Delta_m)$, denote

$$\alpha \doteq \int_K \bar{\theta}_1(x)\, f_1(x)\, dx + \dots + \int_K \bar{\theta}_m(x)\, f_m(x)\, dx \;\in\; I\!R^n\,.$$

Let Δ_m^{ext} be the set of extreme points of Δ_m. According to Lyapunov's theorem, the set

$$\mathscr{A}_\alpha^{ext} \doteq \left\{ \theta \in \mathscr{M}\left(K, \Delta_m^{ext}\right) \;\middle|\; \sum_{i=1}^m \int_K \theta_i(x) f_i(x)\, dx \;=\; \alpha \right\}$$

is nonempty. In the present paper, we aim to provide an alternative proof of this result based on the Baire category method, implying besides that $\mathscr{A}_\alpha^{ext}$ is actually residual in the set

$$\left\{\theta \in \mathscr{M}(K, \Delta_m) \;\middle|\; \sum_{i=1}^{m} \int_K \theta_i(x) f_i(x)\, dx = \alpha\right\}$$

in a "dual" sense.

The equivalence between the range of integral values obtained considering all possible set decompositions and all possible convex combinations of given vector functions plays an important role in optimal control theory, that goes beyond the application to the bang-bang theorem. For instance, it can be used to derive existence theorems for optimal control problems without convexity assumptions (see e.g. [1, 7]).

2 A Dual Approach to Lyapunov's Theorem

For any continuous function $v : K \to I\!R^m$, consider the constrained optimization problem

$$\text{Minimize}_{\theta \in \mathscr{A}_\alpha} \int_K \theta(x) \cdot v(x) dx \tag{1}$$

over the set

$$\mathscr{A}_\alpha \doteq \left\{\theta \in \mathscr{M}(K, \Delta_m) \;\middle|\; \sum_{i=1}^{m} \int_K \theta_i(x) f_i(x)\, dx = \alpha\right\}, \tag{2}$$

where $\theta(x) \cdot v(x) \doteq \sum_{i=1}^{m} \theta_i(x) v_i(x)$ denotes an inner product. It is clear that (1)–(2) admits at least a solution. Indeed, since $\theta_m = 1 - \sum_{i=1}^{m-1} \theta_i$, the problem (1)–(2) is equivalent to

$$\text{Minimize}_{\tilde{\theta} \in \mathscr{B}} \int_K \sum_{i=1}^{m-1} \tilde{\theta}_i(x) \left(v_i(x) - v_m(x)\right) dx \tag{3}$$

over the set

$$\mathscr{B} \doteq \Bigg\{\tilde{\theta} \in \mathbf{L}^\infty(K, I\!R^{m-1}) \;\Bigg|\; \tilde{\theta}_i(x) \geq 0 \;\; \forall i = 1, \ldots, m-1, \; \sum_{i=1}^{m-1} \tilde{\theta}_i(x) \leq 1, \; a.e.\ x \in K,$$
$$\sum_{i=1}^{m-1} \int_K \tilde{\theta}_i(x)\big(f_i(x) - f_m(x)\big)\, dx = \alpha - \int_K f_m(x)\, dx\Bigg\}. \tag{4}$$

Thanks to Alaoglu's theorem, for every sequence $(\tilde{\theta}^n)_{n=1}^{\infty} \subset \mathscr{B}$, there exists a subsequence $(\tilde{\theta}^{n_k})_{k=1}^{\infty}$ converging weakly* to some $\tilde{\theta} \in \mathbf{L}^{\infty}(K, I\!R^{m-1})$ satisfying $\|\tilde{\theta}\|_{\mathbf{L}^{\infty}(K, I\!R^{m-1})} \leq 1$. Hence

$$\lim_{n_k \to +\infty} \int_K \sum_{i=1}^{m-1} [\tilde{\theta}_i^{n_k}(x) - \tilde{\theta}_i(x)] w_i(x)\, dx \;=\; 0 \qquad \forall\, w \in \mathbf{L}^1(K, I\!R^{m-1}) \tag{5}$$

yields

$$\sum_{i=1}^{m-1} \int_K \tilde{\theta}_i(x)\big(f_i(x) - f_m(x)\big)\, dx \;=\; \alpha - \int_K f_m(x)\, dx\,.$$

Since $\sum_{i=1}^{m-1} \tilde{\theta}_i^{n_k}(x) \leq 1$ for a.e. $x \in K$ and $\tilde{\theta}_i^{n_k}(x) \geq 0$ for a.e $x \in K$ and any $i \in \{1, 2, \ldots, m-1\}$, by a contradiction argument one obtains from (5) that $\tilde{\theta}$ satisfies the same properties. Therefore, the set $\mathscr{B}$ is weakly*-compact in $\mathbf{L}^{\infty}(K, I\!R^{m-1})$ and it yields the existence of solutions to (3)–(4).

Let's define

$$\mathscr{V}_{\alpha} \;\doteq\; \big\{ v \in \mathscr{C}(K, I\!R^m) \mid (1) - (2)\ \textit{has a unique solution} \big\}\,. \tag{6}$$

Here, $\mathscr{C}(K, I\!R^m)$ is the space of continuous function on K with values in $I\!R^m$. Our main result is stated as follows.

Theorem 1 *$\mathscr{V}_{\alpha}$ is a residual subset of $\mathscr{C}(K, I\!R^m)$, i.e. it contains the intersection of countably many open dense subsets of $\mathscr{C}(K, I\!R^m)$. Moreover, for any $v \in \mathscr{V}_{\alpha}$, the unique optimal solution θ^* takes values in $Ext\,(\Delta_m)$ almost everywhere in the compact set K.*

The main ingredient in the proof of the above theorem is the following lemma.

Lemma 1 *Let $g : K \to I\!R^n$ be a Lebesgue integrable function. Then the set $\mathscr{W}^g$ of continuous functions $w \in \mathscr{C}(K, I\!R)$ such that*

$$meas\Big(\big\{x \in K \mid w(x) = \lambda \cdot g(x)\big\}\Big) \;=\; 0 \qquad \textit{for all}\ \lambda \in I\!R^n \tag{7}$$

is residual in $\mathscr{C}(K, I\!R)$.

Proof For every positive integer N and every $\varepsilon > 0$, call $\mathscr{W}^g_{\varepsilon,N}$ the set of all $w \in \mathscr{C}(K, I\!R)$ such that

$$\text{meas}\Big(\big\{x \in K \mid w(x) \;=\; \lambda \cdot g(x)\big\}\Big) \;<\; \varepsilon \tag{8}$$

whenever $\lambda \in [-N, N]^n$. The Lemma is proved once we show that, for every ε and N, $\mathscr{W}^g_{\varepsilon,N}$ is open and dense in $\mathscr{C}(K; I\!R)$.

1. We begin by proving that $\mathscr{W}^g_{\varepsilon,N}$ is open. Fix $w \in \mathscr{W}^g_{\varepsilon,N}$. For any $\lambda \in [-N, N]^n$, define

$$\varepsilon_\lambda \doteq \varepsilon - \text{meas}\Big(\big\{x \in K \mid w(x) = \lambda \cdot g(x)\big\}\Big) > 0. \tag{9}$$

Using Lusin's theorem, there exists a continuous function $g_\lambda : K \mapsto I\!R^n$ such that

$$\text{meas}\Big(\big\{x \in K \mid g_\lambda(x) \neq g(x)\big\}\Big) < \varepsilon_\lambda/4. \tag{10}$$

Consider the compact set of $I\!R^n$

$$E_\lambda \doteq \big\{x \in K \mid w(x) = \lambda \cdot g_\lambda(x)\big\}.$$

By the regularity properties of Lebesgue measure, there exists a relatively open set $O_\lambda \subset K$ such that

$$E_\lambda \subseteq O_\lambda \quad \text{and} \quad \text{meas}(O_\lambda \backslash E_\lambda) < \frac{\varepsilon_\lambda}{2}. \tag{11}$$

By the continuity of g_λ and w, one has

$$\min_{x \in K \backslash O_\lambda} \Big| w(x) - \lambda \cdot g_\lambda(x) \Big| \doteq \delta_\lambda > 0.$$

For any function $\tilde{w} \in \mathscr{C}(K, I\!R)$ such that

$$\|\tilde{w} - w\|_\infty = \sup_{x \in K} |\tilde{w}(x) - w(x)| < r_\lambda \doteq \frac{\delta_\lambda}{3 \max\{1, \|g_\lambda\|_\infty\}},$$

it holds

$$\Big|\tilde{w}(x) - \lambda \cdot g_\lambda(x)\Big| > \frac{2}{3}\delta_\lambda \qquad \forall x \in K \setminus O_\lambda.$$

In turn, if $|\tilde{\lambda} - \lambda| < r_\lambda$, this implies

$$\Big|\tilde{w}(x) - \tilde{\lambda} \cdot g_\lambda(x)\Big| > \frac{\delta_\lambda}{3} > 0 \qquad \forall x \in K \backslash O_\lambda$$

and it yields

$$\text{meas}\Big(\big\{x \in K \mid \tilde{w}(x) = \tilde{\lambda} \cdot g_\lambda(x)\big\}\Big) \leq \text{meas}\,(O_\lambda). \tag{12}$$

By (9), (10), (11) and (12), if

$$\|\tilde{w} - w\|_\infty < r_\lambda \qquad \text{and} \qquad |\tilde{\lambda} - \lambda| < r_\lambda, \tag{13}$$

then it holds

$$\begin{aligned}\operatorname{meas}\Big(\big\{x \in K \ \ |\tilde{w}(x) = \tilde{\lambda}\cdot g(x)\big\}\Big) & \\ < \operatorname{meas}\Big(\big\{x \in K \mid \tilde{w}(x) = \tilde{\lambda}\cdot g_\lambda(x)\big\}\Big) + \frac{\varepsilon_\lambda}{4} & \\ \leq \operatorname{meas}(O_\lambda) + \frac{\varepsilon_\lambda}{4} < \operatorname{meas}(E_\lambda) + \frac{3}{4}\varepsilon_\lambda & \\ < \operatorname{meas}\Big(\big\{x \in K \mid w(x) = \lambda\cdot g(x)\big\}\Big) + \frac{1}{4}\varepsilon_\lambda + \frac{3}{4}\varepsilon_\lambda = \varepsilon\,. &\end{aligned} \tag{14}$$

Repeating the above construction, for every $\lambda \in [-N, N]^n$ there exists $r_\lambda > 0$ so that the inequalities (13) imply (14). Since the set $[-N, N]^n$ is compact, we can select a finite family $\{\lambda^1, \dots, \lambda^M\} \subset [-N, N]^n$ such that the corresponding open balls $B(\lambda^k, r_{\lambda^k})$ satisfy

$$[-N, N]^n \subset \bigcup_{k=1}^{M} B(\lambda^k, r_{\lambda^k})\,.$$

Setting $r \doteq \min_{1\leq k\leq M} r_{\lambda^k}$, for every $\tilde{w} \in B(w, r)$ and $\lambda \in [-N, N]^n$ we obtain

$$\operatorname{meas}\Big(\big\{x \in K \mid \tilde{w}(x) = \lambda\cdot g(x)\big\}\Big) < \varepsilon\,.$$

Therefore, $B(w, r) \subseteq \mathscr{W}^g_{\varepsilon,N}$, proving that the set $\mathscr{W}^g_{\varepsilon,N}$ is open in $\mathscr{C}(K, I\!R)$.

2. It remains to prove that each $\mathscr{W}^g_{\varepsilon,N}$ is dense in $\mathscr{C}(K; I\!R)$. Relying on Lusin's theorem, it is not restrictive to assume that g is continuous. Given any $\eta > 0$ and $\tilde{w} \in \mathscr{C}(K, I\!R)$, we will construct a function $w \in \mathscr{W}^g_{\varepsilon,N}$, satisfying

$$\|w - \tilde{w}\|_\infty < \eta\,. \tag{15}$$

For simplicity, without loss of generality we will assume that $K = [0, 1]^d$. Let's choose an integer m sufficiently large so that $m^d \geq n + 1$ and $h \doteq \frac{1}{m}$ satisfies

$$h^d < \frac{\varepsilon}{2n} \tag{16}$$

and

$$(x, x') \in K^2,\ |x - x'| \leq h\sqrt{d} \implies \big|\tilde{w}(x) - \tilde{w}(x')\big| < \frac{\eta}{2}\,. \tag{17}$$

We adopt the following notation: a vector $y \in (I\!R^m)^d$ will be indexed by $y = (y_j)_{j\in\{0,\dots,m-1\}^d}$. For every $\xi \in [0, h]^d$, $\lambda \in [-N, N]^n$ and $y \in (I\!R^m)^d$, define

$$x_{j,\xi} \doteq \xi + h\,j\,, \qquad j \in \{0, \dots, m-1\}^d$$

and

$$J_{\lambda,\xi}(y) \doteq \Big\{j \in \{0, \dots, m-1\}^d \ \Big|\ y_j = \lambda\cdot g(x_{j,\xi})\Big\}\,. \tag{18}$$

We claim that the set

$$Y(\xi) \doteq \left\{ y \in (\mathbb{R}^m)^d \,\middle|\, \# J_{\lambda,\xi}(y) \leq n, \quad \forall \lambda \in [-N, N]^n \right\}$$

is dense in $(\mathbb{R}^m)^d$. Indeed, the complementary of $Y(\xi)$ is contained in the union of a finite family of proper hyperspaces: for every collection of indexes

$$J = \{j_1, \ldots, j_{n+1}\} \subset \{0, \ldots, m-1\}^d,$$

let us define the projection

$$\Pi_J : (\mathbb{R}^m)^d \mapsto \mathbb{R}^{n+1}, \quad \Pi_J(y) \doteq (y_{j_1}, \ldots, y_{j_{n+1}}),$$

and the linear operator

$$A_J : \mathbb{R}^n \mapsto \mathbb{R}^{n+1}, \quad A_J(\lambda) \doteq \left(\lambda \cdot g(x_{\xi,j_1}), \ldots, \lambda \cdot g(x_{\xi,j_{n+1}})\right).$$

Then

$$(\mathbb{R}^m)^d \backslash Y(\xi) \subset \bigcup_{\{J \subset \{0,\ldots,m-1\}^d \,|\, \#J = n+1\}} \left\{ y \in (\mathbb{R}^m)^d \mid \Pi_J(y) \in A_J(\mathbb{R}^n) \right\}.$$

For any $\xi \in [0, h]^d$ and $j \in \{0, \ldots, m-1\}^d$, define

$$\tilde{y}_j(\xi) \doteq \tilde{w}(x_{j,\xi}).$$

By the density of $Y(\xi)$ in $(\mathbb{R}^m)^d$, we can find $y(\xi) \in Y(\xi)$ satisfying

$$\left| y_j(\xi) - \tilde{y}_j(\xi) \right| < \frac{\eta}{2} \qquad \forall j \in \{0, \ldots, m-1\}^d. \tag{19}$$

On the other hand, fixed any $\xi \in [0, h]^d$ and $\lambda \in [-N, N]^n$, there exist $r_\lambda, \delta_\lambda > 0$ such that

$$\inf_{\lambda' \in B(\lambda, r_\lambda)} \left| y_j(\xi) - \lambda' \cdot g(x_{j,\xi}) \right| > \delta_\lambda \qquad \forall j \in \{0, \ldots, m-1\}^d \backslash J_{\lambda,\xi}(y(\xi)).$$

As in the previous step, let $\{\lambda^1, \ldots, \lambda^M\} \subset [-N, N]^n$ be a finite family such that

$$[-N, N]^n \subset \bigcup_{k=1}^{M} B_n\left(\lambda^k, r_{\lambda^k}\right).$$

Set $\delta \doteq \min_{k \in \{1,2,\ldots,M\}} \delta_k$. For any $\lambda \in [-N, N]^n$, there exists an index $k \in \{1, \ldots, M\}$ such that

$$\left|y_j(\xi) - \lambda \cdot g(x_{j,\xi})\right| > \delta \qquad \forall j \in \{0, \dots, m-1\}^d \setminus J_{\xi,\lambda^k}(y(\xi)).$$

Thus, by the uniform continuity of g and the uniformly bound of λ, there exists a neighborhood $\mathcal{N}(\xi)$ of ξ (independent on λ) such that

$$\left|y_j(\xi) - \lambda \cdot g(x_{j,\xi'})\right| > \frac{\delta}{2} \qquad \forall j \in \{0, \dots, m-1\}^d \setminus J_{\xi,\lambda^k}(y(\xi)), \xi' \in \mathcal{N}(\xi).$$

In particular, recalling (18), we obtain that

$$J_{\xi',\lambda}(y(\xi)) \subset J_{\xi,\lambda^k}(y(\xi)) \qquad \forall \xi' \in \mathcal{N}(\xi),$$

and this yields

$$\# J_{\lambda,\xi'}(y(\xi)) \leq n \qquad \forall \lambda \in [-N, N]^n, \ \forall \xi' \in \mathcal{N}(\xi). \tag{20}$$

Cover the set $[0, h[^d$ with finitely many disjoint neighborhoods $\{\mathcal{N}(\xi_k)\}_{k=1,\dots,\ell}$ and define a piecewise constant function $w : [0, 1[^d \mapsto I\!R$ by setting

$$w(x) \doteq y_j(\xi_k) \quad \text{if} \quad x \in \mathcal{N}(\xi_k) + h\,j, \quad k = 1, \dots, \ell, \quad j \in \{0, \dots, m-1\}^d.$$

For any $x \in [0, 1[^d$, let $k \in \{1, \dots, \ell\}$ and $j \in \{0, \dots, m-1\}^d$ be such that $x \in \mathcal{N}(\xi_k) + h\,j$. Then, x and x_{j,ξ_k} belong to $[0, h[^d + h\,j$. Recalling (17) and (19), we have

$$|w(x) - \tilde{w}(x)| \leq |y_j(\xi_k) - \tilde{y}_j(\xi_k)| + |\tilde{w}(x_{\xi_k,j}) - \tilde{w}(x)| < \eta$$

and it yields (15).

Moreover, by (16), (18) and (20), we obtain

$$\begin{aligned}
\text{meas}&\left(\left\{x \in K \ |w(x) = \lambda \cdot g(x)\right\}\right) \\
&= \text{meas}\left(\bigcup_{j\in\{0,\dots,m-1\}^d} \left\{x \in [0, h[^d + h\,j \mid w(x) = \lambda \cdot g(x)\right\}\right) \\
&= \text{meas}\left(\bigcup_{j\in\{0,\dots,m-1\}^d} \bigcup_{k=1}^{\ell} \left\{x \in \mathcal{N}(\xi_k) + h\,j \mid y_j(\xi_k) = \lambda \cdot g(x)\right\}\right) \\
&\leq \sum_{k=1}^{\ell} \text{meas}\left(\bigcup_{j\in\{0,\dots,m-1\}^d} \left\{\xi' \in \mathcal{N}(\xi_k) \mid y_j(\xi_k) = \lambda \cdot g(x_{j,\xi'})\right\}\right) \\
&\leq \sum_{k=1}^{\ell} n \cdot \text{meas}\left(\mathcal{N}(\xi_k)\right) = n\,h^d < \frac{\varepsilon}{2}
\end{aligned}$$

for every $\lambda \in [-N, N]^n$.

Finally, by Lusin's theorem, we then modify w on a set of measure $< \varepsilon/2$ and make it continuous on the entire set K and still satisfying (15). Then $w \in \mathscr{W}^g_{\varepsilon,N} \cap B(\tilde{w}, \eta)$ and the set $\mathscr{W}^g_{\varepsilon,N}$ is dense in $\mathscr{C}(K, I\!R)$. □

We are now going to prove our main theorem.

Proof of Theorem 1. It is divided into 2 steps:
1. Fix $v = (v_1, \ldots, v_m) \in \mathscr{C}(K, I\!R^m)$ and let $\theta^* = (\theta_1^*, \ldots, \theta_m^*)$ be a solution of the optimization problem (1)–(2). We claim that if θ^* is not extremal, then it is not the unique solution of (1)–(2) and there exist two indexes $i_1 \neq i_2 \in \{1, \ldots, m\}$ and a Lagrange multiplier $\lambda = (\lambda_1, \ldots, \lambda_n) \in I\!R^n$ satisfying

$$\text{meas}\Big(\big\{x \in K \mid v_{i_1}(x) - v_{i_2}(x) = \lambda \cdot \big(f_{i_1}(x) - f_{i_2}(x)\big)\big\} \Big) > 0. \tag{21}$$

Indeed, if θ^* is non-extremal then the set

$$K_1 = \big\{x \in K \mid 0 < \theta_i^*(x) < 1 \text{ for some } i \in \{1, \ldots, m\}\big\}$$

has a positive Lebesgue measure. Since $\sum_i^m \theta^*(x) = 1$ for all $x \in K$, we can deduce that there exist two different indexes $i_1, i_2 \in \{1, \ldots, m\}$ such that

$$\text{meas}\big(\big\{x \in K \mid 0 < \theta_i^*(x) < 1\,,\ \forall i \in \{i_1, i_2\}\big\}\big) > 0\,.$$

Observe that

$$\begin{aligned} &\text{meas}\big(\big\{x \in K \ \ |0 < \theta_i^*(x) < 1\,,\ \forall i \in \{i_1, i_2\}\big\}\big) \\ &\quad = \text{meas}\left(\bigcup_{n=3}^{+\infty} \left\{ x \in K \;\middle|\; \frac{1}{n} < \theta_i^*(x) < 1 - \frac{1}{n}\,,\ \forall i \in \{i_1, i_2\} \right\} \right), \end{aligned}$$

there exists $n_0 \geq 3$ such that the set

$$\tilde{K} = \left\{ x \in K \;\middle|\; \frac{1}{n_0} < \theta_i^*(x) < 1 - \frac{1}{n_0}\,,\ \forall i \in \{i_1, i_2\} \right\}$$

has a positive Lebesgue measure.

Consider the auxiliary optimization problem

$$\text{Minimize}_{\xi \in \mathscr{A}_0} \int_{\tilde{K}} \xi(x)\big(v_{i_1}(x) - v_{i_2}(x)\big)\, dx\,, \tag{22}$$

where

$$\mathscr{A}_0 \doteq \left\{ \xi \in \mathscr{M}(\tilde{K}, [-1, 1]) \;\middle|\; \int_{\tilde{K}} \xi(x)\big(f_{i_1}(x) - f_{i_2}(x)\big)\, dx = 0 \right\}. \tag{23}$$

Observe that $\xi^* \equiv 0$ is an optimal solution of (22)–(23). Indeed, for any $\xi \in \mathscr{A}_0$, define the mapping $\tilde{\theta} : K \mapsto I\!R^m$ by

$$\tilde{\theta}(x) \doteq \begin{cases} \theta^*(x) + \frac{1}{n_0}\xi(x)\big(\mathbf{e}_{i_1} - \mathbf{e}_{i_2}\big) & \text{if } x \in \tilde{K} \\ \theta^*(x) & \text{if } x \in K \setminus \tilde{K}\,, \end{cases}$$

where $\{\mathbf{e}_1, \ldots, \mathbf{e}_m\}$ is the canonical basis of $I\!R^m$. Clearly, $\tilde{\theta}$ belongs to $\mathscr{A}_\alpha$. Thus,

$$\int_K \tilde{\theta}(x) \cdot v(x)dx \;\geq\; \int_K \theta^*(x) \cdot v(x)dx$$

and it implies that

$$\int_{\tilde{K}} \xi(x)\big(v_{i_1}(x) - v_{i_2}(x)\big)\, dx \;\geq\; 0\,. \tag{24}$$

Now let's consider the vector subspace Y of $I\!R^n$ generated by

$$\left\{ \int_{\tilde{K}} \xi(x)\big(f_{i_1}(x) - f_{i_2}(x)\big)\, dx \;\middle|\; \xi \in \mathscr{M}(\tilde{K}, [-1, 1]) \right\}$$

and define two convex subsets of $I\!R \times Y$

$$A \doteq \big\{(a_0, 0) \in I\!R \times Y \;\big|\; a_0 < 0\big\}\,,$$

and B the set of elements of the form

$$(b_0, \bar{b}) = \left(\int_{\tilde{K}} \xi(x)\big(v_{i_1}(x) - v_{i_2}(x)\big)\, dx\,,\; \int_{\tilde{K}} \xi(x)\big(f_{i_1}(x) - f_{i_2}(x)\big)\, dx \right),$$

with ξ varying in $\mathscr{M}(\tilde{K}, [-1, 1])$. Recalling (24), one has that $A \cap B = \emptyset$. Thanks to hyperplane separation theorem, there exists $(\lambda_0, \bar{\lambda}) \in \big([0, +\infty) \times Y\big) \setminus \{(0, 0)\}$ such that

$$\lambda_0 a_0 \;\leq\; \lambda_0 b_0 + \bar{\lambda} \cdot \bar{b} \qquad \forall\, a_0 < 0,\, (b_0, \bar{b}) \in B\,.$$

Observe that $\lambda_0 \neq 0$, otherwise we have

$$\bar{\lambda} \cdot \int_{\tilde{K}} \xi(x)\big(f_{i_1}(x) - f_{i_2}(x)\big)\, dx \;\geq\; 0 \qquad \forall \xi \in \mathscr{M}(\tilde{K}, [-1, 1])\,,$$

that is impossible, since $0 \neq \bar{\lambda} \in Y$. Setting $\lambda = -\bar{\lambda}/\lambda_0$, we obtain

$$\int_{\tilde{K}} \xi(x)\big(v_{i_1}(x) - v_{i_2}(x)\big)\, dx - \lambda \cdot \int_{\tilde{K}} \xi(x)\big(f_{i_1}(x) - f_{i_2}(x)\big)\, dx \;\geq\; \lim_{a_0 \to 0-} a_0 \;=\; 0$$

for every $\xi \in \mathscr{M}(\tilde{K}, [-1, 1])$. This yields

$$v_{i_1}(x) - v_{i_2}(x) = \lambda \cdot \big(f_{i_1}(x) - f_{i_2}(x)\big) \qquad \text{a.e. } x \in \tilde{K}$$

and consequently (21).

In order to see that θ^* is not the unique solution of (1)–(2), consider a function $\xi \in \mathscr{A}_0$ such that

$$\text{meas}\left(\left\{x \in \tilde{K} \;\middle|\; \xi(x) \neq 0\right\}\right) > 0.$$

Therefore, the following mappings

$$\tilde{\theta}^{\pm}(x) \doteq \begin{cases} \theta^*(x) \pm \frac{1}{n_0}\xi(x)\big(\mathbf{e}_{i_1} - \mathbf{e}_{i_2}\big) & \text{if } x \in \tilde{K} \\ \theta^*(x) & \text{if } x \in K \setminus \tilde{K} \end{cases}$$

belong to $\mathscr{A}_\alpha$, satisfy $\tilde{\theta}^+ \not\equiv \tilde{\theta}^-$ and

$$\min\left\{\int_K \tilde{\theta}^-(x) \cdot v(x)dx, \int_K \tilde{\theta}^+(x) \cdot v(x)dx\right\} \le \int_K \theta^*(x) \cdot v(x)dx.$$

2. Remark that if the problem (1)–(2) admits two distinct solutions θ^* and θ^{**}, then their convex combination

$$\tilde{\theta} \doteq \frac{\theta^* + \theta^{**}}{2}$$

is still a solution and it is not extremal. Therefore, by the previous step, $\mathscr{V}_\alpha$ contains the set of functions $v = (v_1, \ldots, v_m) \in \mathscr{C}(K, I\!R^m)$ satisfying

$$\text{meas}\Big(\big\{x \in K \mid v_{i_1}(x) - v_{i_2}(x) = \lambda \cdot \big(f_{i_1}(x) - f_{i_2}(x)\big)\big\}\Big) = 0 \quad \forall i_1 \neq i_2,\ \lambda \in I\!R^n.$$

For any Lebesgue integrable function $g : K \to I\!R^n$, define $\mathscr{W}^g$ as in the statement of Lemma 1. We then have

$$\mathscr{V}_\alpha \supset \bigcap_{i_1 \neq i_2 \in \{1,\ldots,m\}} \left\{v = (v_1, \ldots, v_m) \in \mathscr{C}(K, I\!R^m) \;\middle|\; v_{i_1} - v_{i_2} \in \mathscr{W}^{f_{i_1} - f_{i_2}}\right\}.$$

By Lemma 1, the set $\mathscr{W}^{f_{i_1} - f_{i_2}}$ is residual in $\mathscr{C}(K, I\!R)$ for all $i_1 \neq i_2 \in \{1, 2, \ldots, m\}$, i.e., there exists a family of open and dense subsets $\left\{\mathscr{W}_k^{f_{i_1} - f_{i_2}}\right\}_{k \in I\!N}$ of $\mathscr{C}(K, I\!R)$ satisfying

$$\bigcap_{k \in I\!N} \mathscr{W}_k^{f_{i_1} - f_{i_2}} \subset \mathscr{W}^{f_{i_1} - f_{i_2}}.$$

Hence we obtain

$$\mathcal{V}_\alpha \supset \bigcap_{i_1 \neq i_2 \in \{1,\dots,m\}} \left\{ v \in \mathscr{C}(K, I\!R^m) \,\middle|\, v_{i_1} - v_{i_2} \in \bigcap_{k \in I\!N} \mathscr{W}_k^{f_{i_1} - f_{i_2}} \right\}$$
$$\supset \bigcap_{i_1 \neq i_2 \in \{1,\dots,m\},\, k \in I\!N} \left\{ v \in \mathscr{C}(K, I\!R^m) \,\middle|\, v_{i_1} - v_{i_2} \in \mathscr{W}_k^{f_{i_1} - f_{i_2}} \right\}.$$

Moreover, it is not difficult to verify that the sets of the last intersection are open and dense. Therefore we can conclude that $\mathcal{V}_\alpha$ contains the intersection of countably many open dense subsets of $\mathscr{C}(K, I\!R^m)$, i.e. it is residual. □

With similar techniques we can deal with a countable family of integrable functions. Let $(f_i)_{i=1}^\infty$ be a family of Lebesgue integrable functions from $K \subset I\!R^d$ to $I\!R^n$ satisfying

$$\int_K \sup_i \|f_i(x)\| \, dx \; < \; \infty \,, \tag{25}$$

where $\|\cdot\|$ is the norm in $I\!R^n$. Let $(\bar{\theta}_i)_{i=1}^\infty$ be a family of measurable functions from K to $[0, +\infty)$ such that

$$\sum_{i=1}^\infty \bar{\theta}_i(x) = 1 \qquad \forall x \in K \,.$$

We can consider $\bar{\theta} = (\bar{\theta}_i)_{i=1}^\infty$ as an element of the space $\mathbf{L}^\infty(K, \ell^\infty)$, where ℓ^∞ is the space of bounded real sequences. Call

$$\alpha \;\doteq\; \int_K \sum_{i=1}^\infty \bar{\theta}_i(x)\, f_i(x)\, dx \,.$$

Thanks to (25) and dominated convergence, $\alpha \in I\!R^n$. Given $v \in \mathscr{C}(K, \ell^1)$, consider the problem

$$\text{Minimize}_{\theta \in \mathscr{A}_\alpha} \int_K \sum_{i=1}^\infty \theta_i(x) v_i(x) dx \tag{26}$$

over the set

$$\mathscr{A}_\alpha \;\doteq\; \left\{ \theta \in \mathbf{L}^\infty(K, \ell^\infty) \;\middle|\; \theta_i(x) \geq 0 \;\forall i \in I\!N \,, \sum_{i=1}^\infty \theta_i(x) = 1, \; a.e.\ x \in K, \right.$$
$$\left. \int_K \sum_{i=1}^\infty \theta_i(x) f_i(x)\, dx \;=\; \alpha \right\}. \tag{27}$$

This problem admits at least a solution, since it is equivalent to

$$\text{Minimize}_{\tilde{\theta} \in \mathscr{B}} \int_K \sum_{i=1}^\infty \tilde{\theta}_i(x) \big(v_{i+1}(x) - v_1(x)\big) dx$$

over the set

$$\mathscr{B} \doteq \Big\{\tilde\theta \in \mathbf{L}^\infty(K,\ell^\infty) \,\Big|\, \tilde\theta_i(x) \geq 0 \ \forall i \in I\!N\,,\ \sum_{i=1}^{\infty}\tilde\theta_i(x) \leq 1,\ a.e.\ x \in K,$$

$$\sum_{i=1}^{\infty}\int_K \tilde\theta_i(x)\big(f_{i+1}(x) - f_1(x)\big)\,dx \ = \ \alpha - \int_K f_1(x)\,dx\Big\}$$

and $\mathscr{B}$ is weakly*-compact in $\mathbf{L}^\infty(K,\ell^\infty)$.

Theorem 2 *Assume (25). Then the set*

$$\mathscr{V}_\alpha \ \doteq \ \big\{v \in \mathscr{C}(K,\ell^1) \mid (26)\text{--}(27)\ \textit{has a unique solution}\big\}\,. \tag{28}$$

is residual in $\mathscr{C}(K,\ell^1)$. Moreover, for any $v \in \mathscr{V}_\alpha$, the unique optimal solution θ^* verifies $\theta_i^*(x) \in \{0,1\}$ for almost every $x \in K$ and every i.

Proof The proof is similar to the one of Theorem 1. Fix $v \in \mathscr{C}(K,\ell^1)$ and let $\theta^* \in \mathbf{L}^\infty(K,\ell^\infty)$ be a solution of the optimization problem (26)–(27). If θ^* does not verify $\theta_i^*(x) \in \{0,1\}$ for almost every $x \in K$ and every i, then it is possible to show as above that θ^* is not the unique solution of (26)–(27). We claim that there exist two indexes $i_1 \neq i_2$ and $\lambda = (\lambda_1,\ldots,\lambda_n) \in I\!R^n$ satisfying

$$\text{meas}\left(\big\{x \in K \mid v_{i_1}(x) - v_{i_2}(x) = \lambda\cdot\big(f_{i_1}(x) - f_{i_2}(x)\big)\big\}\right) \ > \ 0\,. \tag{29}$$

Indeed, if θ^* is non-extremal, we have

$$\begin{aligned} 0 \ &< \ \text{meas}\big(\big\{x \in K \mid 0 < \theta_i^*(x) < 1 \text{ for some } i\big\}\\ &= \ \text{meas}\left(\bigcup_{I\in I\!N}\bigcup_{n=3}^{+\infty}\left\{x \in K \,\middle|\, \frac{1}{n} < \theta_i^*(x) < 1 - \frac{1}{n},\ \forall i \in \{i_1,i_2\},\ \text{some } i_1 \neq i_2 \leq I\right\}\right). \end{aligned}$$

Consequently, there exist $i_1 \neq i_2$ and $n_0 \geq 3$ such that the set

$$\tilde K \ = \ \left\{x \in K \,\middle|\, \frac{1}{n_0} < \theta_i^*(x) < 1 - \frac{1}{n_0}\,,\ \forall i \in \{i_1,i_2\}\right\}$$

has a positive Lebesgue measure. As in the proof of Theorem 1, one can verify that $\xi^* \equiv 0$ is an optimal solution of the auxiliary problem (22)–(23) and that it satisfies the necessary condition (29) for some Lagrange multiplier $\lambda = (\lambda_1,\ldots,\lambda_n) \in I\!R^n$. Therefore, if we denote by $\mathscr{W}^{f_{i_1}-f_{i_2}}$ is the set of functions $w \in \mathscr{C}(K,I\!R)$ such that

$$\text{meas}\left(\big\{x \in K \mid w(x) = \lambda\cdot(f_{i_1}(x) - f_{i_2}(x))\big\}\right) \ = \ 0 \qquad \text{for all } \lambda \in I\!R^n\,,$$

we obtain

$$\mathscr{V}_\alpha \supset \bigcap_{i_1\neq i_2}\left\{v \in \mathscr{C}(K,\ell^1) \,\middle|\, v_{i_1} - v_{i_2} \in \mathscr{W}^{f_{i_1}-f_{i_2}}\right\}.$$

By Lemma 1, for all $i_1 \neq i_2$ the set $\mathscr{W}^{f_{i_1}-f_{i_2}}$ is residual in $\mathscr{C}(K, I\!R)$, i.e., there exists a family of open and dense subsets $\left\{\mathscr{W}_k^{f_{i_1}-f_{i_2}}\right\}_{k\in I\!N}$ of $\mathscr{C}(K, I\!R)$ satisfying

$$\bigcap_{k\in I\!N} \mathscr{W}_k^{f_{i_1}-f_{i_2}} \subset \mathscr{W}^{f_{i_1}-f_{i_2}}.$$

Hence we obtain

$$\begin{aligned}\mathscr{V}_\alpha &\supset \bigcap_{i_1\neq i_2} \left\{v \in \mathscr{C}(K, \ell^1) \;\middle|\; v_{i_1} - v_{i_2} \in \bigcap_{k\in I\!N} \mathscr{W}_k^{f_{i_1}-f_{i_2}}\right\}\\ &\supset \bigcap_{i_1\neq i_2\,,\,k\in I\!N} \left\{v \in \mathscr{C}(K, \ell^1) \;\middle|\; v_{i_1} - v_{i_2} \in \mathscr{W}_k^{f_{i_1}-f_{i_2}}\right\}.\end{aligned}$$

Consequently, $\mathscr{V}_\alpha$ is residual in $\mathscr{C}(K, \ell^1)$. □

Acknowledgements This work was partially supported by a grant from the Simons Foundation/SFARI (521811, NTK).

References

1. Angell, T.S.: Existence of optimal control without convexity and a bang-bang theorem for linear Volterra equations. J. Optim. Theory Appl. **19**(1), 63–79 (1976)
2. Bressan, A.: Extremal solutions to differential inclusions via Baire category: a dual approach. J. Differ. Equ. **255**, 2392–2399 (2013)
3. Bressan, A., Mazzola, M., Nguyen, Khai T.: The bang-bang theorem via Baire category. A dual approach. NoDEA Nonlinear Differ. Equ. Appl. **23**(4), 23–46 (2016)
4. Cellina, A.: On the differential inclusion $x' \in [-1, 1]$. Atti Accad. Naz. Lincei Rend. Cl. Sci. Fis. Mat. Natur. **69**, 1–6 (1980)
5. Cesari, L.: Optimization - Theory and Applications. Problems with Ordinary Differential Equations. Springer, New York (1983)
6. De Blasi, F.S., Pianigiani, G.: Baire category and boundary value problems for ordinary and partial differential inclusions under Carathéodory assumptions. J. Differ. Equ. **243**, 558–577 (2007)
7. Suryanarayana, M.B.: Existence theorems for optimization problems concerning linear, hyperbolic partial differential equations without convexity conditions. J. Optim. Theory Appl. **19**(1), 47–61 (1976)

Controllability Under Positivity Constraints of Multi-d Wave Equations

Dario Pighin and Enrique Zuazua

Dedicated to Piermarco Cannarsa on the occasion of his 60th birthday

Abstract We consider both the internal and boundary controllability problems for wave equations under non-negativity constraints on the controls. First, we prove the steady state controllability property with nonnegative controls for a general class of wave equations with time-independent coefficients. According to it, the system can be driven from a steady state generated by a strictly positive control to another, by means of nonnegative controls, and provided the time of control is long enough. Secondly, under the added assumption of conservation and coercivity of the energy, controllability is proved between states lying on two distinct trajectories. Our methods are described and developed in an abstract setting, to be applicable to a wide variety of control systems.

Keywords Wave equations · Controllability · Constraints · Quasi-static iteration · Bounded trajectories · Energy conservation

---

D. Pighin (✉) · E. Zuazua
Departamento de Matemáticas, Universidad Autónoma de Madrid,
28049 Madrid, Spain
e-mail: dario.pighin@uam.es

E. Zuazua
DeustoTech, Fundación Deusto, Avda. Universidades, 24,
48007 Bilbao, Basque Country, Spain

Facultad de Ingeniería, Universidad de Deusto, Avda. Universidades, 24,
48007 Bilbao, Basque Country, Spain
e-mail: enrique.zuazua@uam.es

F. Alabau-Boussouira et al. (eds.), *Trends in Control Theory and Partial Differential Equations*, Springer INdAM Series 32,
https://doi.org/10.1007/978-3-030-17949-6_11

1 Introduction

This paper is devoted to the study of the controllability properties of the wave equation, under *positivity* (or nonnegativity) constraints on the *control*.

We address both the case where the control acts in the *interior* of the domain where waves evolve or on its *boundary*.

This problem has been exhaustively considered in the *unconstrained* case but very little is known in the presence of constraints on the control, an issue of primary importance in applications, since whatever the applied context under consideration is, the available controls are always limited. For some of the basic literature on the unconstrained controllability of wave-like equations the reader is referred to: [1, 3–5, 8, 9, 15, 21, 22, 24, 26].

The developments in this paper are motivated by our earlier works on the constrained controllability of heat-like equations ([16, 19]). In that context, due to the well-known comparison principle for parabolic equations, control and state constraints are interlinked. In particular, for the heat equation, nonnegative controls imply that the solution is nonnegative too, when the initial configuration is nonnegative. Therefore, imposing non-negativity constraints on the control ensures that the state satisfies the non-negativity constraint too.

This is no longer true for wave-like equations in which the sign of the control does not determine that of solutions. However, as mentioned above, from a practical viewpoint, it is very natural to consider the problem of imposing control constraints. In this work, to fix ideas, we focus in the particular case of nonnegative controls.

First we address the problem of steady state controllability in which one aims at controlling the solution from a steady configuration to another one. This problem was addressed in [7], in the absence of constraints on the controls for semilinear wave equations. Our main contribution here is to control the system by preserving some constraints on the controls given a priori. And, as we shall see, when the initial and final steady states are associated to positive time-independent control functions, the constrained controllability can be guaranteed to hold if the time-horizon is long enough.

The proof is developed by a step-wise procedure presented in [19] (which differs from the one in [7, 16]), the so-called "stair-case argument", along an arc of steady-states linking the starting and final one. The proof consists on moving recursively from one steady state to the other by means of successive small amplitude controlled trajectories linking successive steady-states. This method and result are presented in a general semigroup setting and it can be successfully implemented for any control system for which controllability holds by means of $L\infty$ controls.

The same recursive approach enables us to prove a state constrained result, under additional dissipativity assumptions. But the time needed for this to hold is even larger than before.

The problem of steady-state controllability is a particular instance of the more general trajectory control problem, in which, given two controlled trajectories of the system, both obtained from nonnegative controls, and one state in each of them (possibly corresponding to two different time-instances) one aims at driving one state into the other one by means of nonnegative constrained controls. This result can also be proved by a similar iterative procedure, but under the added assumption that the system is conservative and its energy coercive so that uncontrolled trajectories are globally bounded.

These results hold for long enough control time horizons. The stepwise procedure we implement needs of a very large control time, much beyond the minimal control time for the control of the wave equation, that is determined by the finite velocity of propagation and the so-called Geometric Control Condition (GCC). It is then natural to introduce the minimal time of control under non-negativity constraints, in both situations above.

There is plenty to be done to understand how these constrained minimal times depends on the data to be controlled. Employing d'Alembert's formula for the one dimensional wave equation, we compute both of them for constant steady states, showing that they coincide with the unconstrained one. In that case we also show that the property of constrained controllability holds in the minimal time too.

Controllability under constraints has already been studied for finite-dimensional models and heat-like equations (see [16, 19]). In both cases it was also proved that controllability by nonnegative controls fails if time is too short, when the initial datum differs from the final target. This fact exhibits a big difference with respect to the unconstrained control problem for these systems, where controllability holds in arbitrary small time in both cases. In the wave-like context addressed in this paper the waiting phenomenon, according to which there is a minimal control time for the constrained problem, is less surprising. But, simultaneously, on the other hand, in some sense, the fact that constraints can be imposed on controls and state seems more striking too.

In [12], authors analysed controllability of the one dimensional wave equation, under the more classical bilateral constraints on the control. Our work is, as far as we know, the first one considering unilateral constraints for wave-like equations.

1.1 Internal Control

Let Ω be a connected bounded open set of $\mathbb{R}^n$, $n \geq 1$, with C^∞ boundary, and let ω and ω_0 be subdomains of Ω such that $\overline{\omega_0} \subset \omega$.

Let $\chi \in C^\infty(\mathbb{R}^n)$ be a smooth function supported in ω such that Range$(\chi) \subseteq [0, 1]$, $\chi\restriction_{\omega_0} \equiv 1$.

We assume further that all derivatives of χ vanish on the boundary of Ω. We will discuss this assumption in Sect. 3.3.

We consider the wave equation controlled from the interior

$$\begin{cases} y_{tt} - \Delta y + cy = u\chi & \text{in } (0,T)\times\Omega \\ y = 0 & \text{on } (0,T)\times\partial\Omega \\ y(0,x) = y_0^0(x),\ \ y_t(0,x) = y_0^1(x) & \text{in } \Omega \end{cases} \tag{1}$$

where $y = y(t,x)$ is the state, while $u = u(t,x)$ is the control whose action is localized on ω by means of multiplication with the smooth cut-off function χ. The coefficient $c = c(x)$ is C^∞ smooth in $\overline{\Omega}$.

It is well known in the literature (e.g. [10, Sect. 7.2]) that, for any initial datum $(y_0^0, y_0^1) \in H_0^1(\Omega)\times L^2(\Omega)$ and for any control $u \in L^2((0,T)\times\omega)$, the above problem admits an unique solution $(y, y_t) \in C^0([0,T]; H_0^1(\Omega)\times L^2(\Omega))$, with $y_{tt} \in L^2(0,T; H^{-1}(\Omega))$.

We assume the *Geometric Control Condition* on (Ω, ω_0, T^*), which basically asserts that all bicharacteristic rays enter in the subdomain ω_0 in time smaller than T^*. This geometric condition is actually equivalent to the property of (unconstrained) *controllability* of the system (see [1, 3]).

1.1.1 Steady State Controllability

The purpose of our first result is to show that, in time large, we can drive (1) from one steady state to another by a *nonnegative* control, assuming the uniform *positivity* of the control defining the steady states.

More precisely, a steady state is a solution to

$$\begin{cases} -\Delta\overline{y} + c\overline{y} = \overline{u}\chi & \text{in } \Omega \\ \overline{y} = 0 & \text{on } \partial\Omega, \end{cases} \tag{2}$$

where $\overline{u} \in L^2(\omega)$ and $\overline{y} \in H^2(\Omega)\cap H_0^1(\Omega)$. Note that, as a consequence of Fredholm Alternative (see [11, Theorem 5.11 page 84]), the existence and uniqueness of the solution of this elliptic problem can be guaranteed whenever zero is not an eigenvalue of $-\Delta + cI : H_0^1(\Omega) \longrightarrow H^{-1}(\Omega)$.

The following result holds:

Theorem 1 (Controllability between steady states) *Take $\overline{y}_0$ and $\overline{y}_1$ in $H^2(\Omega)\cap H_0^1(\Omega)$ steady states associated to L^2-controls $\overline{u}^1$ and $\overline{u}^2$, respectively. Assume further that there exists $\sigma > 0$ such that*

$$\overline{u}^i \geq \sigma, \quad \text{a.e. in } \omega. \tag{3}$$

Then, if T is large enough, there exists $u \in L^2((0,T)\times\omega)$, a control such that

- *the unique solution* (y, y_t) *to the problem* (1) *with initial datum* $(\overline{y}_0, 0)$ *and control* u *verifies* $(y(T, \cdot), y_t(T, \cdot)) = (\overline{y}_1, 0)$;
- $u \geq 0$ *a.e. on* $(0, T) \times \omega$.

Theorem 1 is proved in Sect. 3.1. Inspired by [7], we implement a recursive "staircase" argument to keep the control in a narrow tubular neighborhood of the segment connecting the controls defining the initial and final data. This will guarantee the actual positivity of the control obtained.

1.1.2 Controllability Between Trajectories

The purpose of this section is to extend the above result, under the additional assumption $c(x) > -\lambda_1$, where λ_1 is the first eigenvalue of the Dirichlet Laplacian in Ω. This guarantees that the energy of the system defines a norm

$$\|(y^0, y^1)\|_E^2 = \int_\Omega \left[\|\nabla y^0\|^2 + c\left(y^0\right)^2\right] dx + \int_\Omega (y^1)^2 dx$$

on $H_0^1(\Omega) \times L^2(\Omega)$. Thus, by conservation of the energy, uncontrolled solutions are uniformly bounded for all t.

We assume that both, the initial datum (y_0^0, y_0^1) and the final target (y_1^0, y_1^1), belong to controlled trajectories (see Fig. 1)

$$(y_i^0, y_i^1) \in \left\{(\overline{y}_i(\tau, \cdot), (\overline{y}_i)_t(\tau, \cdot) \mid \tau \in \mathbb{R}\right\}, \tag{4}$$

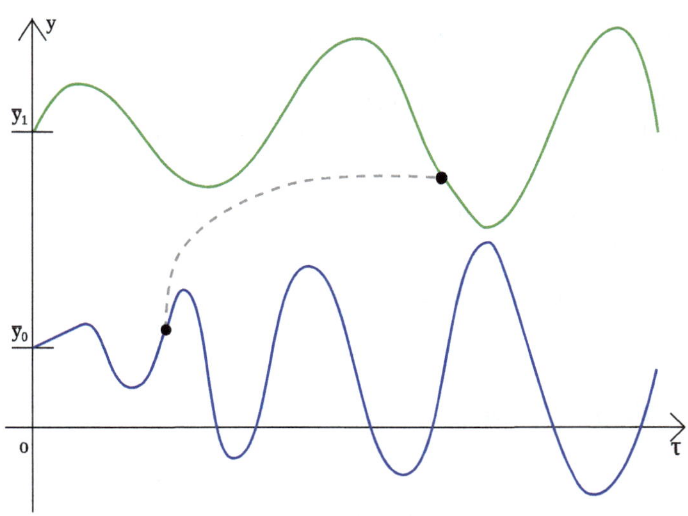

Fig. 1 Controllability between data lying on trajectories

where $(\overline{y}_i, (\overline{y}_i)_t)$ solve (1) with *nonnegative* controls. We suppose that these trajectories are smooth enough, namely

$$(\overline{y}_i, (\overline{y}_i)_t) \in C^{s(n)}(\mathbb{R}; H_0^1(\Omega) \times L^2(\Omega)),$$

with $s(n) = \lfloor n/2 \rfloor + 1$. Hereafter, we denote by $(\overline{y}_0, (\overline{y}_0)_t)$ the initial trajectory, while $(\overline{y}_1, (\overline{y}_1)_t)$ stands for the target one.

Note that the regularity is assumed only in time and not in space. This allows to consider weak steady-state solutions.

We can in particular choose as final target the null state $(y_1^0, y_1^1) = (0, 0)$. It is important to highlight that this is something specific to the wave equation. In the parabolic case (see [16, 19]), this was prevented by the comparison principle, since the zero target cannot be reached in finite time with non-negative controls. But, for the wave equation, the maximum principle does not hold and this obstruction does not apply.

The following result holds

Theorem 2 (Controllability between trajectories) *Suppose $c(x) > -\lambda_1$, for any $x \in \overline{\Omega}$. Let $(\overline{y}_i, (\overline{y}_i)_t) \in C^{s(n)}(\mathbb{R}; H_0^1(\Omega) \times L^2(\Omega))$ be solutions to* (1) *associated to controls $\overline{u}^i \geq 0$ a.e. in $(0, T) \times \omega$, $i = 0, 1$. Take $(y_0^0, y_0^1) = (\overline{y}_0(\tau_0, \cdot), (\overline{y}_0)_t(\tau_0, \cdot))$ and $(y_1^0, y_1^1) = (\overline{y}_1(\tau_1, \cdot), (\overline{y}_1)_t(\tau_1, \cdot))$ for arbitrary values of τ_0 and τ_1. Then, in time $T > 0$ large enough, there exists a control $u \in L^2((0, T) \times \omega)$ such that*

- *the unique solution (y, y_t) to* (1) *with initial datum (y_0^0, y_0^1) verifies the end condition $(y(T, \cdot), y_t(T, \cdot)) = (y_1^0, y_1^1)$;*
- *$u \geq 0$ a.e. in $(0, T) \times \omega$.*

Remark 1 This result is more general than Theorem 1 for two reasons

1. it enables us to link more general data, with nonzero *velocity*, and not only steady states;
2. the control defining the initial and target trajectories is assumed to be only *nonnegative*. This assumption is weaker than the *uniform positivity* one required in Theorem 1.

On the other hand, the present result requires the condition $c(x) > -\lambda_1$ on the potential $c = c(x)$.

We give the proof of Theorem 2 in Sect. 3.2.

1.2 Boundary Control

Let Ω be a connected bounded open set of $\mathbb{R}^n$, $n \geq 1$, with C^∞ boundary, and let Γ_0 and Γ be open subsets of $\partial\Omega$ such that $\overline{\Gamma_0} \subset \Gamma$.

Let $\chi \in C^\infty(\partial\Omega)$ be a smooth function such that $\text{Range}(\chi) \subseteq [0, 1]$, $\text{supp}(\chi) \subset \Gamma$ and $\chi\restriction_{\Gamma_0} \equiv 1$.

We now consider the wave equation controlled on the *boundary*

$$\begin{cases} y_{tt} - \Delta y + cy = 0 & \text{in } (0,T)\times\Omega \\ y = \chi u & \text{on } (0,T)\times\partial\Omega \\ y(0,x) = y_0^0(x),\ y_t(0,x) = y_0^1(x) & \text{in } \Omega \end{cases} \tag{5}$$

where $y = y(t,x)$ is the state, while $u = u(t,x)$ is the boundary control localized on Γ by the cut-off function χ. As before, the space-dependent coefficient c is supposed to be C^∞ regular in $\overline{\Omega}$.

By transposition (see [15]), one can realize that for any initial datum $(y_0^0, y_0^1) \in L^2(\Omega)\times H^{-1}(\Omega)$ and control $u \in L^2((0,T)\times\Gamma)$, the above problem admits an unique solution $(y, y_t) \in C^0([0,T]; L^2(\Omega)\times H^{-1}(\Omega))$.

We assume the *Geometric Control Condition* on (Ω, Γ_0, T^*) which asserts that all generalized bicharacteristics touch the sub-boundary Γ_0 at a non diffractive point in time smaller than T^*. By now, it is well known in the literature that this geometric condition is equivalent to (unconstrained) controllability (see [1, 3]).

1.2.1 Steady State Controllability

As in the context of internal control, our first goal is to show that, in time large, we can drive (5) from one steady state to another, assuming the uniform *positivity* of the controls defining these steady states.

In the present setting a steady state is a time independent solution to (5), namely a solution to

$$\begin{cases} -\Delta\overline{y} + c\overline{y} = 0 & \text{in } \Omega \\ \overline{y} = \chi\overline{u} & \text{on } \partial\Omega. \end{cases} \tag{6}$$

In the present setting, $\overline{u} \in L^2(\partial\Omega)$ and $\overline{y} \in L^2(\Omega)$ solves the above problem in the sense of transposition (see [14, Chap. II, Sect. 4.2] and [13]).

As in the context of internal control, if 0 is not an eigenvalue of $-\Delta + cI : H_0^1(\Omega) \longrightarrow H^{-1}(\Omega)$, for any boundary control $\overline{u} \in L^2(\partial\Omega)$, there exists a unique $\overline{y} \in L^2(\Omega)$ solution to (6) with boundary control $\overline{u}$. This can be proved combining Fredholm Alternative (see [11, Theorem 5.11 page 84]) and transposition techniques [14, Theorem 4.1 page 73].

We prove the following result

Theorem 3 (Steady state controllability). *Let $\overline{y}_i$ be steady states defined by controls $\overline{u}^i$, $i = 0, 1$, so that*

$$\overline{u}^i \geq \sigma, \quad \text{on } \Gamma, \tag{7}$$

with $\sigma > 0$.

Then, if T is large enough, there exists $u \in L^2([0,T]\times\Gamma)$, a control such that

- *the unique solution* (y, y_t) *to* (5) *with initial datum* $(\overline{y}_0, 0)$ *and control* u *verifies* $(y(T, \cdot), y_t(T, \cdot)) = (\overline{y}_1, 0)$*;*
- $u \geq 0$ *on* $(0, T) \times \Gamma$*.*

The proof of the above result can be found in Sect. 4.1. The structure of the proof resembles the one of Theorem 1, with some technical differences due to the different nature of the control.

1.2.2 Controllability Between Trajectories

As in the internal control case, we suppose $c(x) > -\lambda_1$, where λ_1 is the first eigenvalue of the Dirichlet Laplacian in Ω. Then, the generator of the free dynamics is skew-adjoint (see [23, Proposition 3.7.6]), thus generating an unitary group of operators $\{\mathbb{T}_t\}_{t\in\mathbb{R}}$ on $L^2(\Omega) \times H^{-1}(\Omega)$.

Both the initial datum and final target (y_i^0, y_i^1) belong to a smooth trajectory, namely

$$(y_i^0, y_i^1) \in \left\{(\overline{y}_i(\tau, \cdot), (\overline{y}_i)_t(\tau, \cdot)) \mid \tau \in \mathbb{R}\right\}. \tag{8}$$

We assume the *nonnegativity* of the controls $\overline{u}^i$ defining $(\overline{y}_i, (\overline{y}_i)_t)$, for $i = 0, 1$. Hereafter, in the context of boundary control, we take trajectories of class $C^{s(n)}(\mathbb{R}; L^2(\Omega) \times H^{-1}(\Omega))$, with $s(n) = \lfloor n/2 \rfloor + 1$. We set $(\overline{y}_0, (\overline{y}_0)_t)$ to be the initial trajectory and $(\overline{y}_1, (\overline{y}_1)_t)$ be the target one.

Note that, with respect to Theorem 3, we have relaxed the assumptions on the sign of the controls $\overline{u}^i$. Now, they are required to be only *nonnegative* and not uniformly strictly positive.

Theorem 4 (Controllability between trajectories) *Assume* $c(x) > -\lambda_1$*, for any* $x \in \overline{\Omega}$*. Let* $(\overline{y}_i, (\overline{y}_i)_t)$ *be solutions to* (5) *with non-negative controls* $\overline{u}^i$ *respectively. Suppose the trajectories* $(\overline{y}_i, (\overline{y}_i)_t) \in C^{s(n)}([0, T]; L^2(\Omega) \times H^{-1}(\Omega))$*. Pick* $(y_0^0, y_0^1) = (\overline{y}_0(\tau_0, \cdot), (\overline{y}_0)_t(\tau_0, \cdot))$ *and* $(y_1^0, y_1^1) = (\overline{y}_1(\tau_1, \cdot), (\overline{y}_1)_t(\tau_1, \cdot))$*. Then, in time large, we can find a control* $u \in L^2((0, T) \times \Gamma)$ *such that*

- *the solution* (y, y_t) *to* (5) *with initial datum* (y_0^0, y_0^1) *fulfills the final condition* $(y(T, \cdot), y_t(T, \cdot)) = (y_1^0, y_1^1)$*;*
- $u \geq 0$ *a.e. in* $(0, T) \times \Gamma$*.*

The above Theorem is proved in Sect. 4.2. Furthermore, in Sect. 5, we show how Theorem 4 applies in the one dimensional case, providing further information about the minimal time to control and the possibility of controlling the system in the minimal time.

1.2.3 State Constraints

We impose now constraints both on the control and on the state, namely both the *control* and the *state* are required to be nonnegative.

In the parabolic case (see [16, 19]) one can employ the comparison principle to get a state constrained result from a control constrained one. But, now, as we have explained before, the comparison principle is not valid in general for the wave equation. And we cannot rely on comparison to deduce our state constrained result from the control constrained one.

We shall rather apply the "stair-case argument" developed to prove steady state controllability, paying attention to the added need of preserving state constraints as well.

Let λ_1 be the first eigenvalue of the Dirichlet Laplacian. We assume $c > -\lambda_1$ in $\overline{\Omega}$. We also suppose that $\chi \equiv 1$, meaning that the control acts on the whole boundary. We take as initial and final data two steady states y_0^0 and y_1^0 associated to controls $\overline{u}^i \geq \sigma > 0$. Our proof relies on the application of the maximum principle to (6). This ensures that the states $\overline{y}_i \geq \sigma$ once we know $\overline{u}^i \geq \sigma$. For this reason, we need $c > -\lambda_1$ and $\chi \equiv 1$.

Our strategy is the following

- employ the "stair-case argument" used to prove steady state controllability, to keep the control in a narrow tubular neighborhood of the segment connecting $\overline{u}^0$ and $\overline{u}^1$. This can be done by taking the time of control large enough. Since $\overline{u}^i \geq \sigma > 0$, this guarantees the positivity of the control;
- by the continuous dependence of the solution on the data, the controlled trajectory remains also in a narrow neighborhood of the convex combination joining initial and final data. On the other hand, by the maximum principle for the steady problem (6), we have that $y_i^0 \geq \sigma$ in Ω, for $i = 0, 1$. In this way the state y can be assured to remain nonnegative.

Theorem 5 *We assume $c(x) > -\lambda_1$ for any $x \in \overline{\Omega}$ and $\chi \equiv 1$. Let y_0^0 and y_1^0 be solutions to the steady problem*

$$\begin{cases} -\Delta y + cy = 0 & \text{in } \Omega \\ y = \overline{u}^i, & \text{on } \partial\Omega \end{cases} \tag{9}$$

where $\overline{u}^i \geq \sigma$ a.e. on $\partial\Omega$, with $\sigma > 0$. We assume $y_i^0 \in H^{s(n)}(\Omega)$. Then, there exists $\overline{T} > 0$ such that for any $T > \overline{T}$ there exists a control $u \in L^\infty((0, T) \times \partial\Omega)$ such that

- *the unique solution (y, y_t) to (5) with initial datum $(y_0^0, 0)$ and control u is such that $(y(T, \cdot), y_t(T, \cdot)) = (y_1^0, 0)$;*
- *$u \geq 0$ a.e. on $(0, T) \times \partial\Omega$;*
- *$y \geq 0$ a.e. in $(0, T) \times \Omega$.*

The proof of the above Theorem can be found in Sect. 4.3.

Note that the time needed to control the system keeping both the control and the state *nonnegative* is greater (or equal) than the corresponding one with no constraints on the state.

1.3 Orientation

The rest of the paper is organized as follows:

- Section 2: Abstract results;
- Section 3: Internal Control: Proof of Theorems 1 and 2;
- Section 4: Boundary control: Proof of Theorems 3, 4 and 5;
- Section 5: The one dimensional case;
- Section 6: Conclusion and open problems;
- Appendix.

2 Abstract Results

The goal of this section is to provide some results on constrained controllability for some abstract control systems. We apply these results in the context of internal control and boundary control of the wave equation (see Sect. 1).

We begin introducing the abstract control system. Let H and U be two Hilbert spaces endowed with norms $\|\cdot\|_H$ and $\|\cdot\|_U$ respectively. H is called the state space and U the control space. Let $A : D(A) \subset H \longrightarrow H$ be a generator of a C_0-semigroup $(\mathbb{T}_t)_{t\in\mathbb{R}^+}$, with $\mathbb{R}^+ = [0, +\infty)$. The domain of the generator $D(A)$ is endowed with the graph norm $\|x\|^2_{D(A)} = \|x\|^2_H + \|Ax\|^2_H$. We define H_{-1} as the completion of H with respect to the norm $\|\cdot\|_{-1} = \|(\beta I - A)^{-1}(\cdot)\|_H$, with real β such that $(\beta I - A)$ is invertible from H to H with continuous inverse. Adapting the techniques of [23, Proposition 2.10.2], one can check that the definition of H_{-1} is actually independent of the choice of β. By applying the techniques of [23, Proposition 2.10.3], we deduce that A admits a unique bounded extension A from H to H_{-1}. For simplicity, we still denote by A the extension. Hereafter, we write $\mathscr{L}(E, F)$ for the space of all bounded linear operators from a Banach space E to another Banach space F.

Our control system is governed by:

$$\begin{cases} \frac{d}{dt}y(t) = Ay(t) + Bu(t), & t \in (0, \infty), \\ y(0) = y_0, \end{cases} \tag{10}$$

where $y_0 \in H$, $u \in L^2_{loc}([0, +\infty), U)$ is a control function and the control operator $B \in \mathscr{L}(U, H_{-1})$ satisfies the admissibility condition in the following definition (see [23, Definition 4.2.1]).

Definition 1 The control operator $B \in \mathscr{L}(U, H_{-1})$ is said to be admissible if for all $\tau > 0$ we have $\text{Range}(\Phi_\tau) \subset H$, where $\Phi_\tau : L^2((0, +\infty); U) \to H_{-1}$ is defined by:

$$\Phi_\tau u = \int_0^\tau \mathbb{T}_{\tau-r} Bu(r)dr.$$

From now on, we will always assume the control operator to be admissible. One can check that for any $y_0 \in H$ and $u \in L^2_{loc}((0, +\infty); U)$ there exists a unique mild solution $y \in C^0([0, +\infty), H)$ to (10) (see, for instance, [23, Proposition 4.2.5]). We denote by $y(\cdot; y_0, u)$ the unique solution to (10) with initial datum y_0 and control u.

Now, we introduce the following constrained controllability problem

> Let $\mathscr{U}_{ad}$ be a nonempty subset of U. Find a subset E of H so that for each y_0, $y_1 \in E$, there exists $T > 0$ and a control $u \in L^\infty(0, T; U)$ with $u(t) \in \mathscr{U}_{ad}$ for a.e. $t \in (0, T)$, so that $y(T; y_0, u) = y_1$.

We address this controllability problem in the next two subsections, under different assumptions on $\mathscr{U}_{ad}$ and (A, B). In Sect. 2.1, we study the above controllability problem, where the initial and final data are steady states, i.e. solutions to the steady equation:

$$Ay + Bu = 0 \quad \text{for some} \quad u \in U. \tag{11}$$

In Sect. 2.2, we take initial and final data on two different trajectories of (10).

To study the above problem, we need two ingredients, which play a key role in the proofs of Sects. 2.1 and 2.2. First, we introduce the notion of smooth controllability. Before introducing this concept, we fix $s \in \mathbb{N}$ and a Hilbert space V so that

$$V \hookrightarrow U, \tag{12}$$

where $\hookrightarrow$ denotes the continuous embedding. Note that all throughout the remainder of the section, s and V remain fixed.

The concept of smooth controllability is given in the following definition. The notation $y(\cdot; y_0, u)$ stands for the solution of the abstract controlled Eq. (10) with control u and initial data y_0.

Definition 2 The control system (10) is said to be smoothly controllable in time $T_0 > 0$ if for any $y_0 \in D(A^s)$, there exists a control function $v \in L^\infty((0, T_0); V)$ such that

$$y(T_0; y_0, v) = 0$$

and

$$\|v\|_{L^\infty((0,T_0);V)} \leq C\|y_0\|_{D(A^s)}, \tag{13}$$

the constant C being independent of y_0.

Remark 2 (i) In other words, the system is smoothly controllable in time T_0 if for each (regular) initial datum $y_0 \in D(A^s)$, there exists a L^∞-control u with values in the regular space V steering our control system to rest at time T_0.

(ii) The smooth controllability in time T_0 of system (10) is a consequence of the following observability inequality: there exists a constant $C > 0$ such that for any $z \in D(A^*)$

$$\|\mathbb{T}^*_{T_0} z\|_{D(A^s)^*} \leq C \int_0^{T_0} \|i^* B^* \mathbb{T}^*_{T_0 - t} z\|_{V^*} dt,$$

where $D(A^s)^*$ is the dual of $D(A^s)$ and $i : V \hookrightarrow U$ is the inclusion. This inequality, that can often be proved out of classical observability inequalities employing the regularizing properties of the system, provides a way to prove the smooth controllability for system (10). This occurs for parabolic problem enjoying smoothing properties.

(iii) Besides, for some systems (A, B), even if they do not enjoy smoothing properties, there is an alternative way to prove the aforementioned smooth controllability property exploiting the ellipticity properties of the control operator (see [9]).

Under suitable assumptions, the wave system is smoothly controllable (see Lemmas 4 and 5).

The second ingredient is following lemma, which concerns the regularity of the inhomogeneous problem.

Lemma 1 *Fix $k \in \mathbb{N}$ and take $f \in H^k((0,T); H)$ such that*

$$\begin{cases} \frac{d^j}{dt^j} f(0) = 0, & \forall\, j \in \{0, \dots, k\} \\ f(t) = 0, & a.e.\ t \in (\tau, T), \end{cases} \tag{14}$$

with $0 < \tau < T$. Consider y solution to the problem

$$\begin{cases} \frac{d}{dt} y = Ay + f & t \in (0,T) \\ y(0) = 0. \end{cases} \tag{15}$$

Then, $y \in \cap_{j=0}^{k} C^j([\tau, T]; D(A^{k-j}))$ and

$$\sum_{j=0}^{k} \|y\|_{C^j([\tau,T];D(A^{k-j}))} \leq C \|f\|_{H^k((0,T);H)},$$

the constant C depending only on k.

Remark 3 Note that the maximal regularity of the solution is only assured for $t \geq \tau$, after the right hand side term f vanishes.

The proof of this Lemma is given in an Appendix at the end of this paper.

2.1 Steady State Controllability

In this subsection, we study the constrained controllability for some steady states. Recall s and V are given by (12). Before introducing our main result, we suppose:

(H_1) the system (10) is smoothly controllable in time T_0 for some $T_0 > 0$.

(H_2) $\mathcal{U}_{\text{ad}}$ is a closed and convex cone with vertex at 0 and $\text{int}^V(\mathcal{U}_{\text{ad}} \cap V) \neq \varnothing$, where int^V denotes the interior set in the topology of V.

Furthermore, we define the following subset

$$\mathscr{W} = \mathrm{int}^V(\mathscr{U}_{\mathrm{ad}} \cap V) + \mathscr{U}_{\mathrm{ad}}. \tag{16}$$

(Note that, since $\mathscr{U}_{\mathrm{ad}}$ is a convex cone, then $\mathscr{W} \subset \mathscr{U}_{\mathrm{ad}}$.) The main result of this subsection is the following. The solution to (10) with initial datum y_0 and control u is denoted by $y(\cdot; y_0, u)$.

Theorem 6 (Steady state controllability). *Assume* (H_1) *and* (H_2) *hold. Let* $\{(y_i, \overline{u}^i)\}_{i=0}^{1} \subset H \times \mathscr{W}$ *satisfying*

$$Ay_i + B\overline{u}^i = 0, \qquad i = 0, 1.$$

Then there exists $T > T_0$ *and* $u \in L^2(0, T; U)$ *such that*

- $u(t) \in \mathscr{U}_{ad}$ *a.e. in* $(0, T)$*;*
- $y(T; y_0, u) = y_1$.

Remark 4 As we shall see, in the application to the wave equation with positivity constraints:

- for *internal control,* $U = L^2(\omega)$ and $V = H^{s(n)}(\omega)$, with $s = s(n) = \lfloor n/2 \rfloor + 1$;
- for *boundary control,* $U = L^2(\Gamma)$ and $V = H^{s(n)-\frac{1}{2}}(\Gamma)$, where $s(n) = \lfloor n/2 \rfloor + 1$.

$\mathscr{U}_{\mathrm{ad}}$ is the set of nonnegative controls in U. In both cases, $\mathscr{W}$ is nonempty and contains controls u in $L^2(\omega)$ (resp. $L^2(\Gamma)$) such that $u \geq \sigma$, for some $\sigma > 0$. For this to happen, it is essential that $H^{s(n)}(\omega) \hookrightarrow C^0(\overline{\omega})$ (resp. $H^{s(n)-\frac{1}{2}}(\Gamma) \hookrightarrow C^0(\overline{\Gamma})$). This is guaranteed by our special choice of $s = s(n)$. Furthermore, in these special cases:

$$\overline{\mathscr{W}}^U = \mathscr{U}_{\mathrm{ad}},$$

where $\overline{\mathscr{W}}^U$ is the closure of $\mathscr{W}$ in the space U.

In the remainder of the present subsection we prove Theorem 6. The following Lemma is essential for the proof of Theorem 6. Fix $\rho \in C^\infty(\mathbb{R})$ such that

$$\mathrm{Range}(\rho) \subseteq [0, 1], \quad \rho \equiv 1 \quad \text{over } (-\infty, 0] \quad \text{and} \quad \mathrm{supp}(\rho) \subset\subset (-\infty, 1/2). \tag{17}$$

Lemma 2 *Assume that the system* (10) *is smoothly controllable in time* T_0*, for some* $T_0 > 0$*. Let* $(\eta_0, \overline{v}^0) \in H \times U$ *be a steady state, i.e. solution to* (11) *with control* $\overline{v}^0$*. Then, there exists* $w \in L^\infty((1, T_0 + 1); V)$ *such that the control*

$$v(t) = \begin{cases} \rho(t)\overline{v}^0 & \text{in } (0, 1) \\ w & \text{in } (1, T_0 + 1) \end{cases} \tag{18}$$

drives (10) *from* η_0 *to* 0 *in time* $T_0 + 1$*. Furthermore,*

$$\|w\|_{L^\infty((1,T_0+1);V)} \leq C\|\eta_0\|_H. \tag{19}$$

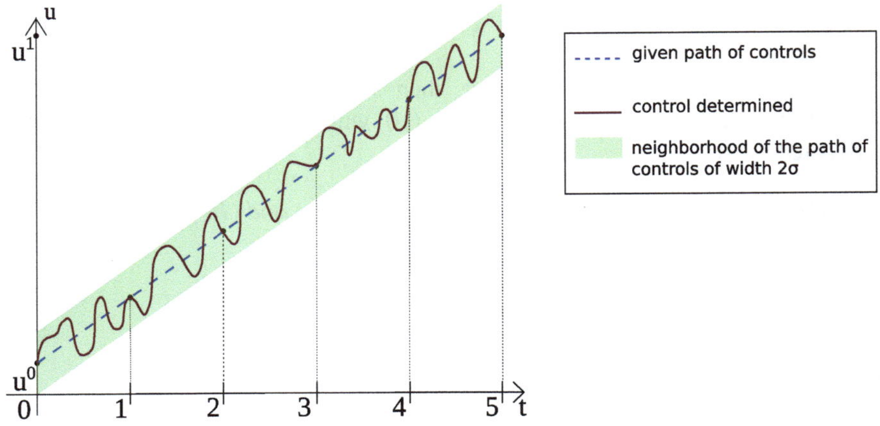

Fig. 2 Stepwise procedure

The proof of the above Lemma can be found in the Appendix.

We prove now Theorem 6, by developing a “stair-case argument” (see Fig. 2).

Proof (Proof of Theorem 6)

Let $\left\{(y_i,\overline{u}^i)\right\}_{i=0}^1$ satisfy

$$Ay_i + B\overline{u}^i = 0 \quad \forall\, i \in \{0,1\}\,. \tag{20}$$

By the definition of $\mathscr{W}$, there exists $\left\{(q^i,z^i)\right\}_{i=0}^1 \subset \mathrm{int}^V(\mathscr{U}_{\mathrm{ad}} \cap V) \times \mathscr{U}_{\mathrm{ad}}$ such that

$$\overline{u}^i = q^i + z^i \quad i = 0,1. \tag{21}$$

Define the segment joining y_0 and y_1

$$\gamma(s) = (1-s)y_0 + sy_1 \quad \forall\, s \in [0,1].$$

For each $s \in [0,1]$, $\gamma(s)$ solves

$$A\gamma(s) + B(q(s)+z(s)) = 0 \quad \forall\, i \in \{0,1\}\,.$$

where $(q(s),z(s)) \in \mathrm{int}^V(\mathscr{U}_{\mathrm{ad}} \cap V) \times \mathscr{U}_{\mathrm{ad}}$ are defined by:

$$q(s) = (1-s)q^0 + sq^1 \quad \text{and} \quad z(s) = (1-s)z^0 + sz^1 \quad \forall\, s \in [0,1].$$

The rest of the proof is divided into two steps.

Step 1 **Show that there exists $\delta > 0$, such that for each $s \in [0,1]$, $q(s) + B^V(0,\delta) \subset \mathrm{int}^V(\mathscr{U}_{\mathrm{ad}} \cap V)$, where $B^V(0,\delta)$ denotes the closed ball in V, centered at 0 and of radius δ.**

Define

$$f(s) = \inf_{y \in V \setminus \text{int}^V(\mathscr{U}_{\text{ad}} \cap V)} \|q(s) - y\|_V, \quad s \in [0, 1]. \tag{22}$$

One can check that f is Lipschitz continuous over the compact interval $[0, 1]$. Then, by Weierstrass' Theorem, we have that

$$\min_{s \in [0,1]} f(s) > 0.$$

Choose $0 < \delta < \min_{s \in [0,1]} f(s)$. Hence, by (22), it follows that, for each $s \in [0, 1]$,

$$q(s) + B^V(0, \delta) \subset \text{int}^V(\mathscr{U}_{\text{ad}} \cap V),$$

as required.

Step 2 **Conclusion**.

Let $C > 0$ be given by Lemma 2. Let $\delta > 0$ be given by Step 1. Choose $N_0 \in \mathbb{N} \setminus \{0\}$ such that

$$N_0 > \frac{2C\|y_0 - y_1\|_H}{\delta}. \tag{23}$$

For each $k \in \{0, \dots, N_0\}$, define:

$$y_k = \left(1 - \frac{k}{N_0}\right) y_0 + \frac{k}{N_0} y_1 \quad \text{and} \quad u_k = \left(1 - \frac{k}{N_0}\right) \overline{u}^0 + \frac{k}{N_0} \overline{u}^1. \tag{24}$$

It is clear that, by (21), for each $k \in \{0, \dots, N_0 - 1\}$,

$$\|y_k - y_{k+1}\|_H = \frac{1}{N_0} \|y_0 - y_1\|_H \quad \text{and} \quad u_k - q\left(\frac{k}{N_0}\right) \in \mathscr{U}_{\text{ad}}. \tag{25}$$

Arbitrarily fix $k \in \{0, \dots, N_0 - 1\}$. Take $\eta_0 = y_k - y_{k+1}$ and $\overline{v}^0 = u_k - u_{k+1}$. Then, we apply Lemma 2, getting a control $w_k \in L^\infty(1, T_0 + 1; V)$ such that

$$y(T_0 + 1; y_k - y_{k+1}, \hat{v}_k) = 0 \tag{26}$$

and

$$\|w_k\|_{L^\infty(1, T_0+1; V)} \leq C\|y_k - y_{k+1}\|_H, \tag{27}$$

where

$$\hat{v}_k(t) = \begin{cases} \rho(t)(u_k - u_{k+1}) & t \in (0, 1] \\ w_k(t) & t \in (1, T_0 + 1). \end{cases} \tag{28}$$

Define

$$v_k(t) = \begin{cases} \rho(t)(u_k - u_{k+1}) + u_{k+1} & t \in (0, 1] \\ w_k(t) + u_{k+1} & t \in (1, T_0 + 1). \end{cases} \tag{29}$$

At the same time, by (20) and (24), we have

$$Ay^{k+1} + Bu_{k+1} = 0 \quad \text{and} \quad y(T_0 + 1; y_{k+1}, u_{k+1}) = y_{k+1}.$$

The above, together with (26), (28) and (29), yields

$$\begin{aligned} y(T_0 + 1; y_k, v_k) &= y(T_0 + 1; y_k - y_{k+1}, \hat{v}_k) + y(T_0 + 1; y_{k+1}, u_{k+1}) \\ &= y_{k+1}. \end{aligned} \tag{30}$$

Next, we claim that

$$v_k(t) \in \mathscr{U}_{\mathrm{ad}} \quad \text{for a.e. } t \in (0, T_0 + 1). \tag{31}$$

To this end, by (16) and since $\mathscr{U}_{\mathrm{ad}}$ is a convex cone, we have

$$\mathscr{W} \quad \text{is convex and} \quad \mathscr{W} \subset \mathscr{U}_{\mathrm{ad}}. \tag{32}$$

By (17), $0 \leq \rho(t) \leq 1$ for all $t \in \mathbb{R}$. Then, by (29) an (32), it follows that, for a.e $t \in (0, 1)$,

$$v_k(t) = \rho(t)u_k + (1 - \rho(t))u_{k+1} \in \rho(t)\mathscr{W} + (1 - \rho(t))\mathscr{W} \subset \mathscr{W} \subset \mathscr{U}_{\mathrm{ad}}.$$

At this stage, to show (31), it remains to prove that

$$v_k(t) \in \mathscr{U}_{\mathrm{ad}} \quad \text{for a.e. } t \in (1, T_0 + 1). \tag{33}$$

Take $t \in (1, T_0 + 1)$. By (27), (25) and (23), we have

$$\|w_k(t)\|_V \leq \frac{C}{N_0}\|y_0 - y_1\|_H \leq \delta/2.$$

From this and Step 1, it follows

$$w_k(t) + q\left(\frac{k+1}{N_0}\right) \in \mathrm{int}^V(\mathscr{U}_{\mathrm{ad}} \cap V).$$

By this, (25), (29) and (16), we get, for a.e. t in $(1, T_0 + 1)$,

$$
\begin{aligned}
v_k(t) &= w_k(t) + u_{k+1} \\
&= w_k(t) + q\left(\frac{k+1}{N_0}\right) + \left(u^{k+1} - q\left(\frac{k+1}{N_0}\right)\right) \\
&\in \mathrm{int}^V(\mathscr{U}_{\mathrm{ad}} \cap V) + \mathscr{U}_{\mathrm{ad}} \\
&= \mathscr{W}.
\end{aligned}
$$

From this and (32), we are led to (33). Therefore, the claim (31) is true.

Finally, define

$$
u(t) = v_k(t - k(T_0+1)),\ \forall\, t \in [k(T_0+1), (k+1)(T_0+1)),\ k \in \{0, \ldots, N_0 - 1\}.
$$

Then, from (30) and (31), the conclusion of this theorem follows. □

In Sects. 3.1 and 4.1, we apply the above Theorem to prove Theorems 1 and 3 respectively. In particular,

- for internal control,

$$
\mathscr{U}_{\mathrm{ad}} = \left\{u \in L^2(\omega) \mid u \geq 0,\ \text{a.e. } \omega\right\};
$$

- for boundary control,

$$
\mathscr{U}_{\mathrm{ad}} = \left\{u \in L^2(\Gamma) \mid u \geq 0,\ \text{a.e. } \Gamma\right\}.
$$

Then, in both cases, $\mathscr{U}_{\mathrm{ad}}$ is closed convex cone with vertex at 0.

Nevertheless, the above techniques can be adapted in a wide variety of contexts.

2.2 *Controllability Between Trajectories*

In this subsection, we study the constrained controllability for some general states lying on trajectories of the system with possibly nonzero time derivative. Recall s and V are given by (12). Before introducing our main result, we assume:

(H_1') the system (10) is smoothly controllable in time T_0 for some $T_0 > 0$.

(H_2') the set $\mathscr{U}_{\mathrm{ad}}$ is a closed and convex and $\mathrm{int}^V(\mathscr{U}_{\mathrm{ad}} \cap V) \neq \varnothing$, where int^V denotes the interior set in the topology of V;

(H_3') the operator A generates a C_0-group $\{\mathbb{T}_t\}_{t\in\mathbb{R}}$ over H and $\|\mathbb{T}_t\|_{\mathscr{L}(H,H)} = 1$ for all $t \in \mathbb{R}$. Furthermore, A is invertible from $D(A)$ to H, with continuous inverse.

The main result of this subsection is the following. The notation $y(\cdot; y_0, u)$ stands for the solution of the abstract controlled Eq. (10) with control u and initial data y_0.

Theorem 7 *Assume (H_1'), (H_2') and (H_3') hold. Let $\overline{y}_i \in C^s(\mathbb{R}; H)$ be solutions to* (10) *with controls $\overline{u}^i \in L^2_{loc}(\mathbb{R}; U)$ for $i = 0, 1$. Assume $\overline{u}^i(t) \in \mathscr{U}_{ad}$ for a.e. $t \in \mathbb{R}$. Let $\tau_0, \tau_1 \in \mathbb{R}$. Then, there exists $T > 0$ and $u \in L^2(0, T; U)$ such that*

- $y(T;\overline{y}_0(\tau_0),u)=\overline{y}_1(\tau_1)$;
- $u(t)\in\mathscr{U}_{ad}$ *for a.e.* $t\in(0,T)$.

Remark 5 (i) Roughly, Theorem 7 addresses the constrained controllability for all initial data y_0 and final target y_1, with $y_0,\ y_1\in E$, where

$$E=\left\{y(\tau)\ \middle|\ \tau\in\mathbb{R},\ y\in C^s(\mathbb{R};H)\ \text{ and }\ \exists\, u\in L^2_{loc}(\mathbb{R};U),\right.$$

$$\left.\text{with }\ u(t)\in\mathscr{U}_{\rm ad}\ \text{ a.e. } t\in\mathbb{R}\ \text{ s.t. }\ \frac{d}{dt}y(t)=Ay(t)+Bu(t),\quad t\in\mathbb{R}\right\}.$$

By Lemma 1, one can check that

$$\left\{y(\tau;0,u)\ \middle|\ \tau\in\mathbb{R},\ u\in C^s(\mathbb{R},\mathscr{U}_{\rm ad}),\ \frac{d^j}{dt^j}u(0)=0,\ j=0,\dots,s\right\}\subset E.$$

Furthermore, we observe that such set E includes some non-steady states.
(ii) There are at least two differences between Theorems 6 and 7. First of all, Theorem 6 studies constrained controllability for some steady states, whereas Theorem 7 can deal with constrained controllability for some non-steady states (see (i) of this remark). Secondly, in Theorem 7 the controls $\overline{u}^i$ $(i=0,1)$ defining the initial datum $\overline{y}^0(\tau_0)$ and final target $\overline{y}^1(\tau_1)$ are required to fulfill the constraint

$$\overline{u}^i(t)\in\mathscr{U}_{\rm ad},\quad \text{a.e. } t\in\mathbb{R},\ i=0,1,$$

while $\overline{u}^i$ in Theorem 6 is required to be in $\mathscr{W}\subsetneq\mathscr{U}_{\rm ad}$. (Then, in Theorem 7 we have weakened the constraints on $\overline{u}^i$. In particular, we are able to apply Theorem 7 to the wave system with nonnegative controls with final target $\overline{y}^1\equiv 0$.)

Before proving Theorem 7, we show a preliminary lemma. Note that such Lemma works with any *contractive* semigroup. In particular, it holds both for wave-like and heat-like systems. A similar result was proved in [17, 20]. For the sake of completeness, we provide the proof of the aforementioned lemma in the Appendix.

Lemma 3 (Null Controllability by small controls) *Assume that A generates a contractive C_0-semigroup $(\mathbb{T}_t)_{t\in\mathbb{R}^+}$ over H. Suppose that (H_1') holds. Let $\varepsilon>0$ and $\eta_0\in D(A^s)$. Then, there exists $\overline{T}=\overline{T}(\varepsilon,\|\eta_0\|_{D(A^s)})>0$ such that, for any $T\geq\overline{T}$, there exists a control $v\in L^\infty((0,T);V)$ such that*

- $y(T;\eta_0,v)=0$;
- $\|v\|_{L^\infty(\mathbb{R}^+;V)}\leq\varepsilon$.

The proof of the Lemma above is given in the Appendix.

We are now ready to prove Theorem 7.

With respect to Theorem 5 we have weakened the constraints on the controls defining the initial and final trajectories. Then, a priori, we have lost the room for oscillations needed in the proof of that Theorem. We shall see how to recover this by

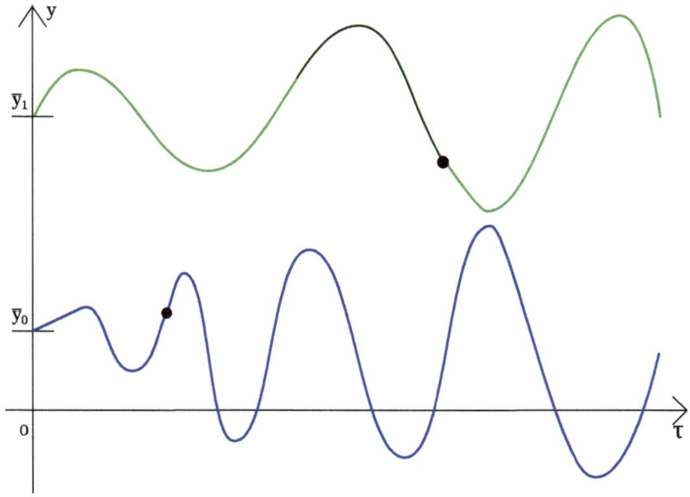

Fig. 3 The two original trajectories. The time τ parameterizing the trajectories is just a parameter independent of the control time t

modifying the initial and final trajectories away from the initial and final data (see Figs. 3, 4 and 5).

Proof (Proof of Theorem 7*)* The main strategy of proof is the following:

(i) we reduce the constrained controllability problem (with initial data $\overline{y}_0(\tau_0)$ and final target $\overline{y}_1(\tau_1)$) to another controllability problem (with initial datum $\hat{y}_0$ and final target 0);
(ii) we solve the latter controllability problem by constructing two controls. The first control is used to improve the regularity of the solution. The second control is small in a regular space and steers the system to rest.

Step 1 **The part (i) of the above strategy**.
For each $T > 0$, we aim to define a new trajectory with the final state $\overline{y}_1(\tau_1)$ as value at time $t = T$. Choose a smooth function $\zeta \in C^\infty(\mathbb{R})$ such that

$$\zeta \equiv 1 \text{ over } \left(-\frac{1}{2}, \frac{1}{2}\right) \quad \text{and} \quad \text{supp}(\zeta) \subset\subset (-1, 1). \tag{34}$$

Take $\sigma \in \text{int}^V(\mathscr{U}_{\text{ad}} \cap V)$. Arbitrarily fix $T > 1$. Define a control

$$\hat{u}_T^1(t) = \zeta(t-T)\overline{u}^1(t-T+\tau_1) + (1-\zeta(t-T))\sigma. \tag{35}$$

We denote by φ_T the unique solution to the problem

$$\begin{cases} \frac{d}{dt}\varphi(t) = A\varphi(t) + B\hat{u}_T^1(t) & t \in \mathbb{R} \\ \varphi(T) = \overline{y}_1(\tau_1). \end{cases} \tag{36}$$

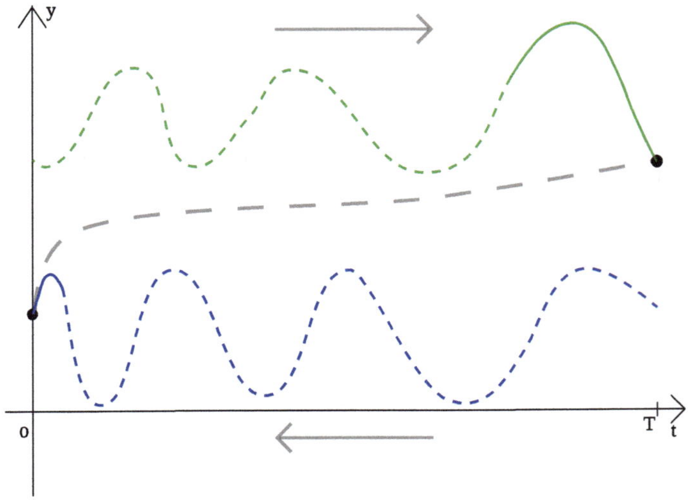

Fig. 4 The new trajectories to be linked, now synchronized with the control time t. Note that (1) we have translated the time parameter defining the trajectories and (2) we have modified them away from the initial and the final data, to apply Lemma 3. The new initial trajectory is represented in blue, while the new final trajectory is drawn in green. The modified part is dashed. Following the notation of the proof of Theorem 7, the new initial trajectory is $y(\cdot; \hat{u}^0, \overline{y}_0(\tau_0))$, while the new final trajectory is φ_T

In what follows, we will construct two controls which send $\overline{y}_0(\tau_0) - \varphi_T(0)$ to 0 in time T, which is part (ii) of our strategy. Recall that ρ is given by (17). We define

$$\hat{u}^0(t) = \rho(t)\overline{u}^0(t + \tau_0) + (1 - \rho(t))\sigma \quad t \in \mathbb{R}.$$

Step 2 **Estimate of** $\|y(1; \overline{y}_0(\tau_0) - \varphi_T(0), \hat{u}^0 - \hat{u}_T^1)\|_{D(A^s)}$
We take the control $(\hat{u}^0 - \hat{u}_T^1)\restriction_{(0,1)}$ to be the first control mentioned in part (ii) of our strategy. In this step, we aim to prove the following regularity estimate associated with this control: there exists a constant $C > 0$ independent of T and σ such that

$$\|y(1; \overline{y}_0(\tau_0) - \varphi_T(0), \hat{u}^0 - \hat{u}_T^1)\|_{D(A^s)} \tag{37}$$
$$\leq C\left[\|\overline{y}_0\|_{C^s([\tau_0,\tau_0+1];H)} + \|\overline{y}_1\|_{C^s([\tau_1-1,\tau_1];H)} + \|\sigma\|_U\right].$$

To begin, we introduce ψ the solution to

$$A\psi + B\sigma = 0. \tag{38}$$

First, we have that

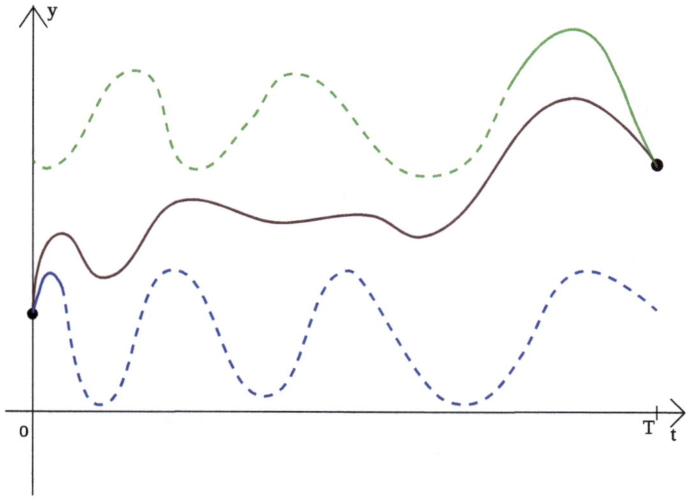

Fig. 5 The new trajectories linked by the controlled trajectory y, pictured in red. As in Fig. 4, the new initial trajectory is drawn in blue, while the new final trajectory is represented in green

$$\begin{aligned} & y(1; \overline{y}(\tau_0) - \varphi_T(0), \hat{u}^0 - \hat{u}_T^1) \\ &= y(1; \overline{y}(\tau_0), \hat{u}^0) - y(1; \varphi_T(0), \hat{u}^1) \\ &= [y(1; \overline{y}(\tau_0), \hat{u}^0) - \psi] - [y(1; \varphi_T(0), \hat{u}_T^1) - \psi] \\ &= y(1; \overline{y}(\tau_0) - \psi, \hat{u}^0 - \sigma) - y(1; \varphi_T(0) - \psi, \hat{u}_T^1 - \sigma). \end{aligned} \tag{39}$$

To estimate (37), we need to compute the norms of the last two terms in (39), in the space $D(A^s)$. We claim that there exists $C_1 > 0$ (independent of T and σ) such that

$$\|y(1; \overline{y}(\tau_0) - \psi, \hat{u}^0 - \sigma)\|_{D(A^s)} \leq C_1 \left(\|\overline{y}_0\|_{C^s([\tau_0, \tau_0+1]; H)} + \|\sigma\|_U \right). \tag{40}$$

To this end, we show that

$$y(t; \overline{y}(\tau_0) - \psi, \hat{u}^0 - \sigma) = \rho(t)(\overline{y}^0(t + \tau_0) - \psi) + \eta_2(t), \quad t \in \mathbb{R}, \tag{41}$$

where η_2 solves

$$\begin{cases} \frac{d}{dt}\eta_2(t) = A\eta_2(t) - \rho'(\overline{y}(t + \tau_0) - \psi) & t \in \mathbb{R} \\ \eta_2(0) = 0. \end{cases} \tag{42}$$

Indeed,

$$\begin{aligned} & \frac{d}{dt} \left[\rho(t)(\overline{y}^0(t + \tau_0) - \psi) + \eta_2(t) \right] \\ &= \rho(t)(A\overline{y}^0(t + \tau_0) + B\overline{u}^0(t + \tau_0)) + \rho'(t)(\overline{y}^0(t + \tau_0) - \psi) \end{aligned}$$

$$
\begin{aligned}
&+ A\eta_2(t) - \rho'(t)(\overline{y}^0(t+\tau_0) - \psi) \\
&= A(\rho(t)\overline{y}^0(t+\tau_0) + \eta_2(t)) + B\left(\rho(t)\overline{u}^0(t+\tau_0)\right) \\
&= A(\rho(t)(\overline{y}^0(t+\tau_0) - \psi) + \eta_2(t)) + \rho(t)A\psi + B\left(\rho(t)\overline{u}^0(t+\tau_0)\right) \\
&= A(\rho(t)(\overline{y}^0(t+\tau_0) - \psi) + \eta_2(t)) + B\left(\rho(t)(\overline{u}^0(t+\tau_0) - \sigma)\right) \\
&= A(\rho(t)(\overline{y}^0(t+\tau_0) - \psi) + \eta_2(t)) + B(\hat{u}^0(t) - \sigma).
\end{aligned} \tag{43}
$$

At the same time, since $\rho(0) = 1$, from (42), it follows that

$$
\rho(t)(\overline{y}^0(t+\tau_0) - \psi) + \eta_2(t)\restriction_{t=0} = \overline{y}^0(\tau_0) - \psi.
$$

From this and (43), we are led to (41).

Next, we will use (41) and (42) to prove (40). To this end, since we assumed $\overline{y}^0 \in C^s(\mathbb{R}; H)$ and ψ is independent of t, we get that

$$
\overline{y}^0(\cdot + \tau_0) - \psi \in C^s(\mathbb{R}; H).
$$

By this, we apply Lemma 1 obtaining the existence of $\hat{C}_1 > 0$ (independent of T and σ) such that

$$
\|\eta_2(1)\|_{D(A^s)} \leq \hat{C}_1 \left(\|\overline{y}^0\|_{C^s([\tau_0,\tau_0+1];H)} + \|\psi\|_H\right). \tag{44}
$$

At the same time, since $\rho(1) = 0$ (see (17)), by (41), we have that

$$
y(1; \overline{y}(T_0) - \psi, \hat{u}^0 - \sigma) = \eta_2(1).
$$

This, together with (44) and (38), yields (40).

At this point, we estimate the norm of the second term in (39) in the space $D(A^s)$, namely we prove the existence of $C_2 > 0$ (independent of T and σ) such that

$$
\|y(1; \varphi_T(0) - \psi, \hat{u}_T^1 - \sigma)\|_{D(A^s)} \leq C_2 \left[\|\overline{y}^1\|_{C^s([\tau_1-1,\tau_1];H)} + \|\sigma\|_U\right]. \tag{45}
$$

To this end, as in the proof of (37), we get that

$$
y(t; \varphi_T(0) - \psi, \hat{u}_T^1 - \sigma) = \zeta(t-T)(\overline{y}^1(t-T+\tau_1) - \psi) + \tilde{\eta}_2(t), \quad t \in \mathbb{R}, \tag{46}
$$

where $\tilde{\eta}_2$ solves

$$
\begin{cases}
\frac{d}{dt}\tilde{\eta}_2(t) = A\tilde{\eta}_2(t) - \zeta'(t-T)(\overline{y}^1(t-T+\tau_1) - \psi) \quad t \in \mathbb{R} \\
\tilde{\eta}_2(T) = 0.
\end{cases} \tag{47}
$$

We will use (46) and (47) to prove (45). Indeed, set

$$
\hat{\eta}(t) = \tilde{\eta}_2(T-t).
$$

By definition of $\hat{\eta}$, we have

$$\begin{cases} \frac{d}{dt}\hat{\eta}(t) = -A\hat{\eta}(t) + \zeta'(-t)(\overline{y}^1(\tau_1 - t) - \psi) & t \in \mathbb{R} \\ \hat{\eta}(0) = 0. \end{cases} \tag{48}$$

Since we have assumed $\overline{y}^1 \in C^s(\mathbb{R}, H)$ and ψ is independent of t (see (38)), we have

$$\overline{y}^1 - \psi \in C^s(\mathbb{R}; H).$$

Recall that $\zeta(t) \equiv 1$ in $\left(-\frac{1}{2}, \frac{1}{2}\right)$ (see (34)). Then, $\zeta'(t) = 0$, for each $t \in \left(-\frac{1}{2}, \frac{1}{2}\right)$. Now, by hypothesis (H_3'), A generates a group of operators. Hence, we can apply Lemma 1 to (48) getting the existence of $\tilde{C}_2 > 0$ (independent of T and σ) such that

$$\|\hat{\eta}(1)\|_{D(A^s)} \leq \tilde{C}_2 \left(\|\overline{y}^1\|_{C^s([\tau_1-1,\tau_1];H)} + \|\psi\|_H\right),$$

whence

$$\|\tilde{\eta}_2(T-1)\|_{D(A^s)} \leq \tilde{C}_2 \left(\|\overline{y}^1\|_{C^s([\tau_1-1,\tau_1];H)} + \|\psi\|_H\right). \tag{49}$$

At the same time, by (H_3') and some computations, we have that

$$\|\mathbb{T}_t\|_{\mathscr{L}(D(A^s),D(A^s))} = 1, \quad \text{for each } t \in \mathbb{R}.$$

Since $\zeta(t-T) = 0$, for each $t \in [0, T-1]$ (see (34)), the above, together with (46) and (47), yields

$$\|y(1; \varphi_T(0) - \psi, \hat{u}_T^1 - \sigma)\|_{D(A^s)} = \|\tilde{\eta}_2(1)\|_{D(A^s)} = \|\tilde{\eta}_2(T-1)\|_{D(A^s)}.$$

This, together with (49) and (38), leads to (45).

Step 3 **Conclusion**.

In this step, we will first construct the second control mentioned in part (ii) of our strategy. Then we put together the first and second controls (mentioned in part (ii)) to get the conclusion.

By (45),

$$\|y(1; \varphi_T(0) - \psi, \hat{u}_T^1 - \sigma)\|_{D(A^s)} \leq C_2 \left[\|\overline{y}^1\|_{C^s([\tau_1-1,\tau_1];H)} + \|\sigma\|_U\right].$$

The above estimate is independent of T. Then for each $T > 0$, by Lemma 3, there exists

$$\overline{T} = \overline{T}(\sigma, \|\overline{y}^0\|_{C^s([\tau_0,\tau_0+1];H)}, \|\overline{y}^1\|_{C^s([\tau_1-1,\tau_1];H)}) > 0$$

and $w_T \in L^\infty(\mathbb{R}^+; V)$ such that

$$\begin{cases} \frac{d}{dt}z(t) = Az(t) + Bw_T(t) & t \in (1, \overline{T}) \\ z(1) = y(1; \overline{y}(\tau_0) - \varphi_T(0), \hat{u}^0 - \hat{u}_T^1), \quad z(\overline{T}) = 0 \end{cases} \tag{50}$$

and

$$\|w_T\|_{L^\infty(1,\overline{T};V)} \leq \frac{1}{2} \inf_{y \in V \setminus \mathrm{int}^V(\mathscr{U}_{\mathrm{ad}} \cap V)} \|\sigma - y\|_V. \tag{51}$$

Note that the last constant is positive, because σ is taken from $\mathrm{int}^V(\mathscr{U}_{\mathrm{ad}})$. Choose $T = \overline{T} + 1$. Define a control:

$$v = \begin{cases} \hat{u}^0(t) & t \in (0,1) \\ w_T(t) + \hat{u}_T^1(t) & t \in (1, \overline{T}) \\ \hat{u}_T^1(t) & t \in (\overline{T}, \overline{T}+1). \end{cases} \tag{52}$$

We aim to show that

$$y(\overline{T}+1; \overline{y}^0(\tau_0), v) = \overline{y}^0(\tau_1) \quad \text{and} \quad v(t) \in \mathscr{U}_{\mathrm{ad}} \text{ a.e. } t \in (1, \overline{T}+1). \tag{53}$$

To this end, by (52), (50) and (36), we get that

$$\begin{aligned} y(\overline{T}+1; \overline{y}^0(\tau_0), v) &= y(\overline{T}+1; \overline{y}^0(\tau_0) - \varphi_T(0), v - \hat{u}_T^1) + y(\overline{T}+1; \varphi_T(0), \hat{u}_T^1) \\ &= \mathbb{T}_1(z_T(\overline{T})) + \varphi_T(\overline{T}+1) \\ &= \overline{y}^1(\tau_1). \end{aligned}$$

This leads to the first conclusion of (53). It remains to show the second condition in (53). Arbitrarily fix $t \in (0,1)$. By (52) and (45), we have

$$\begin{aligned} v(t) &= \rho(t)\overline{u}^0(t+\tau_0) + (1-\rho(t))\sigma \\ &\in \rho(t)\mathscr{U}_{\mathrm{ad}} + (1-\rho(t))\mathscr{U}_{\mathrm{ad}} \subset \mathscr{U}_{\mathrm{ad}}. \end{aligned}$$

Choose also an arbitrary $s \in (1, \overline{T})$. By (52), (51) and (35), we obtain

$$\begin{aligned} v(s) &= w(s) + (1 - \zeta(s - \overline{T} - 1))\sigma + \zeta(s - \overline{T} - 1)\overline{u}^1(s - \overline{T} - 1 + \tau_1) \\ &= w(s) + \sigma \in \mathrm{int}^V(\mathscr{U}_{\mathrm{ad}} \cap V) \subset \mathscr{U}_{\mathrm{ad}}. \end{aligned}$$

Take any $t \in (\overline{T}, \overline{T}+1)$. We find from (52) and (35) that

$$\begin{aligned} v(t) &= \zeta(t - \overline{T} - 1)\overline{u}^1(t - \overline{T} - 1 + \tau_1) + (1 - \zeta(t - \overline{T} - 1))\sigma \\ &\in \zeta(t - \overline{T} - 1)\mathscr{U}_{\mathrm{ad}} + (1 - \zeta(t - \overline{T} - 1))\mathscr{U}_{\mathrm{ad}} \\ &\subset \mathscr{U}_{\mathrm{ad}}. \end{aligned}$$

Therefore, we are led to the second conclusion of (53). This ends the proof. □

3 Internal Control: Proof of Theorems 1 and 2

The present section is organized as follows:

- Section 3.1: proof of Lemma 4 and Theorem 1;
- Section 3.2: proof of Theorem 2;
- Section 3.3: discussion of the issues related to the internal control touching the boundary.

3.1 *Proof of Theorem 1*

We now prove Theorem 1 by employing Theorem 6.

Firstly, we place our control system in the abstract framework introduced in Sect. 2 and we prove that our control system is smoothly controllable (see Definition 2).

The free dynamics is generated by $A : D(A) \subset H \longrightarrow H$, where

$$A = \begin{pmatrix} 0 & I \\ -A_0 & 0 \end{pmatrix}, \qquad \begin{cases} H = H_0^1(\Omega) \times L^2(\Omega) \\ D(A) = \left(H^2(\Omega) \cap H_0^1(\Omega)\right) \times H_0^1(\Omega). \end{cases} \tag{54}$$

where $A_0 = -\Delta + cI : H^2(\Omega) \cap H_0^1(\Omega) \subset L^2(\Omega) \longrightarrow L^2(\Omega)$. The control operator

$$B(v) = \begin{pmatrix} 0 \\ \chi v. \end{pmatrix}$$

defined from $U = L^2(\omega)$ to $H = H_0^1(\Omega) \times L^2(\Omega)$ is bounded, then admissible.

Lemma 4 *In the above framework take $V = H^{s(n)}(\omega)$ and $s = s(n) = \lfloor n/2 \rfloor + 1$. Assume further (Ω, ω_0, T^*) fulfills the Geometric Control Condition. Then, the control system* (1) *is smoothly controllable in any time $T_0 > T^*$.*

The proof of this Lemma can be found in the reference [9, Theorem 5.1].

We are now ready to prove Theorem 1.

Proof (of Theorem 1*)* We choose as set of admissible controls:

$$\mathscr{U}_{\text{ad}} = \left\{u \in L^2(\omega) \mid u \geq 0, \text{ a.e. } \omega\right\}.$$

Then,

$$\bigcup_{\sigma>0} \left\{u \in L^2(\omega) \mid u \geq \sigma, \text{ a.e. } \omega\right\} \subset \mathscr{W}. \tag{55}$$

We highlight that, to prove (55), we need $H^{s(n)}(\omega) \hookrightarrow C^0(\overline{\omega})$. For this reason, we have chosen $s(n) = \lfloor n/2 \rfloor + 1$.

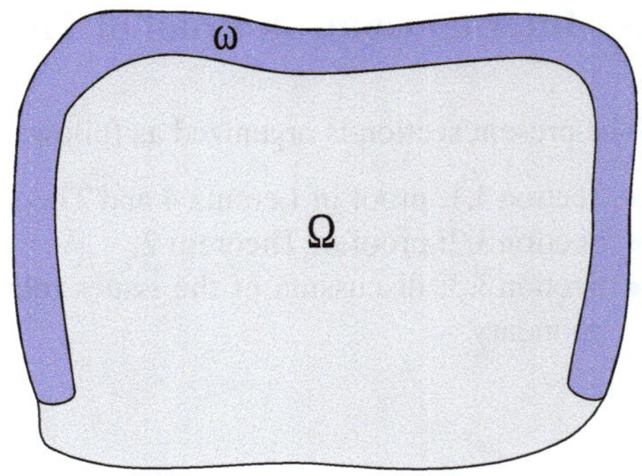

Fig. 6 Controlling from the interior touching the boundary

By Lemma (4), we have that the system is *Smoothly Controllable* with $s = s(n) = \lfloor n/2 \rfloor + 1$ and $V = H^{s(n)}(\omega)$. Then, by Theorem 6 we conclude. □

3.2 *Proof of Theorem 2*

We prove now Theorem 2

Proof (Proof of Theorem 2*).* As we have seen, our system fits the abstract framework. Moreover, we have checked in Lemma 4 that the system is *Smoothly Controllable* with $s(n) = \lfloor n/2 \rfloor + 1$ and $V = H^{s(n)}(\omega)$. Furthermore, $\mathrm{int}^V(\mathscr{U}_{\mathrm{ad}} \cap V) \neq \varnothing$. Indeed, any constant $\sigma > 0$ belongs to $\mathrm{int}^V(\mathscr{U}_{\mathrm{ad}} \cap V)$, since $H^{s(n)}(\omega) \hookrightarrow C^0(\overline{\omega})$. This is guaranteed by our choice of $s(n) = \lfloor n/2 \rfloor + 1$.

Therefore, we are in position to apply Theorem 7 and finish the proof. □

3.3 *Internal Controllability From a Neighborhood of the Boundary*

So far, we have assumed that the control is localized by means of a smooth cut-off function χ so that all its derivatives vanish on the boundary of Ω. This implies that χ must be constant on any connected component of the boundary. This prevents us to localize the internal control in a region touching the boundary only on a subregion, as in Fig. 6.

In this case, as already pointed out in [8], some difficulties in finding regular controls may arise. Indeed, as indicated both in [8] and in [9] a crucial property needs to be verified in order to have controls in $C^0([0, T]; H^s(\omega))$, namely

$$BB^*(D(A^*)^k) \subset D(A^k) \tag{56}$$

for $k = 0, \dots, s$, where we have used the notation of the proof of Theorem 1.

Right now, for any $k \in \mathbb{N}$ we have

$$D(A^k) = \left\{ \begin{pmatrix} \psi_1 \\ \psi_2 \end{pmatrix} \middle| \begin{matrix} \psi_1 \in H^{k+1}(\Omega), \ \Delta^j \psi_1 = 0 \text{ on } \partial\Omega, \quad 0 \le j \le \lfloor k/2 \rfloor \\ \psi_2 \in H^k(\Omega), \ \Delta^j \psi_2 = 0 \text{ on } \partial\Omega, \ 0 \le j \le \lfloor (k+1)/2 \rfloor - 1 \end{matrix} \right\},$$

while

$$D((A^*)^k) = \left\{ \begin{pmatrix} \psi_1 \\ \psi_2 \end{pmatrix} \middle| \begin{matrix} \psi_1 \in H^k(\Omega), \ \Delta^j \psi_1 = 0 \text{ on } \partial\Omega, \ 0 \le j \le \lfloor (k-1)/2 \rfloor \\ \psi_2 \in H^{k-1}(\Omega), \ \Delta^j \psi_2 = 0 \text{ on } \partial\Omega, \ 0 \le j \le \lfloor k/2 \rfloor - 1 \end{matrix} \right\}. \tag{57}$$

Furthermore,

$$BB^* = \begin{pmatrix} 0 & 0 \\ \chi^2 & 0 \end{pmatrix}$$

Then, (56) is verified if and only if for any $\psi \in H^s(\Omega)$ such that

$$(\Delta)^j(\psi) = 0, \quad 0 \le j \le \lfloor (s-1)/2 \rfloor, \quad \text{a.e. on } \partial\Omega$$

the following hold

$$(\Delta)^j(\chi^2 \psi) = 0, \quad 0 \le j \le \lfloor (s-1)/2 \rfloor, \quad \text{a.e. on } \partial\Omega. \tag{58}$$

Choosing χ so that all its normal derivatives vanish on $\partial\Omega$

- in case $s < 5$, we are able to prove (56). Then, by adapting the techniques of [9, Theorem 5.1], we have that our system is *Smoothly Controllable* (Definition 2), with $s(n) = \lfloor n/2 \rfloor + 1$. This enables us to prove Theorem 1 in space dimension $n < 8$.
- in case $s \ge 5$, in (58) the biharmonic operator $(\Delta)^2$ enters into play. By computing it in normal coordinates on the boundary, some terms appear involving the curvature and $\frac{\partial}{\partial \xi_k} \chi \frac{\partial}{\partial \nu} \psi$, where $(\xi_1, \dots, \xi_{n-1})$ are tangent coordinates, while ν is the normal coordinate. In general, these terms do not vanish, unless $\partial\Omega$ is flat. Then, for $n \ge 8$, we are unable to deduce a constrained controllability result in case the internal control is localized along a subregion of $\partial\Omega$.

4 Boundary Control: Proof of Theorems 3, 4 and 5

This section is devoted to boundary control and is organized as follows:

- Section 4.1: proof of Lemma 5 and Theorem 3;
- Section 4.2: proof of Theorem 4;
- Section 4.3: proof of Theorem 5.

4.1 Proof of Theorem 3

We prove Theorem 3.

First of all, we explain how our boundary control system fits the abstract semigroup setting described in Sect. 2. The generator of the free dynamics is:

$$A = \begin{pmatrix} 0 & I \\ -A_0 & 0 \end{pmatrix}, \qquad \begin{cases} H = L^2(\Omega) \times H^{-1}(\Omega) \\ D(A) = H_0^1(\Omega) \times L^2(\Omega), \end{cases} \tag{59}$$

where $A_0 = -\Delta + cI : H_0^1(\Omega) \subset H^{-1}(\Omega) \longrightarrow H^{-1}(\Omega)$. The definition of the control operator is subtler than in the internal control case. Let Δ_0 be the Dirichlet Laplacian. Then, the control operator

$$B(v) = \begin{pmatrix} 0 \\ -\Delta_0 \tilde{z} \end{pmatrix}, \quad \text{where} \begin{cases} -\Delta \tilde{z} = 0 & \text{in } \Omega \\ \tilde{z} = \chi v(\cdot, t) & \text{on } \partial\Omega. \end{cases}$$

defined from $L^2(\Gamma)$ to $H^{-\frac{3}{2}}(\Omega)$. In this case, B is unbounded but admissible (see [15] or [23, proposition 10.9.1 page 349]).

Lemma 5 *In the above framework, set* $V = H^{s(n)-\frac{1}{2}}(\Gamma)$ *and* $s = s(n)$*, with* $s(n) = \lfloor n/2 \rfloor + 1$*. Suppose (GCC) holds for* (Ω, Γ_0, T^*)*. Then, in any time* $T_0 > T^*$*, the control system* (5) *is smoothly controllable in time* T_0.

One can prove the above Lemma, by employing [9, Theorem 5.4].

Proof (Proof of Theorem 3*)* We prove our Theorem, by choosing the set of admissible controls:

$$\mathscr{U}_{ad} = \left\{ u \in L^2(\Gamma) \mid u \geq 0, \text{ a.e. } \Gamma \right\}.$$

Hence,

$$\bigcup_{\sigma>0} \left\{ u \in L^2(\Gamma) \mid u \geq \sigma, \text{ a.e. } \Gamma \right\} \subset \mathscr{W}. \tag{60}$$

Note that, in order to show (60), it is essential that the embedding $H^{s(n)-\frac{1}{2}}(\Gamma) \hookrightarrow C^0(\overline{\Gamma})$ is continuous. This is guaranteed by the choice $s(n) = \lfloor n/2 \rfloor + 1$.

By Lemma 5, we conclude that smooth controllability holds. At this point, it suffices to apply Theorem 6 to conclude. □

4.2 Proof of Theorem 4

We prove now Theorem 4.

Proof (Proof of Theorem 4) We have explained above how our control system (5) fits the abstract framework presented in Sect. 2. Furthermore, by Lemma 5, the system is *Smoothly Controllable* with $s(n) = \lfloor n/2 \rfloor + 1$ and $V = H^{s(n)-\frac{1}{2}}(\Gamma)$. Moreover, the set $\text{int}^V(\mathscr{U}_{\text{ad}} \cap V)$ is non empty, since all constants $\sigma > 0$ belong to it. This is consequence of the continuity of $H^{s(n)-\frac{1}{2}}(\Gamma) \hookrightarrow C^0(\overline{\Gamma})$, valid for $s(n) = \lfloor n/2 \rfloor + 1$. The result holds as a consequence of Theorem 7. □

4.3 State Constraints. Proof of Theorem 5

We conclude this section proving Theorem 5 about state constraints. The following result is needed.

Lemma 6 *Let $s \in \mathbb{N}^*$ and $T > T^*$. Take a steady state solution η_0 associated to the control $v^0 \in H^{s-\frac{1}{2}}(\Gamma)$. Then, there exists $v \in \cap_{j=0}^{s} C^j([0, T]; H^{s-\frac{1}{2}-j}(\Gamma))$ such that the unique solution (η, η_t) to (5) with initial datum $(\eta_0, 0)$ and control v is such that $(\eta(T, \cdot), \eta_t(T, \cdot)) = (0, 0)$. Furthermore,*

$$\sum_{j=0}^{s} \|v\|_{C^j([0,T]; H^{s-\frac{1}{2}-j}(\Gamma))} \leq C(T) \|v^0\|_{H^{s-\frac{1}{2}}(\Gamma)}, \tag{61}$$

the constant C being independent of η_0 and v^0. Finally, if $s = s(n) = \lfloor n/2 \rfloor + 1$, then the control $v \in C^0([0, T] \times \overline{\Gamma})$ and

$$\|v\|_{C^0([0,T]\times\overline{\Gamma})} \leq C(T) \|v^0\|_{H^{s(n)-\frac{1}{2}}(\Gamma)}. \tag{62}$$

The above Lemma can be proved by using the techniques of Lemma 2. We now prove our Theorem about *state constraints*.

Proof (of Theorem 5)

Step 1 **Consequences of Lemma** 6.

Let $T_0 > T^*$, T^* being the critical time given by the *Geometric Control Condition*. By Lemma 6, for any $\varepsilon > 0$, there exists $\delta_\varepsilon > 0$ such that for any pair of steady states y_0 and y_1 defined by regular controls $\overline{u}^i \in H^{s(n)-\frac{1}{2}}(\Gamma)$, such that:

$$\|\overline{u}^1 - \overline{u}^0\|_{H^{s(n)-\frac{1}{2}}(\Gamma)} < \delta_\varepsilon \tag{63}$$

we can find a control u driving (10) from y_0 to y_1 in time T_0 and verifying

$$\sum_{j=0}^{s(n)} \|u - \overline{u}^1\|_{C^j([0,T_0];H^{s(n)-\frac{1}{2}-j}(\Gamma))} < \varepsilon, \tag{64}$$

where $\overline{u}^1$ is the control defining y_1. Moreover, if (y, y_t) is the unique solution to (5) with initial datum $(y_0, 0)$ and control u, we have

$$\|y - y_1\|_{C^0([0,T_0]\times\overline{\Omega})} \leq C\|y - y_1\|_{C^0([0,T_0];H^{s(n)}(\Omega))}$$

$$\leq C\sum_{j=0}^{s(n)} \|u - \overline{u}^1\|_{C^j([0,T_0];H^{s(n)-\frac{1}{2}-j}(\Gamma))} \leq C\varepsilon,$$

where we have used the boundedness of the inclusion $H^{s(n)}(\Omega) \hookrightarrow C^0(\overline{\Omega})$ and the continuous dependence of the data

.

Step 2 **Stepwise procedure and conclusion.**
We consider the convex combination $\gamma(s) = (1-s)y_0 + sy_1$. Then, let

$$z_k = \gamma\left(\frac{k}{\overline{n}}\right), \quad k = 0, \dots, \overline{n}$$

be a finite sequence of steady states defined by the control $\overline{u}_k = \frac{\overline{n}-k}{\overline{n}}\overline{u}^0 + \frac{k}{\overline{n}}\overline{u}^1$. Let $\delta > 0$. By taking $\overline{n}$ sufficiently large,

$$\|\overline{u}_k - \overline{u}_{k-1}\|_{H^{s(n)-\frac{1}{2}}(\Gamma)} < \delta. \tag{65}$$

By the above reasonings, choosing δ small enough, for any $1 \leq k \leq \overline{n}$, we can find a control u^k joining the steady states z_{k-1} and z_k in time T_0, with

$$\|y^k - z_k\|_{C^0([0,T_0]\times\overline{\Omega})} \leq \sigma,$$

where $(y^k, (y^k)_t)$ is the solution to (5) with initial datum z_{k-1} and control u^k. Hence,

$$y^k = y^k - z_k + z_k \geq -\sigma + \sigma = 0, \quad \text{on } (0, T_0) \times \Omega, \tag{66}$$

where we have used the maximum principle for elliptic equations (see [2]) to assert that $z^k \geq \sigma$ because $u_k \geq \sigma$.

By taking the traces in (66), we have $u^k \geq 0$ for $1 \leq k \leq \overline{n}$.

In conclusion, the control $u : (0, \overline{n}T_0) \longrightarrow H^{s(n)-\frac{1}{2}}(\Gamma)$ defined as $u(t) = u_k(t - (k-1)T_0)$ for $t \in ((k-1)T_0, kT_0)$ is the required one. This finishes the proof. □

5 The One Dimensional Wave Equation

We consider the one dimensional wave equation, controlled from the boundary

$$\begin{cases} y_{tt} - y_{xx} = 0 & (t,x) \in (0,T) \times (0,1) \\ y(t,0) = u_0(t), \ y(t,1) = u_1(t) & t \in (0,T) \\ y(0,x) = y_0^0(x), \ y_t(0,x) = y_0^1(x). & x \in (0,1) \end{cases} \tag{67}$$

As in the general case, by transposition (see [15]), for any initial datum $(y_0^0, y_0^1) \in L^2(0,1) \times H^{-1}(0,1)$ and controls $u_i \in L^2(0,T)$, the above problem admits an unique solution $(y, y_t) \in C^0([0,T]; L^2(0,1) \times H^{-1}(0,1))$.

We show how Theorem 4 reads in this one-dimensional setting, in the special case where both the initial trajectory $(\overline{y}_0, (\overline{y}_0)_t)$ and the final one $(\overline{y}_1, (\overline{y}_1)_t)$ are constant (independent of x) steady states.

We determine explicitly a pair of *nonnegative* controls steering (67) from one positive constant to the other. The controlled solution remains *nonnegative*.

In this special case, we show further that

- the minimal controllability time is the same, regardless whether we impose the positivity constraint on the control or not;
- constrained controllability holds in the minimal time.

The minimal controllability time for (67) is defined as follows.

Let $(y_0^0, y_0^1) \in L^2(0,1) \times H^{-1}(0,1)$ be an initial datum and $(y_1^0, y_1^1) \in L^2(0,1) \times H^{-1}(0,1)$ be a final target. Then the minimal controllability time without constraints is defined as follows:

$$T_{\min} \stackrel{\text{def}}{=} \inf\left\{T > 0 \,\middle|\, \exists u_i \in L^2(0,T), \ (y(T,\cdot), y_t(T,\cdot)) = (y_1^0, y_1^1)\right\}. \tag{68}$$

Similarly, the minimal time under *positivity* constraints on the *control* is defined as:

$$T_{\min}^c \stackrel{\text{def}}{=} \inf\left\{T > 0 \,\middle|\, \exists u_i \in L^2(0,T)^+, \ (y(T,\cdot), y_t(T,\cdot)) = (y_1^0, y_1^1)\right\}. \tag{69}$$

Finally, we introduce the minimal time with constraints on the state and and the control:

$$T_{\min}^s \stackrel{\text{def}}{=} \inf\left\{T > 0 \,\middle|\, \exists u_i \in L^2(0,T)^+, \ (y(T,\cdot), y_t(T,\cdot)) = (y_1^0, y_1^1), \ y \geq 0\right\}. \tag{70}$$

The problem of controllability of the one-dimensional wave equation under bilateral constraints on the control has been studied in [12]. In the next Proposition, we concentrate on unilateral constraints and we compute explicitly the minimal time for the specific data considered.

Proposition 1 *Let* $(y_0^0, 0)$ *be the initial datum and* $(y_1^0, 0)$ *be the final target, with* $y_0^0 \in \mathbb{R}^+$ *and* $y_1^0 \in \mathbb{R}^+$. *Then,*

1. *for any time $T > 1$, there exists two nonnegative controls*

$$u_0(t) = \begin{cases} y_0^0 & t \in [0, 1) \\ (y_1^0 - y_0^0)\frac{t-1}{T-1} + y_0^0 & t \in (1, T] \end{cases} \tag{71}$$

$$u_1(t) = \begin{cases} (y_1^0 - y_0^0)\frac{t}{T-1} + y_0^0 & t \in [0, T-1) \\ y_1^0 & t \in [T-1, T] \end{cases} \tag{72}$$

driving (67) *from* $(y_0^0, 0)$ *to* $(y_1^0, 0)$ *in time* T. *Moreover, the corresponding solution remains nonnegative, i.e.*

$$y(t, x) \geq 0, \quad \forall (t, x) \in [0, T] \times [0, 1].$$

2. $T^s_{min} = T^c_{min} = T_{min} = 1$;
3. *the nonnegative controls* $\hat{u}_0 \equiv y_0^0$ *and* $\hat{u}_1 \equiv y_1^0$ *in* $L^2(0, 1)$ *steers* (67) *from* $(y_0^0, 0)$ *to* $(y_1^0, 0)$ *in the minimal time. Furthermore, the corresponding solution* $y \geq 0$ *a.e. in* $(0, 1) \times (0, 1)$;
4. *the controls in the minimal time are not unique. In particular, for any* $\lambda \in [0, 1]$, $\hat{u}_\lambda^0 = (1 - \lambda)y_0^0 + \lambda y_1^0$ *and* $\hat{u}_\lambda^1 = (1 - \lambda)y_1^0 + \lambda y_0^0$ *drives* (67) *from* $(y_0^0, 0)$ *to* $(y_1^0, 0)$ *in the minimal time.*

Proof We proceed in several steps.

Step 1. **Proof of the constrained controllability in time $T > 1$.**
By D'Alembert's formula, the solution (y, y_t) to (67) with initial datum $(y_0^0, 0)$ and controls u_i defined in (71) and (72), reads as

$$y(t, x) = f(x + t), \quad (t, x) \in [0, T] \times [0, 1],$$

where

$$f(\xi) = \begin{cases} y_0^0 & \xi \in [0, 1) \\ (y_1^0 - y_0^0)\frac{\xi-1}{T-1} + y_0^0 & \xi \in [1, T) \\ y_1^0. & \xi \in [T, T+1]. \end{cases}$$

This finishes the proof of (1.).
Step 2 **Computation of the minimal time.**
In any time $T > 1$, controllability under state and control constraints holds. Then, $T_{\min} \leq T^c_{\min} \leq T^s_{\min} \leq 1$.

It remains to prove that $T_{\min} \geq 1$. This can be obtained by adapting the techniques of [18, Proposition 4.1].

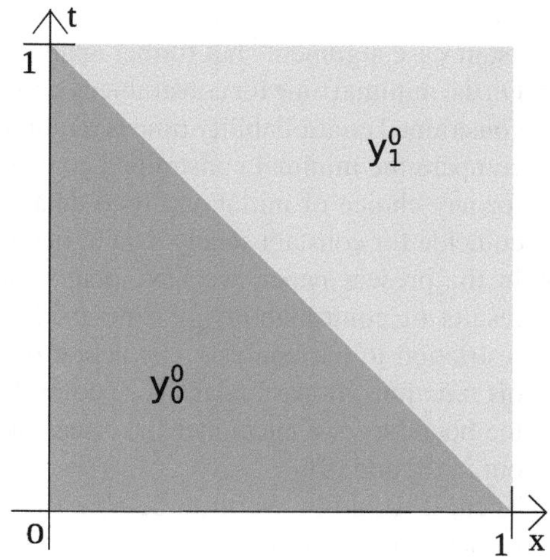

Fig. 7 Level sets of the solution to (67) with initial datum $(y_0^0, 0)$ and controls $\hat{u}^i$. In the darker region the solution takes value y_0^0, while in the complement it coincides with y_1^0

Step 3 **Controllability in the minimal time.**
One can check (see Fig. 7) that the unique solution to (67) with initial datum $(y_0^0, 0)$ and controls $\hat{u}^i$ is

$$y(t,x) = \begin{cases} y_0^0 & t+x<1 \\ y_1^0 & t+x>1 \end{cases} \tag{73}$$

This concludes the argument. □

6 Conclusions and Open Problems

In this paper we have analyzed the controllability of the wave equation under *positivity* constraints on the control and on the state.

1. In the general case (without assuming that the energy defines a norm), we have shown how to steer the wave equation from one steady state to another in time large, provided that both steady states are defined by positive controls, away from zero;
2. in case the energy defines a norm, we have generalized the above result to data lying on trajectories. Furthermore, the controls defining the trajectory are supposed to be only nonnegative, thus allowing us to take as target $(y_1^0, y_1^1) = (0, 0)$.

We present now some open problems, which as long as we know, have not been treated in the literature so far.

- Further analysis of controllability of the wave under *state* constraints. As pointed out in [16, 19], in the case of parabolic equations a state constrained result follows from a control constrained one by means of the comparison principle. For the wave equation, such principle does not hold. We have proved Theorem 5, using a "stair-case argument" but further analysis is required.

- On the minimal time for constrained controllability. Further analysis of the minimal constrained controllability time is required. In particular, it would be interesting to compare the minimal constrained controllability time and the unconstrained one for any choice of initial and final data. As we have seen in Proposition 1, they coincide for constant steady data in one space dimension.
- In the present paper, we have determined nonnegative controls by employing results of controllability of smooth data by smooth controls. This imposes a restriction to our analysis: the action of the control is localized by smooth cut-off functions. In particular, when controlling (1) from an interior subset touching the boundary, we encounter the issues discussed in Sect. 3.3 and already pointed out in [8] and [9].
  Then, it would be worth to be able to build nonnegative controls without using smooth controllability.
- Derive the Optimality System (OS) for the controllability of the wave by nonnegative controls.
- Extend our results to the semilinear setting, by employing the analysis carried out in [4, Theorem 1.3], [5, 6, 25].
- Extend the results to more general classes of potentials c. For instance, one could assume c to be bounded, instead of C^∞ smooth.

Acknowledgements This research was supported by the Advanced Grant DyCon (Dynamical Control) of the European Research Council Executive Agency (ERC), the MTM2014-52347 and MTM2017-92996 Grants of the MINECO (Spain) and the ICON project of the French ANR-16-ACHN-0014. The authors thank Yubiao Zhang for his helpful revision of the paper.

Appendix

Regularity results

In what follows, H is a real Hilbert space and $A : D(A) \subset H \longrightarrow H$ is a generator of a C^0-semigroup.

Lemma 7 *Let $k \in \mathbb{N}$. Take $y \in C^k([0,T]; H) \cap H^{k+1}((0,T); H_{-1})$ solution to the homogeneous equation:*

$$\frac{d}{dt}y = Ay, \quad t \in (0,T). \tag{74}$$

Then, $y \in \cap_{j=0}^{k} C^j([0,T]; D(A^{k-j}))$ and

$$\sum_{j=0}^{k} \|y\|_{C^j([0,T];D(A^{k-j}))} \leq C(k)\|y\|_{C^k([0,T];H)},$$

the constant $C(k)$ depending only on k.

The proof of the above Lemma can be done by using the Eq. (74) (see [2]).
We prove now Lemma 1.

Proof (Proof of Lemma 1*) Step 1* **Time regularity**
By induction on $j = 0, \ldots, k$, we prove that $y \in C^j([0, T]; H)$ and

$$\|y\|_{C^j([0,T];H)} \leq C\|f\|_{H^j((0,T);H)}.$$

For $j = 0$, the validity of the assertion is a consequence of classical semigroup theory (e.g. [23, Proposition 4.2.5] with control space $U = H$ and control operator $B = Id_H$). Assume now that the result hold up to $j - 1$. Then, let w solution to

$$\begin{cases} \frac{d}{dt}w = Aw + f' & t \in (0, T) \\ w(0) = 0. \end{cases} \tag{75}$$

By induction assumption, $w \in C^{j-1}([0, T]; H)$ and the corresponding estimate holds. Then, $\tilde{y}(t) = \int_0^t w(\sigma)d\sigma \in C^j([0, T]; H)$ and

$$\|\tilde{y}\|_{C^j([0,T];H)} \leq C\|f\|_{H^j((0,T);H)}.$$

Then, it remains to show that $y = \tilde{y}$. Now, for any $t \in [0, T]$

$$\tilde{y}(t) - \tilde{y}(0) = \int_0^t [w(\sigma) - w(0)]d\sigma = \int_0^t \int_0^\sigma [Aw(\xi) + f'(\xi)]d\xi d\sigma$$
$$= \int_0^t [A\tilde{y}(\sigma) + f(\sigma)]d\sigma.$$

By uniqueness of solution to (15), we have $y = \tilde{y}$. This finishes the first step.
Step 2 **Conclusion**
We start observing that y solves

$$y_t = Ay, \qquad t \in (\tau, T).$$

Then, by classical semigroup arguments (see [2, Chapter 7]), we conclude. □

Proof of Lemma 2
We give the proof of Lemma 2.

Proof (Proof of Lemma 2*)* Let v be given by (18). The proof is made of two steps.
Step 1 **Show that** $y(1; \eta_0, v) \in D(A^s)$, **with** s **given by** (12)
We apply Lemma 1 with $y = y(\cdot; \eta_0, \rho\overline{v}^0) - \rho\eta_0$ and $f = \rho'\eta_0$, getting

$$y(1; \eta_0, \rho\overline{v}^0) - \rho\eta_0 \in D(A^s).$$

Since $\rho\eta_0 = 0$ over $(\delta, 1)$, for some $\delta \in (0, 1)$, we have that

$$y(1; \eta_0, \rho\overline{v}^0) \in D(A^s).$$

Step 2 **Conclusion**
Since $y(1; \eta_0, \rho\overline{v}^0) \in D(A^s)$, we are in position to apply the smooth controllability (see Definition 2) and determine $w \in L^\infty((1, T_0 + 1); V)$ steering the solution to (10) from $y(1; \eta_0, v)$ at time $t = 1$ to 0 at time $t = T_0 + 1$.
Hence, the desired control v reads as (18).

Finally, by similar reasonings the estimate (19) follows. This ends the proof of this Lemma. □

Proof of Lemma 3
We prove now Lemma 3.

Proof (Proof of Lemma 3*)* We split the proof in two steps.
Step 1 **Proof of the inequality** $\|\mathbb{T}_t\|_{\mathscr{L}(D(A^s),D(A^s))} \leq 1$ **with** $t \in \mathbb{R}^+$
Recall that

$$\|x\|^2_{D(A^s)} = \sum_{j=0}^{s} \|A^j x\|^2_H \quad \forall\, x \in D(A^s).$$

Now, for any $x \in D(A^s)$ and $t \in \mathbb{R}^+$, we have

$$\|A^j \mathbb{T}_t x\|_H = \|\mathbb{T}_t A^j x\|_H \leq \|A^j x\|_H \quad \forall\, j = 0, \ldots, s.$$

This yields $\|\mathbb{T}_t\|_{\mathscr{L}(D(A^s),D(A^s))} \leq 1$ for any $t \in \mathbb{R}^+$.
Step 2 **Conclusion.**
Let $C > 0$ be given by (2). Take

$$N > \frac{C\|\eta_0\|_{D(A^s)}}{\varepsilon}. \tag{76}$$

Arbitrarily fix $k \in \{0, \ldots, N-1\}$. Consider the following equation

$$\begin{cases} \frac{d}{dt}y(t) = Ay(t) + B\chi_{(kT_0,(k+1)T_0)}(t)u_k(t) & t \in \mathbb{R}^+ \\ y(0) = \frac{1}{N}\eta_0, \end{cases} \tag{77}$$

where $\chi_{(kT_0,(k+1)T_0)}$ is the characteristic function of the set $(kT_0, (k+1)T_0)$ and $u_k \in L^2(\mathbb{R}^+, V)$. From step 1 and (76), we have that

$$\|y(kT_0; (1/N)\eta_0, 0)\|_{D(A^s)} \leq (1/N)\|\eta_0\|_{D(A^s)} \leq \varepsilon. \tag{78}$$

Then, we apply smooth controllability (given by (H_1')) to find some control $\hat{u}_k \in L^\infty(\mathbb{R}^+; V)$ so that the solution to (77) with control $u_k = \hat{u}_k$ satisfies

$$y((k+1)T_0; (1/T_0)\eta_0, \chi_{(kT_0,(k+1)T_0)}\hat{u}_k) = 0 \quad \text{and} \quad \|\hat{u}_k\|_{L^\infty((kT_0,(k+1)T_0);V)} \le \varepsilon. \tag{79}$$

Now, we define:

$$v(t) = \sum_{k=0}^{N-1} \chi_{(kT_0,(k+1)T_0)}(t)u_k(t) \quad t \in \mathbb{R}^+. \tag{80}$$

Then, from (79) and (80), we know

$$y(NT_0; \eta_0, v) = 0 \quad \text{and} \quad \|v\|_{L^\infty((0,NT_0);V)} \le \varepsilon.$$

This leads to the conclusion where $\overline{T} = NT_0$. □

References

1. Bardos, C., Lebeau, G., Rauch, J.: Sharp sufficient conditions for the observation, control, and stabilization of waves from the boundary. SIAM J. Control. Optim. **30**, 1024–1065 (1992)
2. Brezis, H.: Functional Analysis, Sobolev Spaces and Partial Differential Equations. Universitext. Springer, New York (2010)
3. Burq, N., Gérard, P.: Condition nécessaire et suffisante pour la contrôlabilité exacte des ondes. Comptes Rendus l'Académie Sci.-Ser. I-Math. **325**, 749–752 (1997) (Elsevier)
4. Cannarsa, P., Komornik, V., Loreti, P.: Well posedness and control of semilinear wave equations with iterated logarithms. ESAIM: Control, Optim. Calc. Var. **4**, 37–56 (1999) (EDP Sciences)
5. Cannarsa, P., Komornik, V., Loreti, P.: Controllability of semilinear wave equations with infinitely iterated logarithms. Control Cybern. **28**, 449–461 (1999) (EDP Sciences)
6. Cannarsa, P., Komornik, V., Loreti, P.: One-sided and internal controllability of semilinear wave equations with infinitely iterated logarithms. Discret. Contin. Dyn. Syst. **8**, 745–756 (2002)
7. Coron, J.-M., Trélat, E.: Global steady-state stabilization and controllability of 1D semilinear wave equations. Commun. Contemp. Math. **8**, 535–567 (2006)
8. Dehman, B., Lebeau, G.: Analysis of the HUM control operator and exact controllability for semilinear waves in uniform time. SIAM J. Control Optim. **48**, 521–550 (1995)
9. Ervedoza, Sylvain, Zuazua, Enrique: A systematic method for building smooth controls for smooth data. Discrete Contin. Dyn. Syst. Ser. B **14**, 1375–1401 (2010)
10. Evans, L.C.: Partial Differential Equations. Graduate Studies in Mathematics. American Mathematical Society (2010)
11. Gilbarg, D., Trudinger, N.S.: Elliptic Partial Differential Equations of Second Order. Springer, Berlin (2001)
12. Gugat, M., Grimm, V.: Optimal boundary control of the wave equation with pointwise control constraints. Comput. Optim. Appl. **49**, 123–147 (2011) (Springer)
13. Lions, J.L., Magenes, E.: Problmes aux Limites Non Homognes et Applications. Grundlehren der mathematischen Wissenschaften. Springer, Berlin (1968)
14. Lions, J.L.: Optimal Control of Systems Governed by Partial Differential Equations. Grundlehren der mathematischen Wissenschaften. Springer, Berlin (1971)
15. Lions, J.L.: Exact controllability, stabilization and perturbations for distributed systems. SIAM Rev. **30**, 1–68 (1988)
16. Lohéac, J., Trélat, E., Zuazua, E.: Minimal controllability time for the heat equation under unilateral state or control constraints. Math. Model. Methods Appl. Sci. **27**, 1587–1644 (2017)
17. Lohéac, J., Tucsnak, M.: Maximum principle and bang-bang property of time optimal controls for Schrodinger-type systems. SIAM J. Control Optim. **51**, 4016–4038 (2013)

18. Lohéac, J., Zuazua, Z.: Norm saturating property of time optimal controls for wave-type equations. IFAC-PapersOnLine **49**, 37–42 (2016) (Elsevier)
19. Pighin, D., Zuazua, E.: Controllability under positivity constraints of semilinear heat equations (2017). arXiv preprint arXiv:1711.07678
20. Phung, K.D., Wang, G., Zhang, X.: On the existence of time optimal controls for linear evolution equations. Discret. Contin. Dyn. Syst. Ser. B **8**, 925 (2007) (CiteSeerX)
21. Russell, D.L.: A unified boundary controllability theory for hyperbolic and parabolic partial differential equations. Stud. Appl. Math. **52**, 189–211 (1973) (Wiley Online Library)
22. Russell, D.L.: Controllability and stabilizability theory for linear partial differential equations: recent progress and open questions. Siam Rev. **20** (1978)
23. Tucsnak, M., Weiss, G.: Observation and Control for Operator Semigroups. Springer Science & Business Media, Berlin (2009)
24. Zuazua, E.: Exact controllability for the semilinear wave equation. J. Mathématiques Pures Appliquées **69**, 1–31 (1990) (Elsevier)
25. Zuazua, E.: Exact controllability for semilinear wave equations in one space dimension. Ann. l'Institut Henri Poincare (C) Non Linear Anal., 109–129 (1993) (Elsevier)
26. Zhang, X., Zuazua, E.: Exact controllability of the semi-linear wave equation. Unsolved Problems in Mathematical Systems and Control Theory, p. 173. Princeton University Press, New Jersey (2004)

Asymptotic Analysis of a Cucker–Smale System with Leadership and Distributed Delay

Cristina Pignotti and Irene Reche Vallejo

Abstract We extend the analysis developed in Pignotti and Reche Vallejo (J Math Anal Appl 464:1313–1332, 2018) [34] in order to prove convergence to consensus results for a Cucker–Smale type model with hierarchical leadership and distributed delay. Flocking estimates are obtained for a general interaction potential with divergent tail. We analyze also the model when the ultimate leader can change its velocity. In this case we give a flocking result under suitable conditions on the leader's acceleration.

Keywords Cucker–Smale model · Flocking · Time delay

1 Introduction

The celebrated Cucker–Smale model has been introduced in [14, 15] as a model for flocking, namely for phenomena where autonomous agents reach a consensus based on limited environmental information. Let us consider $N \in \mathbb{N}$ agents and let $(x_i(t), v_i(t)) \in \mathbb{R}^{2d}$, $i = 1, \ldots, N$, be their phase-space coordinates. As usual $x_i(t)$ denotes the position of the ith agent and $v_i(t)$ the velocity. The Cucker–Smale model reads, for $t > 0$,

$$\begin{aligned} \dot{x}_i(t) &= v_i(t), \\ \dot{v}_i(t) &= \sum_{j=1}^{N} \psi_{ij}(t)(v_j(t) - v_i(t)), \qquad i = 1, \ldots, N, \end{aligned} \tag{1.1}$$

C. Pignotti (✉) · I. Reche Vallejo
Dipartimento di Ingegneria e Scienze dell'Informazione e Matematica, Università di L'Aquila, Via Vetoio, Loc. Coppito, 67010 L'Aquila, Italy
e-mail: pignotti@univaq.it

I. Reche Vallejo
e-mail: irenereche92@gmail.com

F. Alabau-Boussouira et al. (eds.), *Trends in Control Theory and Partial Differential Equations*, Springer INdAM Series 32,
https://doi.org/10.1007/978-3-030-17949-6_12

where the communication rates $\psi_{ij}(t)$ are of the form

$$\psi_{ij}(t) = \psi(|x_i(t) - x_j(t)|), \tag{1.2}$$

being $\psi : [0, +\infty) \to (0, +\infty)$ a suitable non-increasing potential functional.

Definition 1.1 We say that a solution of (1.1) converges to consensus (or flocking) if

$$\sup_{t>0} |x_i(t) - x_j(t)| < +\infty \quad \text{and} \quad \lim_{t\to+\infty} |v_i(t) - v_j(t)| = 0, \quad \forall\, i, j = 1, \dots, N. \tag{1.3}$$

The potential function considered by Cucker and Smale in [14, 15] is $\psi(s) = \frac{1}{(1+s^2)^\beta}$ with $\beta \geqslant 0$. They proved that there is unconditional convergence to flocking whenever $\beta < 1/2$. In the case $\beta \geqslant 1/2$, they obtained a conditional flocking result, namely convergence to flocking under appropriate assumptions on the initial data. Actually, unconditional flocking can be obtained also for $\beta = 1/2$ [20].

The extension of the flocking result to cover the case of non-symmetric communication rates is due to Motsch and Tadmor [30]. Other variants and generalizations have been proposed, e.g. more general interaction potentials, cone-vision constraints, leadership [10, 12, 21, 29, 31, 36, 38, 40], stochastic terms [13, 18, 19], pedestrian crowds [11, 23], infinite-dimensional kinetic models [1, 2, 4, 7, 17, 22, 37] and control models [3, 5, 6, 33, 39].

Here, we consider the Cucker–Smale system with hierarchical leadership introduced by Shen [36]. In this model the agents are ordered in a specific way, depending on which other agents they are leaders of or led by. This reflects natural situations, e.g. in animals groups, where some agents are more influential than the others. We also add a distributed delay term (cf. [32]), namely we assume that the agent i adjusts its velocity depending on the information received from other agents on a time interval $[t - \tau, t]$. Indeed, it is natural to assume that there is a time delay in the information's transmission from an agent to the others. The case of CS-model with hierarchical leadership and a pointwise time delay has been recently studied by the authors [34]. Other models with (pointwise) time delay, without leadership, have been considered in [8, 9, 28, 35], while for other extensions of Shen's results, without delay, we refer to [16, 24–27].

In order to present our model, we first recall some definitions from [36].

Definition 1.2 The **leader set** $\mathcal{L}(i)$ of an agent i in a flock $[1, 2, \dots, N]$ is the subgroup of agents that directly influence agent i.

The Cucker–Smale system considered by Shen is then, for all $i \in \{1, \dots, N\}$ and $t > 0$,

$$\begin{aligned} \frac{dx_i}{dt} &= v_i, \\ \frac{dv_i}{dt} &= \sum_{j\in\mathcal{L}(i)} \psi_{ij}(t)(v_j - v_i). \end{aligned} \tag{1.4}$$

The interaction potential ψ_{ij}, for $j \in \mathcal{L}(i)$, was analogous to the one of Cucker and Smale's papers, namely

$$\psi_{ij}(t) = \frac{1}{(1 + |x_i(t) - x_j(t)|^2)^\beta}, \quad j \in \mathcal{L}(i).$$

Note that if $j \notin \mathcal{L}(i)$ then the agent j does not influence the dynamics of the agent i; we say $\psi_{ij} = 0$ if $j \notin \mathcal{L}(i)$. For such a model Shen proved convergence to consensus for $\beta < 1/2$.

Definition 1.3 A flock $[1, \ldots, N]$ is an **HL-flock**, namely a flock under hierarchical leadership, if the agents can be ordered in such a way that:

1. if $j \in \mathcal{L}(i)$ then $j < i$, and
2. for all $i > 1$, $\mathcal{L}(i) \neq \emptyset$.

Definition 1.4 For each agent $i = 1, \ldots, N$, we define the m-th level leaders of i as

$$\mathcal{L}^0(i) = \{i\}, \quad \mathcal{L}^1(i) = \mathcal{L}(i), \quad \mathcal{L}^2(i) = \mathcal{L}(\mathcal{L}(i)), \quad \ldots, \quad \mathcal{L}^m = \mathcal{L}(\mathcal{L}^{m-1}(i)), \quad \ldots$$

for $m \in \mathbb{N}$, and denote the set of all leaders of the agent i, direct or indirect, as

$$[\mathcal{L}](i) = \mathcal{L}^0(i) \cup \mathcal{L}^1(i) \cup \ldots$$

For a fixed positive time τ and for every $t > 0$, our system is the following:

$$\begin{aligned} \frac{dx_i}{dt}(t) &= v_i(t), \\ \frac{dv_i}{dt}(t) &= \sum_{j \in \mathcal{L}(i)} \int_{t-\tau}^{t} \mu(t-s)\psi_{ij}(s)[v_j(s) - v_i(t)]\, ds, \end{aligned} \tag{1.5}$$

for all $i \in \{1, \ldots, N\}$, with initial conditions, for $s \in [-\tau, 0]$,

$$\begin{aligned} x_i(s) &= x_i^0(s), \\ v_i(s) &= v_i^0(s), \end{aligned} \tag{1.6}$$

for some continuous functions x_i^0 and v_i^0, $i = 1, \ldots, N$. The communication rates are

$$\psi_{ij}(t) = \psi(|x_i(t) - x_j(t)|), \quad j \in \mathcal{L}(i),$$

for some non-increasing, positive, continuous interaction potential ψ. For further uses we define

$$\psi_{ij} = 0, \quad j \notin \mathcal{L}(i).$$

The weight function $\mu : [0, \tau] \to \mathbb{R}$ is assumed to be bounded and non–negative, with

$$\int_0^\tau \mu(s)ds = \mu_0 > 0. \tag{1.7}$$

We will prove a flocking result under the assumption

$$\int_0^{+\infty} \psi(s)ds = +\infty\,. \tag{1.8}$$

Then, our result extends and generalizes the one of Shen. Note that in [34] we have proved a flocking result in the case of a pointwise time delay. We can formally obtain the model studied in [34] if the weight $\mu(\cdot)$ is a Dirac delta function centered at $t = \tau$.

The paper is organized as follows. In Sect. 2 we give some preliminary properties of system (1.5), in particular we prove the positivity and boundedness properties for the velocities. In Sect. 3 we will prove the flocking result for the system (1.5). Finally, in Sect. 4 we will consider the model under hierarchical leadership and a free-will leader and we will prove flocking estimates under suitable growth assumptions on the acceleration of the free-will leader.

2 Preliminary Properties

Before proving our main result, namely the convergence to consensus theorem, we need some general properties of the Cucker–Smale model (1.5), such as the positivity property and the boundedness of the velocities. The following propositions extend analogous results of [36].

Proposition 2.1 *Let us consider the system of scalar equations*

$$\begin{aligned} &\frac{du_i}{dt}(t) = \sum_{j\in\mathcal{L}(i)} \int_{t-\tau}^t \mu(t-s)\psi_{ij}(s)[u_j(s) - u_i(t)]\,ds, \quad i = 1, \dots, N,\ t > 0,\\ &u_i(s) = u_i^0(s), \quad i = 1, \dots, N,\ s \in [-\tau, 0], \end{aligned} \tag{2.1}$$

where $u_i^0(\cdot)$, $i = 1, \dots, N$, are continuous functions. If $u_i^0(s) \geqslant 0$ for all $i = 1, \dots, N$, and all $s \in [-\tau, 0]$, then $u_i(t) \geqslant 0$ for all i and $t > 0$.

Proof Observe that if an agent j is in the leader set $[\mathcal{L}](i)$ of the agent i, then it is not influenced by agents outside of $[\mathcal{L}](i)$. Thus, it is sufficient to prove the statement for the system (2.1) restricted to the agents in $[\mathcal{L}](i)$, for each $i = 1, \dots, N$.

We then proceed by induction. Consider the first agent, i.e. agent 1. By definition of an HL-flock, $\mathcal{L}(1) = \emptyset$, which gives

$$\frac{du_1}{dt} = 0 \text{ and so } \quad u_1(t) = u_1(0) = u_1^0(0) \geqslant 0, \quad \forall\, t \geqslant 0. \tag{2.2}$$

Using (2.2), the equation for the agent 2 becomes

$$\frac{du_2}{dt}(t) = \int_{t-\tau}^{t} \mu(t-s)\psi_{21}(s)[u_1(s) - u_2(t)]ds = (u_1(0) - u_2(t)) \int_{t-\tau}^{t} \mu(t-s)\psi_{21}(s)ds\,.$$

Arguing by contradiction, we assume that $u_2(\bar{t}) < 0$ for some $\bar{t} > 0$. Then, let us denote

$$t^* = \inf\{t > 0 \mid u_2(s) < 0 \text{ for } s \in (t, \bar{t})\}.$$

Hence, by definition of t^*, $u_2(t^*) = 0$ and $u_2(s) < 0$ for $s \in (t^*, \bar{t})$. So, using again (2.2),

$$\frac{du_2}{dt}(t) = (u_1(0) - u_2(t)) \int_{t-\tau}^{t} \mu(t-s)\psi_{21}(s)ds \geqslant 0, \quad t \in [t^*, \bar{t}),$$

which is in contradiction with $u_2(t) < 0$ for $t \in (t^*, \bar{t})$ and $u_2(t^*) = 0$. This ensures that $u_2(t) \geqslant 0$ for all $t \geq 0$.

Now, as the induction hypothesis, assume that $u_i(t) \geqslant 0$ for all $t > 0$ and for all $i \in \{1, \ldots, k-1\}$.

The equation for agent k is

$$\frac{du_k}{dt}(t) = \sum_{j \in \mathcal{L}(k)} \int_{t-\tau}^{t} \mu(t-s)\psi_{kj}(s)[u_j(s) - u_k(t)]ds\,, \quad t > 0.$$

As in the first step, let us assume by contradiction that $u_k(\bar{t}) < 0$ for some $\bar{t} > 0$ and let us denote

$$t^* = \inf\{t > 0 \mid u_k(s) < 0 \text{ for } s \in (t, \bar{t})\}.$$

Then, $u_k(t^*) = 0$ and $u_k(s) < 0$ for $s \in (t^*, \bar{t})$. We can use the induction hypothesis on the agents $j \in \mathcal{L}(k) \subseteq \{1, \ldots, k-1\}$, so

$$\frac{du_k}{dt}(t) = \sum_{j \in \mathcal{L}(k)} \int_{t-\tau}^{t} \mu(t-s)\psi_{kj}(s)[u_j(s) - u_k(t)]ds \geqslant 0, \quad t \in [t^*, \bar{t})\,,$$

which gives a contradiction.

Therefore, we have proved that $u_i(t) \geqslant 0$ for all $i \in \{1, \ldots, N\}$. ∎

As in the undelayed case (see Theorem 4.2 of [36]) we can now deduce from the previous proposition the boundedness result for the velocities.

Proposition 2.2 *Let Ω be a convex and compact domain in $\mathbb{R}^d$ and let (x_i, v_i) be a solution of system (1.5). If $v_i(s) \in \Omega$ for all $i = 1, \dots, N$ and $s \in [-\tau, 0]$, then $v_i(t) \in \Omega$ for all $i = 1, \dots, N$ and $t > 0$. In particular, if Ω is the ball with center 0 and radius*

$$D_0 = \max_{1 \leqslant i \leqslant N} \max_{s \in [-\tau, 0]} |v_i(s)|, \tag{2.3}$$

then $|v_i(t)| \leqslant D_0$ for all $t > 0$ and $i = 1, \dots, N$.

3 Convergence to Consensus

Here we will prove the announced flocking result for the CS-model under hierarchical leadership with distributed delay (1.5). Our proof extends to the model at hand the one in [34], with pointwise delay. We need a preliminary lemma.

Lemma 3.1 *Let (x, v) be a trajectory in the phase-space, namely $\frac{dx}{dt}(t) = v(t)$ for $t \geq 0$. Assume that*

$$\frac{d|v|}{dt}(t) \leq -d_0 \psi(|x(t)| + M)|v(t)| + ce^{-bt} \quad \forall\, t \geq t_0, \tag{3.1}$$

for some non-negative constants M, c, t_0 and $b, d_0 > 0$, where $\psi : [0, +\infty) \to (0, +\infty)$ is a continuous function satisfying (1.8). Then, there exists a suitable positive constant C such that

$$|x(t)| \leq C, \quad t \geq 0.$$

Proof Let us consider the functionals (cfr. [20, 34])

$$\mathcal{F}_{\pm}(t) = |v(t)| \pm d_0 \varphi(|x(t)| + M), \tag{3.2}$$

where φ is a primitive of ψ, namely $\varphi'(s) = \psi(s)$, $s \in (0, +\infty)$.

From (3.1) we deduce

$$\begin{aligned} \frac{d\mathcal{F}_{\pm}}{dt}(t) &= \frac{d|v|}{dt}(t) \pm d_0 \psi(|x(t)| + M) \frac{d|x|}{dt}(t) \\ &\leqslant -d_0 \psi(|x(t)| + M)|v(t)| \pm d_0 \psi(|x(t)| + M) \frac{d|x|}{dt}(t) + ce^{-bt} \\ &= d_0 \psi(|x(t)| + M) \left(\pm \frac{d|x|}{dt}(t) - |v(t)| \right) + ce^{-bt} \leqslant ce^{-bt}, \quad t \geq t_0, \end{aligned} \tag{3.3}$$

where we have used

$$\left| \frac{d|x(t)|}{dt} \right| \leq |v(t)|. \tag{3.4}$$

Now, integrating (3.3) on the time interval $[t_0, t]$, we obtain

$$\mathcal{F}_{\pm}(t) - \mathcal{F}_{\pm}(t_0) \leqslant c \int_{t_0}^{t} e^{-bs}\, ds = \frac{c}{b}(e^{-bt_0} - e^{-bt}) \leqslant \frac{c}{b},$$

which implies

$$|v(t)| - |v(t_0)| \leqslant \pm d_0 \left(\varphi\left(|x(t_0)| + M\right) - \varphi\left(|x(t)| + M\right)\right) + \frac{c}{b},$$

namely

$$|v(t)| - |v(t_0)| \leqslant -d_0 \left| \int_{|x(t_0)|+M}^{|x(t)|+M} \psi(s)\, ds \right| + \frac{c}{b}. \tag{3.5}$$

In particular, from (3.5), we deduce

$$|v(t_0)| + \frac{c}{b} \geqslant d_0 \left| \int_{|x(t_0)|+M}^{|x(t)|+M} \psi(s)\, ds \right|. \tag{3.6}$$

Then, assumption (1.8) ensures the existence of a constant $x_M > 0$ such that

$$|v(t_0)| + \frac{c}{b} = d_0 \int_{|x(t_0)|+M}^{x_M} \psi(s)\, ds,$$

which, together with (3.6), implies

$$|x(t)| \leqslant C, \quad \forall\, t \geq 0,$$

being ψ is a non-negative function. ■

Theorem 3.2 *Let (x_i, v_i), $i = 1, \ldots, N$, be a solution of the Cucker–Smale system under hierarchical leadership with distributed delay (1.5) with initial conditions (1.6). Assume that the potential function ψ satisfies (1.8). Then,*

$$|v_i(t) - v_j(t)| = O(e^{-Bt}), \quad \forall\, i, j = 1, \ldots, N, \tag{3.7}$$

for a suitable constant $B > 0$ depending only on the initial configuration and the parameters of the system.

Proof We will use induction on the number of agents in the flock. Consider first a flock of 2 agents [1, 2]. Recall that, by definition of an HL-flock, $\mathcal{L}(2) \neq \emptyset$, i.e. $\psi_{21} > 0$. Moreover, $\psi_{12} = 0$. Then,

$$\frac{dv_1}{dt} = 0 \quad \Rightarrow \quad v_1(t) = v_1(0), \quad \forall\, t > 0, \tag{3.8}$$

and

$$\frac{dv_2}{dt}(t) = \int_{t-\tau}^{t} \mu(t-s)\psi_{21}(s)[v_1(s) - v_2(t)]ds \tag{3.9}$$
$$= (v_1(0) - v_2(t)) \int_{t-\tau}^{t} \mu(t-s)\psi_{21}(s)ds, \quad t \geq \tau.$$

We now denote

$$y_2(t) = x_2(t) - x_1(t) \quad \text{and} \quad w_2(t) = v_2(t) - v_1(t). \tag{3.10}$$

Then, from (3.9), we obtain

$$\frac{dw_2}{dt}(t) = \frac{dv_2}{dt}(t) - \frac{dv_1}{dt}(t) = \int_{t-\tau}^{t} \mu(t-s)\psi_{21}(s)[v_1(s) - v_2(t)]ds, \quad t \geq \tau, \tag{3.11}$$

and thus, using also (3.8),

$$\frac{1}{2}\frac{d|w_2|^2}{dt}(t) = -|w_2(t)|^2 \int_{t-\tau}^{t} \mu(t-s)\psi_{21}(s)ds\,,$$

which implies

$$\frac{d|w_2|}{dt}(t) \leqslant -|w_2(t)| \int_{t-\tau}^{t} \mu(t-s)\psi\left(|x_2(s) - x_1(s)|\right) ds\,, \quad t \geq \tau\,. \tag{3.12}$$

Therefore, from (3.12), we deduce that $|w_2(t)|$ is decreasing in time for $t \geq \tau$. Now, observe that for $t > \tau$ and $s \in [t-\tau, t]$, we have

$$x_1(s) - x_2(s) = x_1(t) - x_2(t) + \int_{t}^{s} (x_1 - x_2)'(\sigma)\, d\sigma$$
$$= x_1(t) - x_2(t) + \int_{s}^{t} w_2(\sigma)\, d\sigma,$$

which gives, recalling Proposition 2.2,

$$|x_1(s) - x_2(s)| \leqslant |x_1(t) - x_2(t)| + 2D_0\tau = |y_2(t)| + 2D_0\tau\,, \quad t \geq \tau\,, \tag{3.13}$$

with $y_2(t)$, $w_2(t)$ defined in (3.10) and D_0 the bound on the initial velocities defined in (2.3).

Using this inequality in (3.12) and recalling that the potential function ψ is not increasing, we obtain

$$\frac{d|w_2|}{dt}(t) \leqslant -|w_2(t)| \int_{t-\tau}^{t} \mu(t-s)\psi(|y_2(t)| + 2\tau D_0)ds \tag{3.14}$$
$$= -\mu_0|w_2(t)|\psi(|y_2(t)| + 2\tau D_0)\,, \quad t \geq \tau\,,$$

where μ_0 is the positive constant in (1.7). Then, the pair state-velocity (y_2, w_2) satisfies the inequality (3.1) with $t_0 = \tau$, $d = \mu_0$, $M = 2\tau D_0$ and $c = 0$. Therefore, we can apply Lemma 3.1 obtaining $|y_2(t)| \leq C_2$ for some positive constant C_2. So, for a suitable constant y_M^2,

$$|y_2(t)| + 2\tau D_0 \leqslant y_M^2, \quad t \geq \tau. \tag{3.15}$$

Now, from (3.14) and (3.15) we deduce

$$\frac{d|w_2(t)|}{dt} \leqslant -\mu_0 \psi(y_M^2)|w_2(t)|, \quad t \geq \tau,$$

and the Gronwall inequality implies

$$|w_2(t)| \leqslant e^{-\mu_0 \psi(y_M^2)(t-\tau)}|w_2(\tau)|, \quad t \geq \tau. \tag{3.16}$$

In order to complete our inductive step we will need also estimates on the distances $|v_i(s) - v_j(t)|$ and $|v_i(s) - v_j(s)|$ for $j = 1, 2$ and $s \in [t-\tau, t]$.

Now, since $v_1(t)$ is constant for $t \geq \tau$, we easily deduce

$$|v_1(s) - v_2(t)| = |v_1(t) - v_2(t)| = O(e^{-\psi(y_M^2)t}). \tag{3.17}$$

Observe also that, for $s \in [t-\tau, t]$,

$$\begin{aligned} |v_2(s) - v_2(t)| &= \left| \int_s^t v_2{}'(\sigma)\, d\sigma \right| = \left| \int_s^t \int_{\sigma-\tau}^{\sigma} \mu(\sigma - r)\psi_{21}(r)[v_1(r) - v_2(\sigma)]\, dr\, d\sigma \right| \\ &\leqslant c \int_s^t e^{-\psi(y_M^2)\sigma}\, d\sigma \leqslant c\tau e^{-\psi(y_M^2)(t-\tau)} = c\tau e^{\psi(y_M^2)\tau} e^{-\psi(y_M^2)t} = O(e^{-\psi(y_M^2)t}). \end{aligned} \tag{3.18}$$

Since

$$|v_2(s) - v_1(t)| \leqslant |v_2(s) - v_2(t)| + |v_2(t) - v_1(t)|, \tag{3.19}$$

from previous estimates we thus obtain

$$|v_2(s) - v_1(t)| = O(e^{-\psi(y_M^2)t}), \quad t > \tau, \ s \in [t-\tau, t]. \tag{3.20}$$

Moreover, of course, $|v_1(s) - v_1(t)| = O(e^{-\psi(y_M^2)t})$, being $v_1(t)$ constant for $t \geq \tau$.

We assume now, by induction, that analogous exponential estimates are satisfied for a flock of $l-1$ agents $[1, \ldots, l-1]$ with $l > 2$, i.e. there exists some constant $b > 0$ such that, $\forall\, i, j \in \{1, \ldots, l-1\}$,

$$|v_i(t) - v_j(t)| = O(e^{-bt}), \tag{3.21}$$

$$|v_i(s) - v_j(t)| = O(e^{-bt}), \quad t > \tau, \ s \in [t-\tau, t]. \tag{3.22}$$

Then, we want to prove that such estimates hold true also for a flock with $l > 2$ agents $[1, \dots, l]$. This will complete the proof. For this aim, define the average position and velocity of the leaders of agent l,

$$\hat{x}_l = \frac{1}{d_l} \sum_{i \in \mathcal{L}(l)} x_i(t) \quad \text{and} \quad \hat{v}_l = \frac{1}{d_l} \sum_{i \in \mathcal{L}(l)} v_i(t), \quad d_l = \#\mathcal{L}(l). \tag{3.23}$$

Also, define

$$y_l(t) = x_l(t) - \hat{x}_l(t) \quad \text{and} \quad w_l(t) = v_l(t) - \hat{v}_l(t). \tag{3.24}$$

Then,

$$\frac{dw_l}{dt}(t) = \frac{dv_l}{dt}(t) - \frac{d\hat{v}_l}{dt}(t) = \sum_{j \in \mathcal{L}(l)} \int_{t-\tau}^{t} \mu(t-s)\psi_{lj}(s)[v_j(s) - v_l(t)]ds - \frac{d\hat{v}_l}{dt}(t)\,. \tag{3.25}$$

By adding and subtracting $\sum_{j \in \mathcal{L}(l)} \int_{t-\tau}^{t} \mu(t-s)\psi_{lj}(s)ds\ \hat{v}_l(t)$ in (3.25) we get

$$\frac{dw_l}{dt} = -w_l(t) \sum_{j \in \mathcal{L}(l)} \int_{t-\tau}^{t} \mu(t-s)\psi_{lj}(s)ds + \sum_{j \in \mathcal{L}(l)} \int_{t-\tau}^{t} \mu(t-s)\psi_{lj}(s)[v_j(s) - \hat{v}_l(t)]ds - \frac{d\hat{v}_l}{dt}. \tag{3.26}$$

Using the induction hypothesis (3.22), since $\mathcal{L}(i), \mathcal{L}(l) \subseteq [1, \dots, l-1]$,

$$\frac{d\hat{v}_l}{dt} = \frac{1}{d_l} \sum_{i \in \mathcal{L}(l)} \frac{dv_i}{dt} = \frac{1}{d_l} \sum_{i \in \mathcal{L}(l)} \sum_{j \in \mathcal{L}(i)} \int_{t-\tau}^{t} \mu(t-s)\psi_{ij}(s)[v_j(s) - v_i(t)]\, ds = O(e^{-bt}). \tag{3.27}$$

Using again the induction hypothesis (3.22),

$$\begin{aligned} &\sum_{j \in \mathcal{L}(l)} \int_{t-\tau}^{t} \mu(t-s)\psi_{lj}(s)[v_j(s) - \hat{v}_l(t)]\, ds \\ &\qquad = \frac{1}{d_l} \sum_{j \in \mathcal{L}(l)} \int_{t-\tau}^{t} \mu(t-s)\psi_{lj}(s)\Big(\sum_{i \in \mathcal{L}(l)} [v_j(s) - v_i(t)]\Big)\, ds = O(e^{-bt}). \end{aligned} \tag{3.28}$$

So, identity (3.26) can be rewritten as

$$\frac{dw_l}{dt}(t) = -w_l(t) \sum_{j \in \mathcal{L}(l)} \int_{t-\tau}^{t} \mu(t-s)\psi_{lj}(s)\, ds + O(e^{-bt}), \quad t \geq \tau\,. \tag{3.29}$$

with

$$\psi_{lj}(s) = \psi(|x_l(s) - x_j(s)|).$$

Observe that for every $j \in \mathcal{L}(l)$ it results

$$\begin{aligned} |x_l(s) - x_j(s)| &\le |x_l(s) - \hat{x}_l(s)| + |x_j(s) - \hat{x}_l(s)| \\ &\leqslant |y_l(s)| + M_l, \end{aligned} \tag{3.30}$$

for some positive M_l, due to the induction's assumption. Then, (3.29) gives

$$\frac{d|w_l|}{dt}(t) \leqslant -d_l |w_l(t)| \int_{t-\tau}^{t} \mu(t-s)\psi\left(|y_l(s)| + M_l\right) \, ds + ce^{-bt}, \quad t \ge \tau. \tag{3.31}$$

Now, note that from Proposition 2.2, $|v_i(t)| \leqslant D_0$ for all i and for all $t > 0$, which implies

$$|w_l(t)| \leqslant \frac{1}{d_l} \sum_{j \in \mathcal{L}(l)} |v_j(t) - v_l(t)| \leqslant \frac{1}{d_l} \sum_{j \in \mathcal{L}(l)} 2D_0 = 2D_0.$$

Then,

$$|y_l(s)| \leqslant |y_l(t)| + 2\tau D_0, \quad t \ge \tau, \; s \in [t-\tau, t], \tag{3.32}$$

which used in (3.31), recalling that ψ in not increasing, yields

$$\frac{d|w_l|}{dt}(t) \leqslant -d_l \mu_0 \psi\left(|y_l(t)| + 2\tau D_0 + M_l\right) |w_l(t)| + ce^{-bt}. \tag{3.33}$$

We can then apply Lemma 3.1 to the pair state-velocity (y_l, w_l) to conclude that $|y_l(t)| \le C_l$ for some positive constant C_l. So, for a suitable constant y_M^l,

$$|y_l(t)| + 2\tau D_0 + M_l \le y_M^l, \quad t \ge \tau.$$

Using the above estimate in (3.32) we then obtain

$$\frac{d|w_l|}{dt} \leqslant -d_l \mu_0 \psi(y_M^l) |w_l(t)| + ce^{-bt},$$

and therefore, from the Gronwall's inequality we deduce,

$$|w_l(t)| \le Ce^{-B^l t}, \tag{3.34}$$

for suitable positive constants C, B^l.

Thus, from (4.5) and the induction hypothesis (3.21), for every $j \in \mathcal{L}(l)$, we have

$$|v_l(t) - v_j(t)| \le |v_l(t) - \hat{v}_l(t)| + |\hat{v}_l(t) - v_j(t)| = O(e^{-Bt}). \tag{3.35}$$

Now, to complete the induction argument, we only have to prove that, for all $t > 0$ and $i, j \in \{1, \dots, l\}$,

$$|v_i(s) - v_j(t)| = O(e^{-Bt}), \tag{3.36}$$

for a suitable positive constant B.

If $i, j \in \{1, \ldots, l-1\}$, then (4.14) is true by (3.22). Let us consider the case $i \in \{1, \ldots, l-1\}$ and $j = l$. Then,

$$|v_i(s) - v_l(t)| \leqslant |v_i(s) - v_i(t)| + |v_i(t) - v_l(t)| = O(e^{-Bt}),$$

by (3.22) and (4.13), for a suitable B.

Consider now $i = j = l$. Then, using previous estimates we see that

$$\begin{aligned} |v_l(s) - v_l(t)| &= \left| \int_s^t v_l'(\sigma)\, d\sigma \right| = \left| \int_s^t \sum_{k \in \mathcal{L}(l)} \int_{\sigma-\tau}^{\sigma} \mu(\sigma - r)\psi_{lj}(r)\,(v_k(r) - v_l(\sigma))\, dr\, d\sigma \right| \\ &\leqslant \bar{c} \int_s^t e^{-B\sigma}\, d\sigma \leq \bar{c}\tau e^{-B(t-\tau)} = \bar{c}\tau e^{B\tau} e^{-Bt} = O(e^{-Bt}). \end{aligned} \tag{3.37}$$

Also for the last case, where $j \in \{1, \ldots, l-1\}$ and $i = l$, using (4.15) we have

$$|v_l(s) - v_j(t)| \leqslant |v_l(s) - v_l(t)| + |v_l(t) - v_j(t)| = O(e^{-Bt}),$$

by the previous case and (4.13). Then, we have proved that (4.14) is satisfied for all $i, j \in \{1, \ldots, l\}$ and this concludes the proof of the theorem. ■

4 The Case of Free-Will Leader

It may happen that the leader of the flock, instead of moving at a constant velocity, takes off or changes its rate in order to avoid a danger, for instance due to the presence of predator species. Thus, it is important to consider this situation in the mathematical model.

The Cucker–Smale model with a free-will leader is, then,

$$\begin{aligned} \frac{dx_1}{dt}(t) &= v_1(t), \\ \frac{dv_1}{dt}(t) &= f(t), \end{aligned} \tag{4.1}$$

where $f : [0, +\infty) \to \mathbb{R}^d$ is a continuous integrable function, that is,

$$\|f\|_1 = \int_0^{+\infty} |f(t)|\, dt < +\infty\,, \tag{4.2}$$

for the motion of the free-will leader, and the Cucker–Smale model under hierarchical leadership and distributed delay, as in the previous sections, for the other agents, namely

$$\frac{dx_i}{dt}(t) = v_i(t),$$
$$\frac{dv_i}{dt}(t) = \sum_{j\in\mathcal{L}(i)} \int_{t-\tau}^{t} \mu(t-s)\psi_{ij}(s)[v_j(s) - v_i(t)]\,ds, \tag{4.3}$$

for all $i \in \{2, \dots, N\}$. The initial data are assigned, as usual, on the time interval $[-\tau, 0]$, i.e.

$$x_i(s) = x_i^0(s),$$
$$v_i(s) = v_i^0(s), \tag{4.4}$$

for some continuous functions x_i^0 and v_i^0, for $i = 1, \dots, N$.

The flocking result below extends the one proved by Shen [36] for the undelayed case. The case with pointwise delay has been studied in [34]. Here, we consider a more general acceleration function with respect to [34, 36], for the free-will leader. Indeed we assume

$$|f(t)| = o((1+t)^{1-N}) \quad \text{and} \quad t^{N-2}|f(t)| \in L^1(0, +\infty) \tag{4.5}$$

instead of

$$|f(t)| = O((1+t)^{-\mu}), \quad \mu > N-1\,. \tag{4.6}$$

Then, for instance, f can be in the form

$$f(t) = \frac{C}{(1+t)^{\mu}}, \ \mu > N-1,$$

as in [34, 36], but also

$$f(t) = \frac{C}{(1+t)^{N-1}\ln^2(2+t)}\,.$$

Note that, from (4.5) it results

$$t^k|f(t)| = o((1+t)^{1-N+k}), \quad \forall\, k = 1, \dots, N-1\,. \tag{4.7}$$

In order to prove our flocking result, we will need the following lemma, which is a generalization of Lemma 3.1 above.

Lemma 4.1 *Let (x, v) be a trajectory in the phase-space, namely $\frac{dx}{dt}(t) = v(t)$ for $t \geq 0$. Assume that*

$$\frac{d|v|}{dt}(t) \leq -d_0\psi(|x(t)| + M)|v(t)| + g(t) \quad \forall\, t \geq t_0, \tag{4.8}$$

for some non-negative constants M, t_0, a constant $d_0 > 0$ and a continuous and integrable function $g : [t_0, +\infty) \to (0, +\infty)$, where $\psi : [0, +\infty) \to (0, +\infty)$ is a *continuous function satisfying (1.8). Then, there exists a suitable positive constant C such that*

$$|x(t)| \leq C, \quad t \geq 0.$$

Proof Let us consider the functionals $\mathcal{F}_{\pm}$ introduced in (3.2) with d_0, M, ψ as in the statement. From (4.8) we deduce

$$\begin{aligned}\frac{d\mathcal{F}_{\pm}}{dt}(t) &= \frac{d|v|}{dt}(t) \pm d_0\psi(|x(t)|+M)\frac{d|x|}{dt}(t)\\ &\leqslant -d_0\psi(|x(t)|+M)|v(t)| \pm d_0\psi(|x(t)|+M)\frac{d|x|}{dt}(t) + g(t)\\ &= d_0\psi(|x(t)|+M)\left(\pm\frac{d|x|}{dt}(t) - |v(t)|\right) + g(t) \leqslant g(t), \quad t \geq t_0,\end{aligned} \tag{4.9}$$

where we have used inequality (3.4).

Now, we integrate (4.9) on the time interval $[t_0, t]$, obtaining

$$\mathcal{F}_{\pm}(t) - \mathcal{F}_{\pm}(t_0) \leqslant \|g\|_{L^1(t_0,+\infty)},$$

which gives

$$|v(t)| \leqslant \pm d_0\left(\varphi\left(|x(t_0)|+M\right) - \varphi\left(|x(t)|+M\right)\right) + |v(t_0)| + \|g\|_{L^1(t_0,+\infty)},$$

namely

$$|v(t)| \leqslant -d_0\left|\int_{|x(t_0)|+M}^{|x(t)|+M}\psi(s)\,ds\right| + |v(t_0)| + \|g\|_{L^1(t_0,+\infty)}. \tag{4.10}$$

Therefore, from (4.10), we have

$$|v(t_0)| + \|g\|_{L^1(t_0,+\infty)} \geqslant d_0\left|\int_{|x(t_0)|+M}^{|x(t)|+M}\psi(s)\,ds\right|. \tag{4.11}$$

The assumption (1.8) ensures then the existence of a constant $x_M > 0$ such that

$$|v(t_0)| + \|g\|_{L^1(t_0,+\infty)} = d_0\int_{|x(t_0)|+M}^{x_M}\psi(s)\,ds,$$

which, together with (4.11), implies $|x(t)| \leqslant C, \quad \forall\, t \geq 0$. ■

Theorem 4.2 *Let (x_i, v_i), $i = 1, \ldots, N$, be a solution of the Cucker–Smale system under hierarchical leadership with delay (4.1)–(4.3) with initial conditions (4.4). Assume that (1.8) is satisfied and that the acceleration of the free-will leader satisfies (4.5). Then, it results*

$$|v_i(t) - v_j(t)| \to 0, \quad \text{for } t \to +\infty, \quad \forall\, i, j = 1, \ldots, N. \tag{4.12}$$

Proof As in the previous convergence to consensus result, we argue by induction. First, we look at the first agent, i.e. the free-will leader. Equation (4.1) gives

$$v_1(t) = v_1(0) + \int_0^t f(s)\,ds\,,$$

and so, from (4.2),

$$|v_1(t)| \le |v_1(0)| + \|f\|_1 = C_1\,, \quad \forall\, t \ge 0\,. \tag{4.13}$$

Now, let us consider the 2-flock. As before, let us denote

$$w_2(t) = v_2(t) - v_1(t) \quad \text{and} \quad y_2(t) = x_2(t) - x_1(t), \quad t \ge 0\,.$$

From (4.1) and (4.3)

$$\begin{aligned} \frac{dw_2}{dt}(t) &= \frac{dv_2}{dt}(t) - \frac{dv_1}{dt}(t) = \int_{t-\tau}^t \mu(t-s)\psi_{21}(s)[v_1(s) - v_2(t)]\,ds - f(t) \\ &= (v_1(t) - v_2(t)) \int_{t-\tau}^t \mu(t-s)\psi_{21}(s)\,ds - \int_{t-\tau}^t \mu(t-s)\psi_{21}(s)[v_1(t) - v_1(s)]\,ds - f(t) \\ &= -w_2(t) \int_{t-\tau}^t \mu(t-s)\psi_{21}(s)\,ds - \int_{t-\tau}^t \mu(t-s)\psi_{21}(s) \int_s^t f(\sigma)\,d\sigma\,ds - f(t)\,, \quad t \ge \tau\,. \end{aligned} \tag{4.14}$$

Now, from (4.5), it results

$$\begin{aligned} &\left| \int_{t-\tau}^t \mu(t-s)\psi_{21}(s) \int_s^t f(\sigma)\,d\sigma\,ds \right| + |f(t)| \\ &\qquad \le \tau\mu_0 \max_{s\in[0,+\infty)} \psi(s) \int_{t-\tau}^t |f(s)|\,ds + |f(t)| = O(|f|)\,. \end{aligned} \tag{4.15}$$

Then, from (4.14) and (4.15) we obtain

$$\frac{d|w_2|}{dt}(t) \le -|w_2(t)| \int_{t-\tau}^t \mu(t-s)\psi_{21}(s)\,ds + \tilde{f}(t)\,, \quad t \ge \tau\,. \tag{4.16}$$

where

$$\tilde{f}(t) := \tau\mu_0 \max_{s\in[0,+\infty)} \psi(s) \int_{t-\tau}^t |f(s)|\,ds + |f(t)| = O(|f|)\,. \tag{4.17}$$

Therefore,

$$|w_2(t)| \le |w_2(\tau)| + \int_\tau^{+\infty} \tilde{f}(t)\,dt \le D_2, \quad \forall\, t \ge \tau\,, \tag{4.18}$$

for some constant $D_2 > 0$. Since

$$y_2(s) = y_2(t) + \int_t^s w_2(\sigma)\, d\sigma\,,$$

from (4.18) we have

$$|y_2(s)| \le |y_2(t)| + \tau D_2\,, \quad \forall\, s \in [t-\tau, t]\,. \tag{4.19}$$

From (4.16) and (4.19), we then deduce

$$\frac{d|w_2|}{dt}(t) \le -\mu_0 \psi(|y_2(t)| + \tau D_2)|w_2(t)| + \tilde{f}(t)\,, \quad t \ge \tau\,. \tag{4.20}$$

Then, we can apply Lemma 4.1 to the pair (y_2, w_2) with $d = \mu_0$, $M = \tau D_2$ and $g = \tilde{f}$, obtaining that

$$|y_2(t)| + \tau D_2 \leqslant y_R^2, \quad t \ge 0\,, \tag{4.21}$$

for a suitable positive constant y_R^2 . So, from (4.20) and (4.21) we have

$$\frac{d|w_2|}{dt}(t) \leqslant -\psi(y_R^2)|w_2(t)| + \tilde{f}(t), \quad t \ge \tau\,,$$

and thus, for every $T > \tau$, applying Gronwall's lemma we deduce

$$\begin{aligned} |w_2(T)| &\le e^{-\psi(y_R^2)\frac{T}{2}}|w_2(T/2)| + \int_{\frac{T}{2}}^{T} e^{-\psi(y_R^2)(T-t)}\,\tilde{f}(t)\,dt \\ &\le e^{-\psi(x_R^2)\frac{T}{2}}D_2 + \int_{\frac{T}{2}}^{T} \tilde{f}(t)\,dt \le e^{-\psi(x_R^2)\frac{T}{2}}D_2 + \tilde{f}_2(T)\,, \end{aligned} \tag{4.22}$$

where, recalling (4.5), $\tilde{f}_2$, is a suitable function satisfying

$$\tilde{f}_2(t) = O(t|f|) = o((1+t)^{2-N})\,. \tag{4.23}$$

Thus,

$$|v_2(t) - v_1(t)| = o((1+t)^{2-N})\,. \tag{4.24}$$

Note also that

$$|v_1(t-\tau) - v_1(t)| \le \int_{t-\tau}^{t} |f(t)|\,dt = O(|f|)\,, \tag{4.25}$$

and then

$$\begin{aligned} |v_2(t-\tau) - v_2(t)| &\le |v_2(t-\tau) - v_1(t-\tau)| \\ &+ |v_1(t-\tau) - v_1(t)| + |v_1(t) - v_2(t)| = o((1+t)^{2-N})\,. \end{aligned} \tag{4.26}$$

Therefore, (4.24)-(4.26) imply

$$|v_i(t-\tau) - v_j(t)| = O(\tilde{f}_2) = o((1+t)^{2-N}), \quad \text{for } i, j \in \{1, 2\}. \tag{4.27}$$

Now, as induction hypothesis, assume that for a flock of $l-1$ agents $[1, \dots, l-1]$ with $2 < l \leq N$, we have

$$|v_i(t) - v_j(t)| = O(t^{l-2}|f|) = o((1+t)^{l-1-N}), \tag{4.28}$$

$$|v_i(t-\tau) - v_j(t)| = O(t^{l-2}|f|) = o((1+t)^{l-1-N}), \tag{4.29}$$

for all $i, j \in \{1, \dots, l-1\}$.

Then, we want to prove the same kind of estimates for a flock with l agents. This will complete our theorem.

As before, we will use the average position and velocity of the leaders of agent l, introduced in (3.23) and let y_l, w_l be defined as in (3.24). Then, as before we can write

$$\frac{dw_l}{dt} = -w_l(t) \sum_{j \in \mathcal{L}(l)} \int_{t-\tau}^{t} \mu(t-s)\psi_{lj}(s)ds + \sum_{j \in \mathcal{L}(l)} \int_{t-\tau}^{t} \mu(t-s)\psi_{lj}(s)[v_j(s) - \hat{v}_l(t)]ds - \frac{d\hat{v}_l}{dt}. \tag{4.30}$$

Using the induction hypothesis (4.29), since $\mathcal{L}(i), \mathcal{L}(l) \subseteq [1, \dots, l-1]$,

$$\frac{d\hat{v}_l}{dt} = \frac{1}{d_l} \sum_{i \in \mathcal{L}(l)} \frac{dv_i}{dt} = \chi_{1 \in \mathcal{L}(l)} \frac{1}{d_l} f(t) + \frac{1}{d_l} \sum_{i \in \mathcal{L}(l) \setminus \{1\}} \frac{dv_i}{dt} = O(t^{l-2}|f|) = o((1+t)^{l-1-N}). \tag{4.31}$$

From the induction hypotheses (4.29) we deduce also

$$\begin{aligned} &\sum_{j \in \mathcal{L}(l)} \int_{t-\tau}^{t} \mu(t-s)\psi_{lj}(s)[v_j(s) - \hat{v}_l(t)]\, ds \\ &\quad = \frac{1}{d_l} \sum_{j \in \mathcal{L}(l)} \int_{t-\tau}^{t} \mu(t-s)\psi_{lj}(s)\Big(\sum_{i \in \mathcal{L}(l)} [v_j(s) - v_i(t)]\Big)\, ds = O(t^{l-2}|f|) = o((1+t)^{l-1-N}). \end{aligned} \tag{4.32}$$

Then, identity (4.30) can be rewritten as

$$\frac{dw_l}{dt}(t) = -w_l(t) \sum_{j \in \mathcal{L}(l)} \int_{t-\tau}^{t} \mu(t-s)\psi_{lj}(s)\, ds + O(t^{l-2}|f|), \quad t \geq \tau. \tag{4.33}$$

As before one can now observe that for every $j \in \mathcal{L}(l)$ it results

$$\begin{aligned} |x_l(s) - x_j(s)| &\leq |x_l(s) - \hat{x}_l(s)| + |x_j(s) - \hat{x}_l(s)| \\ &\leqslant |y_l(s)| + R_l, \end{aligned} \tag{4.34}$$

for some positive R_l, due to the induction's assumption. Thus, (4.33) implies

$$\frac{d|w_l|}{dt}(t) \leqslant -d_l|w_l(t)| \int_{t-\tau}^{t} \mu(t-s)\psi\left(|y_l(s)| + R_l\right) \, ds + O(t^{l-2}|f|), \quad t \geq \tau. \tag{4.35}$$

Note that (4.35) implies

$$\frac{d|w_l|}{dt} \leqslant O(t^{l-2}|f|). \tag{4.36}$$

So, recalling the assumptions (4.5) on the acceleration f of the free-will leader, we deduce

$$|w_l(t)| \leq |w_l(\tau)| + \int_{\tau}^{+\infty} O(t^{l-2}|f|) \, dt \leq C_l. \tag{4.37}$$

Then,

$$|x^l(t-\tau)| \leqslant |x^l(t)| + \int_{t-\tau}^{t} |v^l(s)| \, ds \leq |x^l(t)| + C_l\tau, \quad t \geq \tau, \tag{4.38}$$

which, used in (4.35), gives

$$\frac{d|w_l|}{dt}(t) \leqslant -d_l\mu_0\psi\left(|y_l(t)| + 2\tau C_l + R_l\right)|w_l(t)| + O(t^{l-2}|f|). \tag{4.39}$$

We can then apply Lemma 4.1 to the pair state-velocity (y_l, w_l) and conclude that $|y_l(t)| \leq C_l$ for some positive constant C_l. So, for a suitable constant y_M^l,

$$|y_l(t)| + 2\tau C_l + R_l \leq y_M^l, \quad t \geq \tau.$$

Using the above estimate in (4.39) we then obtain

$$\frac{d|w_l|}{dt} \leqslant -d_l\mu_0\psi(y_M^l)|w_l(t)| + O(t^{l-2}|f|).$$

Thus, we can apply the Gronwall's lemma analogously to the 2−flock case obtaining

$$|v^l(t)| = O(t^{l-1}|f|) = o(t^{l-N}). \tag{4.40}$$

Then, from (4.40) and the induction hypothesis (4.28), for every $j \in \mathcal{L}(l)$, we have

$$|v_l(t) - v_j(t)| \leq |v_l(t) - \hat{v}_l(t)| + |\hat{v}_l(t) - v_j(t)| = O(t^{l-1}|f|) = o(t^{l-N}). \tag{4.41}$$

Now, it remains to prove that, for all $i, j \in \{1, \ldots, l\}$,

$$|v_i(t-\tau) - v_j(t)| = O(t^{l-1}|f|) = o(|f|^{l-N}). \tag{4.42}$$

If $i, j \in \{1, \ldots, l-1\}$, then (4.42) is true by (4.29). Consider the case $i \in \{1, \ldots, l-1\}$ and $j = l$. Then,

$$|v_i(t-\tau)-v_l(t)| \leqslant |v_i(t-\tau)-v_i(t)| + |v_i(t)-v_l(t)| = O(t^{l-1}|f|) = o(|f|^{l-N}),$$

by (4.29) and (4.41).

For the case $i = j = l$, using the previous estimates, we obtain

$$\begin{aligned}|v_l(s)-v_l(t)| &= \left|\int_s^t v_l'(\sigma)\,d\sigma\right| = \left|\int_s^t \sum_{k\in\mathcal{L}(l)} \int_{\sigma-\tau}^{\sigma} \mu(\sigma-r)\psi_{lk}(r)\,(v_k(r)-v_l(\sigma))\,dr\,d\sigma\right| \\ &\leqslant C\int_s^t O(\sigma^{l-1}|f(\sigma)|)\,d\sigma = O(t^{l-1}|f|), \quad s\in[t-\tau,t].\end{aligned} \tag{4.43}$$

Also for the last case, where $j \in \{1,\dots,l-1\}$ and $i = l$, using (4.41) and (4.43) we obtain

$$|v_l(t-\tau)-v_j(t)| \leqslant |v_l(t-\tau)-v_l(t)| + |v_l(t)-v_j(t)| = O(t^{l-1}|f|) = o(|f|^{l-N}).$$

Therefore, (4.42) is satisfied for all $i, j \in \{1,\dots,l\}$ and so the theorem is proved. ■

Remark 4.3 Note that our generalization concerning the acceleration function f of the free-will leader is suitable also for the problem without delay considered by Shen [36] and for the problem with pointwise delay studied by the authors [34]. Therefore, our flocking estimates (4.13) could be obtained, under the same assumptions on f, for the problem with free-will leader studied in [36] and the more general one considered in [34].

Acknowledgements The research of the first author was partially supported by the GNAMPA 2017 project *Comportamento asintotico e controllo di equazioni di evoluzione non lineari* and the GNAMPA 2018 project *Analisi e controllo di modelli differenziali non lineari* (INdAM).

References

1. Albi, G., Herty, M., Pareschi, L.: Kinetic description of optimal control problems and applications to opinion consensus. Commun. Math. Sci. **13**, 1407–1429 (2015)
2. Bellomo, N., Herrero, M.A., Tosin, A.: On the dynamics of social conflict: Looking for the Black Swan. Kinet. Relat. Models **6**, 459–479 (2013)
3. Borzì, A., Wongkaew, S.: Modeling and control through leadership of a refined flocking system. Math. Models Methods Appl. Sci. **25**, 255–282 (2015)
4. Canuto, C., Fagnani, F., Tilli, P.: An Eulerian approach to the analysis of Krause's consensus models. SIAM J. Control Optim. **50**, 243–265 (2012)
5. Caponigro, M., Fornasier, M., Piccoli, B., Trélat, E.: Sparse stabilization and optimal control of the Cucker-Smale model. Math. Cont. Relat. Fields **3**, 447–466 (2013)
6. Caponigro, M., Fornasier, M., Piccoli, B., Trélat, E.: Sparse stabilization and control of alignment models. Math. Models Methods Appl. Sci. **25**, 521–564 (2015)
7. Carrillo, J.A., Fornasier, M., Rosado, J., Toscani, G.: Asymptotic flocking dynamics for the kinetic Cucker-Smale model. SIAM J. Math. Anal. **42**, 218–236 (2010)
8. Choi, Y.P., Haskovec, J.: Cucker-Smale model with normalized communication weights and time delay. Kinet. Relat. Models **10**, 1011–1033 (2017)

9. Choi, Y.P., Li, Z.: Emergent behavior of Cucker-Smale flocking particles with heterogeneous time delays. Appl. Math. Lett. **86**, 49–56 (2018)
10. Couzin, I., Krause, J., Franks, N., Levin, S.: Effective leadership and decision making in animal groups on the move. Nature **433**, 513–516 (2005)
11. Cristiani, E., Piccoli, B., Tosin, A.: Multiscale modeling of granular flows with application to crowd dynamics. Multiscale Model. Simul. **9**, 155–182 (2011)
12. Cucker, F., Dong, J.G.: A general collision-avoiding flocking framework. IEEE Trans. Automat. Cont. **56**, 1124–1129 (2011)
13. Cucker, F., Mordecki, E.: Flocking in noisy environments. J. Math. Pures Appl. **89**, 278–296 (2008)
14. Cucker, F., Smale, S.: Emergent behaviour in flocks. IEEE Trans. Autom. Control **52**, 852–862 (2007)
15. Cucker, F., Smale, S.: On the mathematics of emergence. Jpn. J. Math. **2**, 197–227 (2007)
16. Dalmao, F., Mordecki, E.: Cucker-smale flocking under hierarchical leadership and random interactions. SIAM J. Appl. Math. **71**, 1307–1316 (2011)
17. Degond, P., Motsch, S.: Continuum limit of self-driven particles with orientation interaction. Math. Models Methods Appl. Sci. **18**, 1193–1215 (2008)
18. Erban, R., Haskovec, J., Sun, Y.: On cucker-smale model with noise and delay. SIAM J. Appl. Math. **76**(4), 1535–1557 (2016)
19. Ha, S.Y., Lee, K., Levy, D.: Emergence of time-asymptotic flocking in a stochastic Cucker-Smale system. Commun. Math. Sci. **7**, 453–469 (2009)
20. Ha, S.Y., Liu, J.G.: A simple proof of the Cucker-Smale flocking dynamics and mean-field limit. Commun. Math. Sci. **7**, 297–325 (2009)
21. Ha, S.Y., Slemrod, M.A.: Flocking dynamics of singularly perturbed oscillator chain and the cucker-smale system. J. Dyn. Diff. Equ. **22**, 325–330 (2010)
22. Ha, S.Y., Tadmor, E.: From particle to kinetic and hydrodynamic descriptions of flocking. Kinet. Relat. Models **1**, 415–435 (2008)
23. Lemercier, S., Jelic, A., Kulpa, R., Hua, J., Fehrenbach, J., Degond, P., Appert Rolland, C., Donikian, S., Pettré, J.: Realistic following behaviors for crowd simulation. Comput. Graph. Forum **31**, 489–498 (2012)
24. Li, Z.: Effectual leadershipin flocks with hierarchy and individual preference. Discret. Contin. Dyn. Syst. **34**, 3683–3702 (2014)
25. Li, Z., Ha, S.Y., Xue, X.: Emergent phenomena in an ensemble of Cucker-Smale particles under joint rooted leadership. Math. Models Methods Appl. Sci. **24**, 1389–1419 (2014)
26. Li, Z., Xue, X.: Cucker-Smale flocking under rooted leadership with fixed and switching topologies. SIAM J. Appl. Math. **70**, 3156–3174 (2010)
27. Li, C.H., Yang, S.Y.: A new discrete Cucker-Smale flocking model under hierarchical leadership. Discret. Contin. Dyn. Syst. Ser. B **21**, 2587–2599 (2016)
28. Liu, Y., Wu, J.: Flocking and asymptotic velocity of the Cucker-Smale model with processing delay. J. Math. Anal. Appl. **415**, 53–61 (2014)
29. Mecholsky, N., Ott, E., Antonsen, T.M.: Obstacle and predator avoidance in a model for flocking. Phys. D **239**, 988–996 (2010)
30. Motsch, S., Tadmor, E.: A new model for self-organized dynamics and its flocking behavior. J. Stat. Phys. **144**, 923–947 (2011)
31. Motsch, S., Tadmor, E.: Heterophilious dynamics enhances consensus. SIAM Rev. **56**, 577–621 (2014)
32. Nicaise, S., Pignotti, C.: Stabilization of the wave equation with boundary or internal distributed delay. Differ. Integral Equ. **21**, 935–958 (2008)
33. Piccoli, B., Rossi, F., Trélat, E.: Control to flocking of the kinetic Cucker-Smale model. SIAM J. Math. Anal. **47**, 4685–4719 (2015)
34. Pignotti, C., Reche Vallejo, I.: Flocking estimates for the Cucker–Smale model with time lag and hierarchical leadership. J. Math. Anal. Appl. **464**, 1313–1332 (2018)
35. Pignotti, C., Trélat, E.: Convergence to consensus of the general finite-dimensional Cucker-Smale model with time-varying delays. Commun. Math. Sci. (to appear). arXiv:1707.05020

36. Shen, J.: Cucker-Smale flocking under hierarchical leadership. SIAM J. Appl. Math. **68**, 694–719 (2007/08)
37. Toscani, G.: Kinetic models of opinion formation. Commun. Math. Sci. **4**, 481–496 (2006)
38. Vicsek, T., Czirok, A., Ben Jacob, E., Cohen, I., Shochet, O.: Novel type of phase transition in a system of self-driven particles. Phys. Rev. Lett. **75**, 1226–1229 (1995)
39. Wongkaew, S., Caponigro, M., Borzì, A.: On the control through leadership of the Hegselmann-Krause opinion formation model. Math. Models Methods Appl. Sci. **25**, 565–585 (2015)
40. Yates, C., Erban, R., Escudero, C., Couzin, L., Buhl, J., Kevrekidis, L., Maini, P., Sumpter, D.: Inherent noise can facilitate coherence in collective swarm motion. Proc. Natl. Acad. Sci. USA **106**, 5464–5469 (2009)

Global Non-negative Approximate Controllability of Parabolic Equations with Singular Potentials

Judith Vancostenoble

Abstract In this work, we consider the linear $1-d$ heat equation with some singular potential (typically the so-called inverse square potential). We investigate the global approximate controllability via a multiplicative (or bilinear) control. Provided that the singular potential is not *super-critical*, we prove that any non-zero and non-negative initial state in L^2 can be steered into any neighborhood of any non-negative target in L^2 using some static bilinear control in L^∞. Besides the corresponding solution remains non-negative at all times.

Keywords Bilinear control · Multiplicative control · Parabolic equation · Singular potential

1 Introduction and Main Results

1.1 Introduction

In this paper, we analyze controllability properties for parabolic equations with singular potential. Typically, we consider the following linear $1-D$ heat equation with an *inverse square potential* (that arises for example in the context of combustion theory or quantum mechanics):

$$\begin{cases} u_t - u_{xx} - \dfrac{\mu}{x^2} u = 0 & x \in (0,1),\ t > 0, \\ u(0,t) = 0 = u(1,t) & t > 0, \\ u(x,0) = u_0(x) & x \in (0,1), \end{cases} \tag{1.1}$$

J. Vancostenoble (✉)
Institut de Mathématiques de Toulouse, UMR CNRS 5219, Université Paul Sabatier Toulouse III, 118 route de Narbonne, 31 062 Toulouse Cedex 4, France
e-mail: vancoste@math.univ-toulouse.fr

F. Alabau-Boussouira et al. (eds.), *Trends in Control Theory and Partial Differential Equations*, Springer INdAM Series 32,
https://doi.org/10.1007/978-3-030-17949-6_13

where $u_0 \in L^2(0,1)$ and μ is a real parameter. *We concentrate on the above typical problem to simplify the presentation. However notice that this work covers more general cases that are mentioned later in Sect.* 1.4.

Since the pioneering works by Baras and Goldstein [2, 3], we know that inverse square potentials generate interesting phenomena. In particular, existence/blow-up of positive solutions is determined by the value of μ with respect to the constant $\mu^\star = 1/4$ appearing in the Hardy inequality [18, 25]:

$$\forall z \in H_0^1(0,1), \qquad \frac{1}{4}\int_0^1 \frac{z^2}{x^2}\,dx \le \int_0^1 z_x^2\,dx. \tag{1.2}$$

When $\mu < 1/4$, the operator $z \mapsto -z_{xx} - \mu x^{-2} z$ generates a coercive quadratic form in $H_0^1(0,1)$. This allows showing the well-posedness in the classical variational setting of the linear heat equation with smooth coefficients, that is: for any $u_0 \in L^2(0,1)$, there exists a unique solution $u \in \mathcal{C}([0,+\infty[; L^2(0,1)) \cap L^2(0,+\infty; H_0^1(0,1))$.

For the critical value $\mu = 1/4$, the space $H_0^1(0,1)$ has to be slightly enlarged as shown in [30] but a similar result of well-posedness occurs. (See Sect. 2 for details).

Finally, when $\mu > 1/4$, the problem is ill-posed (due to possible instantaneous blow-up) as proved in [2].

Recently, the null controllability properties of (1.1) began to be studied. For any $\mu \le 1/4$, it has been proved in [29] that such equations can be controlled (in any time $T > 0$) by a *locally distributed control*: $\forall \mu \le 1/4$, $\forall u_0 \in L^2(0,1)$, $\forall T > 0$, $\forall 0 \le a < b \le 1$, there exists $h \in L^2((0,1)\times(0,T))$ such that the solution of

$$\begin{cases} u_t - u_{xx} - \dfrac{\mu}{x^2}u = h(x,t)\chi_{(a,b)}(x) & x \in (0,1),\ t \in (0,T),\\ u(0,t) = 0 = u(1,t) & t \in (0,T),\\ u(x,0) = u_0(x) & x \in (0,1), \end{cases} \tag{1.3}$$

satisfies $u(\cdot,T) \equiv 0$. On the contrary, when $\mu > \mu_\star = 1/4$, the property fails as shown in [12].

After these first results, several other works followed extending them in various situations. See for instance [4, 10, 24, 27, 28]. In (1.3), $h\chi_{[a,b)}$ represents a locally distributed control that enters the model as an *additive term* describing the effect of some external force or source on the process at hand. However this is not always realistic to act on the system in such a way.

In the present work, we are interested in studying the effect of other kind of controls on problem (1.1). In the spirit of the works by Khapalov [19–22], we aim to consider a *multiplicative* (also called *bilinear*) control. This means that the control enters now as a *multiplicative coefficient* in the equation (see Sect. 1.2). The advantages of such controls mainly rely on the fact that, instead of being some external action on the system, they may represent changes of parameters of the considered process. We refer to [22, Chap. 1] for a list of situations for which additive controls do not seem realistic whereas multiplicative ones provide a precious alternative. Moreover, multiplicative controls also allow to steer non-negative initial states to non-negative targets *preserving the non-negativity of the solutions during the process*. Even though this last property is naturally expected in many concrete situations, this was not guaranteed when dealing with additive controls!

Let us finally mention several other contributions to multiplicative controls of parabolic pde's: we refer for instance the reader to [5–8, 14, 26] and the references therein.

1.2 Description of the Multiplicative Control Problem

Let $T > 0$ and let us consider the following Dirichlet boundary problem:

$$\begin{cases} u_t - u_{xx} - \dfrac{\mu}{x^2}u = \alpha(x,t)u & (x,t) \in Q_T := (0,1) \times (0,T), \\ u(0,t) = 0 = u(1,t) & t \in (0,T), \\ u(x,0) = u_0(x) & x \in (0,1). \end{cases} \tag{1.4}$$

Here $\alpha \in L^\infty(Q_T)$ is a control function of multiplicative/bilinear form. Our goal is to study the global approximate controllability properties of system (1.4). So following Khapalov (see [22, Chap. 2, Definition 2.1]), we use the following notion:

Definition 1.1 System (1.4) is non-negatively globally approximately controllable in $L^2(0,1)$ if, for any $\varepsilon > 0$ and for every non-negative $u_0, u_d \in L^2(0,1)$ with $u_0 \not\equiv 0$, there exist some $T = T(\varepsilon, u_0, u_d)$ and some bilinear control $\alpha(x,t) \in L^\infty(Q_T)$ such that the corresponding solution of (1.4) satisfies

$$\|u(\cdot,T) - u_d\|_{L^2(0,1)} \leq \varepsilon.$$

Besides, we say that the bilinear control is *static* if $\alpha = \alpha(x) \in L^\infty(0,1)$.

1.3 Main Result

Now we are ready to state our main result concerning the case of the inverse square potential (proved later in Sect. 3):

Theorem 1.1 *Assume that $\mu \leq 1/4$. Then system* (1.4) *is non-negatively globally approximately controllable in $L^2(0,1)$ by means of static controls $\alpha = \alpha(x)$ in $L^\infty(0,1)$. Moreover, the corresponding solution to* (1.4) *remains non-negative at all times.*

1.4 Other Results

1.4.1 Larger Class of Data

With no change in the proof of Theorem 1.1, one can actually state a result that concerns a larger class of data (see the proof in Sect. 4.1):

Theorem 1.2 *For any* $u_0, u_d \in L^2(0,1)$ *such that*

$$\langle u_0, u_d \rangle_{L^2(0,1)} > 0 \text{ and } u_d \geq 0,$$

for avery $\varepsilon > 0$*, there exist some* $T = T(\varepsilon, u_0, u_d)$ *and some static bilinear control* $\alpha = \alpha(x)$ *in* $L^\infty(0,1)$ *such that the corresponding solution of* (1.4) *satisfies*

$$\|u(\cdot, T) - u_d\|_{L^2(0,1)} \leq \varepsilon.$$

1.4.2 General Form of Singular Potential

We considered previously the typical case of the inverse square potential $V(x) = \mu/x^2$ with $\mu \leq \mu_\star = 1/4$. Now we turn to more general singular potentials. Let V be a locally integrable function defined on $(0,1)$ and assume that

$$0 \leq V(x) \leq \frac{\mu}{x^2} \qquad \text{with some} \quad \mu \leq \mu_\star = \frac{1}{4}. \tag{1.5}$$

For $T > 0$, we now consider the following Dirichlet boundary problem:

$$\begin{cases} u_t - u_{xx} - V(x)u = \alpha(x,t)u & (x,t) \in Q_T := (0,1) \times (0,T), \\ u(0,t) = 0 = u(1,t) & t \in (0,T), \\ u(x,0) = u_0(x) & x \in (0,1). \end{cases} \tag{1.6}$$

And we prove (see Sect. 4.1):

Theorem 1.3 *Assume that* $V(x)$ *satisfies* (1.5)*. Then system* (1.6) *is non-negatively globally approximately controllable in* $L^2(0,1)$ *by means of static controls* $\alpha = \alpha(x)$ *in* $L^\infty(0,1)$*. Moreover, the corresponding solution to* (1.6) *remains non-negative at all times.*

1.5 Perpectives

1.5.1 Degenerate/Singular Heat Equation

This present work complements [5, 6, 14] where the case of the heat equation with some degenerate diffusion coefficient

$$u_t - (x^\alpha u_x)_x = 0, \qquad x \in (0, 1), \tag{1.7}$$

was investigated. Here $\alpha \geq 0$ represents the order of degeneracy of the diffusion coefficient that may vanish at $x = 0$. Associated to suitable boundary conditions, this problem is well-posed. In the first studies of its controllability properties, it has been proved that (1.7) is controllable via *additive* control if an only if $\alpha < 2$, see [9]. In [5, 6, 14], the authors prove that, still assuming $\alpha < 2$, it can also be controlled via *multiplicative* controls.

Let us now consider the following degenerate/singular equation:

$$u_t - (x^\alpha u_x)_x - \frac{\mu}{x^\beta} u = 0, \qquad x \in (0, 1). \tag{1.8}$$

It has been proved in [27] that, provided that the parameters α, μ, β satisfies some precise (and optimal) conditions, problem (1.8) is controllable via *additive* controls. (See also [16, 17] to various extensions). We expect that it should also be controllable via *multiplicative* controls under the same conditions on the parameters.

1.5.2 Semilinear Heat Equation with Singular Potential

Another perspective is the study of the null controllability by multiplicative control of semilinear perturbations of the present singular heat equation. In [20], Khapalov studied the case of the classical heat equation with a semilinear term whereas the case of some degenerate heat equation has been studied by Floridia in [14]. We expect that it would be possible to get similar results as in [14, 20] in the case of the heat equation with a singular potential.

1.5.3 Nonnegative Controllability in Small Time

In the present work, following the strategy introduced by Khapalov in [20], we got a result of controllability in *large* time. In the case of the classical heat equation, Khapalov also developed a second approach ("a qualitative approach" presented in [21]) that allows him to get a result of controllability in *small* time. An open and interesting question would be to obtain a similar result of controllability in *small* time in the case of the heat equation with a singular potential. For now, the question remains open. The proofs in [21] are strongly based on specific properties of the classical heat operator (regularity of the solutions) which are no more true when the operator is perturbed by a singular potential. So some new argument has to be found to treat the singular case.

2 Functional Setting and Preliminaries

2.1 Functional Framework

For any $\mu \leq 1/4$, we define

$$H_0^{1,\mu}(0,1) := \{z \in L^2(0,1) \cap H^1_{loc}((0,1]) \mid z(0) = 0 = z(1)$$
$$\text{and } \int_0^1 \left(z_x^2 - \frac{\mu}{x^2} z^2\right) dx < +\infty\}.$$

In the case of a *sub-critical parameter* $\mu < 1/4$, thanks to Hardy inequality (1.2), it is easy to see that $H_0^{1,\mu}(0,1) = H_0^1(0,1)$. On the contrary, for *the critical value* $\mu = \mu_\star = 1/4$, the space is enlarged (see [30] for a precise description of this space):

$$H_0^1(0,1) \underset{\neq}{\subset} H_0^{1,\mu=1/4}(0,1) =: H^\star(0,1). \tag{2.1}$$

Next we prove

Lemma 2.1 *Let $\mu \leq 1/4$ be given. Then $H_0^{1,\mu}(0,1) \hookrightarrow L^2(0,1)$ with compact embedding.*

Proof For any $\mu < 1/4$, $H_0^{1,\mu}(0,1) = H_0^1(0,1)$ so the result is well-known. Consider now $\mu = 1/4$. Deriving some improved Hardy-Poincaré inequalities (see [30, Theorem 2.2]), Vázquez and Zuazua noticed that

$$H^\star(0,1) \hookrightarrow W_0^{1,q}(0,1) \text{ if } 1 \leq q < 2.$$

Then, for $0 \leq s < 1$, we use the fact that $W_0^{1,q}(0,1)$ is compactly embedded in $H_0^s(0,1)$ for suitable $q = q(s)$ close enough to 2. It follows that

$$H^\star(0,1) \hookrightarrow H_0^s(0,1) \text{ with compact embedding if } 0 \leq s < 1. \tag{2.2}$$

Finally, we conclude using the fact that $H_0^s(0,1)$ is compactly embedded in $L^2(0,1)$. □

For any $\mu \leq 1/4$ and $z \in H_0^{1,\mu}(0,1)$, we consider the positive and negative parts of z respectively defined by

$$z^+(x) := \max(z(x), 0), \qquad x \in (0,1),$$
$$z^-(x) := -\min(0, z(x)), \qquad x \in (0,1),$$

so that $z = z^+ - z^-$. We will need the following result of regularity for z^+ and z^-:

Lemma 2.2 *Let $\mu \le 1/4$ and consider $z \in H_0^{1,\mu}(0,1)$. Then for any $1 \le q < 2$, z^+ and z^- belong to $W^{1,q}(0,1)$. Moreover*

$$(z^+)_x = \begin{cases} z_x & \text{in } \{x \in (0,1) \mid z(x) > 0\}, \\ 0 & \text{in } \{x \in (0,1) \mid z(x) \le 0\}, \end{cases} \tag{2.3}$$

and

$$(z^-)_x = \begin{cases} 0 & \text{in } \{x \in (0,1) \mid z(x) \ge 0\}, \\ -z_x & \text{in } \{x \in (0,1) \mid z(x) < 0\}. \end{cases} \tag{2.4}$$

Proof Consider $z \in H_0^{1,\mu}(0,1)$ with $\mu \le 1/4$. From (2.1) and (2.2), we deduce that, for any $1 \le q < 2$, z belongs to $W^{1,q}(0,1)$. Next, using Theorem 5.1 in appendix, one deduce that $z^+, z^- \in W^{1,q}(0,1)$ and (2.3) and (2.4) hold true. □

2.2 *The Unperturbed Operator*

Let us describe here the *unperturbed operator* corresponding to the heat equation with inverse square potential. We define it in the following way:

$$\begin{cases} D(A_0) := \left\{ z \in H^2_{loc}((0,1]) \cap H_0^{1,\mu}(0,1) \mid z_{xx} + \dfrac{\mu}{x^2} z \in L^2(0,1) \right\}, \\ A_0 z := z_{xx} + \dfrac{\mu}{x^2} z. \end{cases} \tag{2.5}$$

In this context, A_0 is a closed, self-adjoint, dissipative operator with dense domain in $L^2(0,1)$ (see [30]). Therefore A_0 is the infinitesimal generator of a C_0-semigroup of contractions in $L^2(0,1)$.

Moreover, from the spectral theory for self-adjoint operators with compact inverse (see [30]), we have:

Lemma 2.3 *Assume $\mu \le 1/4$. There exists an nondecreasing sequence $(\bar{\lambda}_k)_{k \ge 1}$, $\bar{\lambda}_k \to +\infty$ as $k \to +\infty$, such that the eigenvalues of A_0 are given by $(-\bar{\lambda}_k)_{k \ge 1}$ and have finite multiplicity. Besides the corresponding eigenfunctions $\{\bar{\omega}_k\}_{k \ge 1}$ form a complete orthonormal system in $L^2(0,1)$.*

Concerning, the eigenfunction associated to the first eigenvalue, we have the following result:

Lemma 2.4 *Assume $\mu \leq 1/4$. The first eigenvalue $-\bar{\lambda}_1$ is simple and the corresponding eigenfunction $\bar{\omega}_1$ satisfies*

$$\bar{\omega}_1(x) > 0 \textit{ for all } x \in (0,1) \quad \textit{or} \quad \bar{\omega}_1(x) < 0 \textit{ for all } x \in (0,1).$$

Proof The expression of the eigenfunctions of A_0 has been computed in [24] and we recall it in Proposition 5.1 in appendix. So, using Proposition 5.1, the first normalized eigenfunction is

$$\bar{\omega}_1 = \pm \frac{1}{|J'_\nu(j_{\nu,1})|} \sqrt{x} J_\nu(j_{\nu,1}x), \qquad x \in (0,1),$$

and since $j_{\nu,1}$ is the first positive zero of J_ν, this function does not vanish in $(0, 1)$. □

2.3 Perturbed Operators

Next, for any $\alpha = \alpha(x) \in L^\infty(0,1)$ given, we consider now the *perturbed* operator

$$A := A_0 + \alpha I \text{ with domain } D(A) := D(A_0).$$

Then one can prove

Proposition 2.1 *Let $\mu \leq 1/4$ and $\alpha \in L^\infty(0,1)$ be given. Then the above operator $(A, D(A))$ satisfies*

- *$D(A)$ is compactly embedded and dense in $L^2(0,1)$.*
- *$A : D(A) \to L^2(0,1)$ is the infinitesimal generator of a strongly continuous semigroup e^{tA} of bounded linear operators on $L^2(0,1)$.*

Problem (1.4) can be rewritten in the Hilbert space $L^2(0, 1)$ in the following way

$$\begin{cases} u'(t) = Au(t), & t \in (0,T), \\ u(0) = u_0. \end{cases} \tag{2.6}$$

In the following, we will simply denote by $\| \cdot \|$ the norm in $L^2(0, 1)$ and by $\langle \cdot, \cdot \rangle$ the scalar product in $L^2(0, 1)$. We recall (see [1]) that a weak solution of (2.6) is a function $u \in \mathcal{C}^0([0,T]; L^2(0,1))$ such that, for every $v \in D(A^\star)$, the function $v \mapsto \langle u(t), v \rangle$ is absolutely continuous on $[0, T]$ and

$$\frac{d}{dt}\langle u(t), v \rangle = \langle u(t), A^\star v \rangle, \qquad \text{a.e. } t \in (0,T).$$

Theorem 2.1 *Let $\mu \leq 1/4$ be given. For every $\alpha \in L^\infty(0,1)$ and for every $u_0 \in L^2(0,1)$, there exists a unique weak solution*

$$u \in \mathcal{B}(0,T) := \mathcal{C}^0([0,T]; L^2(0,1)) \cap L^2(0,T; H_0^{1,\mu}(0,1))$$

to (1.4), *which coincides with $e^{tA}u_0$.*

Besides, once again from the spectral theory for self-adjoint operators with compact inverse, we have:

Lemma 2.5 *Assume $\mu \leq 1/4$ and $\alpha \in L^\infty(0,1)$. There exists an nondecreasing sequence $(\lambda_k)_{k\geq 1}$, $\lambda_k \to +\infty$ as $k \to +\infty$, such that the eigenvalues of A are given by $(-\lambda_k)_{k\geq 1}$ and have finite multiplicity. Besides the corresponding eigenfunctions $\{\omega_k\}_{k\geq 1}$ form a complete orthonormal system in $L^2(0,1)$. Moreover, the first eigenvalue of A is given by:*

$$\lambda_1 = \inf_{z\in D(A)\setminus\{0\}} \frac{-\langle Az, z\rangle}{\|z\|^2}.$$

2.4 Maximum Principle

For perturbed operators of the form $A = A_0 + \alpha I$, we will also need the following result:

Lemma 2.6 *Let $\mu \geq 1/4$ be given. Let $T > 0$, $\alpha \in L^\infty(0,1)$ and $u_0 \in L^2(0,1)$ such that $u_0 \geq 0$ in $(0,1)$. Consider $u \in \mathcal{B}(0,T)$ be the corresponding solution of* (1.4). *Then $u \geq 0$ in Q_T.*

Proof Consider $u \in \mathcal{B}(0,T)$ the solution of (1.4) and let us prove that $u^- \equiv 0$ in Q_T (which suffices to show that $u \geq 0$ in Q_T). Multiplying the equation by u^- and integrating on $(0,1)$, we get

$$\int_0^1 \Big(u_t u^- - u_{xx}u^- - \frac{\mu}{x^2}uu^- - \alpha uu^-\Big)dx = 0.$$

Using $u = u^+ - u^-$, we compute each term:

$$\int_0^1 u_t u^- dx = \int_0^1 (u^+ - u^-)_t u^- dx = -\int_0^1 (u^-)_t u^- dx = -\frac{1}{2}\frac{d}{dt}\int_0^1 \big(u^-\big)^2 dx,$$

$$\int_0^1 u_{xx}u^- dx = -\int_0^1 u_x (u^-)_x dx = -\int_0^1 (u^+ - u^-)_x (u^-)_x dx = \int_0^1 \Big((u^-)_x\Big)^2 dx,$$

$$\int_0^1 \frac{\mu}{x^2} uu^- dx = \int_0^1 \frac{\mu}{x^2}(u^+ - u^-)u^- dx = -\int_0^1 \frac{\mu}{x^2}\Big(u^-\Big)^2 dx,$$

$$\int_0^1 \alpha u u^- dx = \int_0^1 \alpha(u^+ - u^-)u^- dx = -\int_0^1 \alpha(u^-)^2 dx.$$

We deduce

$$-\frac{1}{2}\frac{d}{dt}\int_0^1 \left(u^-\right)^2 dx - \int_0^1 \left((u^-)_x\right)^2 dx + \int_0^1 \frac{\mu}{x^2}(u^-)^2 dx + \int_0^1 \alpha(u^-)^2 dx = 0.$$

Hence

$$\frac{d}{dt}\int_0^1 \left(u^-\right)^2 dx = 2\int_0^1 \alpha(u^-)^2 dx - 2\int_0^1 \left[\left((u^-)_x\right)^2 - \frac{\mu}{x^2}(u^-)^2\right] dx.$$

From Hardy inequality (1.2) and the fact that $\mu \leq 1/4$, we have

$$\int_0^1 \frac{\mu}{x^2}(u^-)^2 dx \leq \frac{1}{4}\int_0^1 \frac{(u^-)^2}{x^2} dx \leq \int_0^1 \left((u^-)_x\right)^2 dx,$$

so

$$-2\int_0^1 \left[\left((u^-)_x\right)^2 - \frac{\mu}{x^2}(u^-)^2\right] dx \leq 0.$$

Then we deduce

$$\frac{d}{dt}\int_0^1 \left(u^-\right)^2 dx \leq 2\int_0^1 \alpha(u^-)^2 dx \leq 2\|\alpha\|_{L^\infty(0,1)}\int_0^1 (u^-)^2 dx.$$

Using Gronwall's inequality, it follows that

$$\forall t \in (0,T), \qquad \int_0^1 u^-(x,t)^2 dx \leq \int_0^1 u^-(x,0)^2 dx \, e^{2\|\alpha\|_{L^\infty(0,1)} t}.$$

But $u_0 \geq 0$ so $u^-(x,0) = 0$. This implies that $u^-(x,t) \equiv 0$ for $(x,t) \in Q_T$. □

2.5 Specific Perturbed Operator

In this paragraph, we consider now some special perturbed operator that will be used later in order to exhibit a suitable bilinear control. We prove:

Lemma 2.7 *Let $u \in D(A_0)$ be given such that $u > 0$ on $(0,1)$ and such that*

$$\frac{u_{xx}}{u} + \frac{\mu}{x^2} \in L^\infty(0,1).$$

Next consider the operator

$$A := A_0 + \alpha_\star I \text{ with domain } D(A) := D(A_0),$$

and where $\alpha_\star$ *is defined by*

$$\alpha_\star(x) := -\frac{u_{xx}}{u} - \frac{\mu}{x^2} \text{ for } x \in (0, 1).$$

Let $(-\lambda_k)_{k\geq 1}$ *and* $\{\omega_k\}_{k\geq 1}$ *be the eigenvalues and eigenfunctions of A given in Lemma 2.5. Then the first eigenvalue* $-\lambda_1$ *is simple and its value is* $\lambda_1 = 0$*. Moreover, the corresponding normalized eigenfunction* ω_1 *satisfies*

$$|\omega_1| = \frac{u}{\|u\|}.$$

Besides, ω_1 *is the only element of* $\{\omega_k\}_{k\geq 1}$ *that does not change sign on* $(0, 1)$.

Proof Let us compute

$$A\frac{u}{\|u\|} = \frac{u_{xx}}{\|u\|} + \frac{\mu}{x^2}\frac{u}{\|u\|} + \alpha_\star(x)\frac{u}{\|u\|} = 0.$$

It follows that $u/\|u\|$ is an eigenfunction (with norm 1) of A associated to the eigenvalue $\lambda = 0$. Hence there exists $k_\star \geq 1$ such that $\lambda_{k_\star} = 0$ and

$$\omega_{k_\star} = \frac{u}{\|u\|} > 0 \text{ or } \omega_{k_\star} = -\frac{u}{\|u\|} < 0.$$

By orthogonality of the family $\{\omega_k\}_{k\geq 1}$, we have

$$\forall l \neq k_\star, \qquad \int_0^1 \omega_{k_\star}(x)\omega_l(x)dx = 0.$$

Consequently, $\omega_{k_\star}$ is the only element of $\{\omega_k\}_{k\geq 1}$ that does not change sign in $(0, 1)$.

Let us now prove that $k_\star = 1$, that is $\lambda_1 = 0$. Since $-\lambda = 0$ is an eigenvalue and since the sequence $(-\lambda_k)_{k\geq 1}$ is decreasing, we have $-\lambda_1 \geq 0$ that is $\lambda_1 \leq 0$. So it is sufficient to show that $\lambda_1 \geq 0$.

We use the characterization of the first eigenvalue of A:

$$\lambda_1 = \inf_{z\in D(A)\setminus\{0\}} \frac{-\langle Az, z\rangle}{\|z\|^2}.$$

For any $z \in D(A)$, we compute

$$\langle Az, z\rangle = \int_0^1 \Big(z_{xx} + \frac{\mu}{x^2}z + \alpha^\star z\Big)zdx = \int_0^1 \Big(z_{xx}z + \frac{\mu}{x^2}z^2 - \frac{u_{xx}}{u}z^2 - \frac{\mu}{x^2}z^2\Big)dx$$

$$
\begin{aligned}
&= \int_0^1 z_{xx} z \, dx - \int_0^1 u_{xx}\left(\frac{z^2}{u}\right) dx = - \int_0^1 z_x^2 dx + \int_0^1 u_x \left(\frac{z^2}{u}\right)_x dx \\
&= - \int_0^1 z_x^2 dx + 2\int_0^1 \left(\frac{u_x}{u} z\right) z_x dx - \int_0^1 u_x^2 \frac{z^2}{u^2} dx \\
&\leq - \int_0^1 z_x^2 dx + \int_0^1 \left(\frac{u_x}{u} z\right)^2 dx + \int_0^1 z_x^2 dx - \int_0^1 u_x^2 \frac{z^2}{u^2} dx = 0.
\end{aligned}
$$

It follows that $\langle Az, z\rangle \leq 0$ for any $z \in D(A)$ which implies that $\lambda_1 \geq 0$.

It remains to prove that λ_1 is simple. Observe that

$$
\forall z \in D(A), \quad \frac{-\langle Az, z\rangle}{\|z\|^2} = Q(z)
$$

where Q is the quadratic form defined by

$$
\forall z \in H_0^{1,\mu}(0,1), \quad Q(z) := \frac{1}{\|z\|^2} \int_0^1 \left(z_x^2 - \frac{\mu}{x^2} z^2 - \alpha_\star(x) z^2\right) dx.
$$

Another characterization of the first eigenvalue $-A$ is

$$
\lambda_1 = \inf_{z \in H_0^{1,\mu}(0,1)\setminus\{0\}} Q(z).
$$

Any eigenfunction ω of A associated to λ_1 is a minimizer of Q. Reciprocally, by standard arguments of the calculus of variations, any minimizer ω of Q is an eigenfunction of A corresponding to λ_1.

We argue by contradiction assuming that $\lambda_2 = \lambda_1$ so that ω_2 is another eigenfunction of A associated to λ_1. It follows that ω_2 is a minimizer of Q. By Lemma 2.2, it is easy to show that $Q(|\omega_2|) = Q(\omega_2)$ so $|\omega_2|$ is also a minimizer of Q. Therefore $|\omega_2|$ is an eigenfunction associated to λ_1. So we get

$$
\begin{cases}
A|\omega_2| = \lambda_1 |\omega_2| = 0 \text{ in } (0,1), \\
|\omega_2|(x=0) = 0 = |\omega_2|(x=1),
\end{cases}
$$

i.e.

$$
\begin{cases}
-(|\omega_2|)_{xx} = \frac{\mu}{|x|^2}|\omega_2| + \alpha_\star(x)|\omega_2| \text{ in } (0,1), \\
|\omega_2|(x=0) = 0 = |\omega_2|(x=1).
\end{cases}
$$

We deduce

$$
\begin{cases}
-(|\omega_2|)_{xx} + \|\alpha_\star\|_{L^\infty(0,1)}|\omega_2| = \frac{\mu}{|x|^2}|\omega_2| + (\|\alpha_\star\|_{L^\infty(0,1)} + \alpha_\star(x))|\omega_2| \geq 0 \text{ in } (0,1), \\
|\omega_2|(x=0) = 0 = |\omega_2|(x=1).
\end{cases}
$$

Since $|\omega_2| \not\equiv 0$, by strong maximum principle, we have $|\omega_2| > 0$ in $(0, 1)$. It follows that $\omega_2 > 0$ in $(0, 1)$ or $\omega_2 < 0$ in $(0, 1)$. This contradicts the fact that ω_1 is the only element of $\{\omega_k\}_{k\geq 1}$ that does not change sign in $(0, 1)$. □

3 Proof of Theorem 1.1

In this section, we prove Theorem 1.1. The proof is divided into the following 3 steps: first of all, we show that it is sufficient to consider some well-chosen subset of targets; secondly, for any u_d in the previous subset, one exhibit some $\alpha_\star = \alpha_\star(x)$ such that u_d is simply co-linear the first positive eigenfunction of the perturbed operator $A_0 + \alpha_\star I$; finally, one construct a small perturbation $\alpha = \alpha_\star + \delta$ of $\alpha_\star$ that solves the question at hand.

3.1 Step 1

In a first step, we show that it is sufficient to prove the result for the following set of non-negative target states u_d:

$$u_d \in D(A_0), \qquad u_d > 0 \text{ in } (0, 1) \text{ such that } \frac{u_{d,xx}}{u_d} + \frac{\mu}{x^2} \in L^\infty(0, 1). \tag{3.1}$$

Indeed let us consider u_d as in Theorem 1.1, that is u_d satisfying $u_d \in L^2(0, 1)$ and $u_d \geq 0$ in $(0, 1)$. Let us fix $\varepsilon > 0$. Using a regularization by convolution, one can find a function u_d^ε such that

$$u_d^\varepsilon \in \mathcal{C}^\infty([0, 1]), \qquad u_d^\varepsilon > 0 \text{ in } (0, 1) \text{ such that } \|u_d - u_d^\varepsilon\| \leq \frac{\varepsilon}{2}. \tag{3.2}$$

Let us denote $\bar{\omega}_1$ the first positive eigenfunction (corresponding to the eigenvalue $-\bar{\lambda}_1$) of A_0 with norm 1 that we introduced in Lemma 2.4. Of course, $\bar{\omega}_1$ belongs to $D(A_0)$ and is a solution of the following Sturm–Liouville problem

$$\begin{cases} \bar{\omega}_{1,xx} + \dfrac{\mu}{x^2}\bar{\omega}_1 + \bar{\lambda}_1\bar{\omega}_1 = 0, & x \in (0, 1), \\ \bar{\omega}_1(0) = 0 = \bar{\omega}_1(1). \end{cases} \tag{3.3}$$

Consider some cut-off function as follows: for $\sigma > 0$ small, $\xi_\sigma \in \mathcal{C}^\infty([0, 1])$ is such that

$$\begin{cases} 0 \leq \xi_\sigma(x) \leq 1, & x \in [0, 1], \\ \xi_\sigma(x) = 1, & x \in [0, \sigma/2], \\ \xi_\sigma(x) = 0, & x \in [\sigma, 1]. \end{cases} \tag{3.4}$$

And define

$$\bar{u}_d^\varepsilon(x) := \xi_\sigma(x)\bar{\omega}_1(x) + (1 - \xi_\sigma(x))u_d^\varepsilon(x).$$

Since $\bar{u}_d^\varepsilon = \bar{\omega}_1$ on $[0, \sigma/2]$, one can easily check that $\bar{u}_d^\varepsilon \in D(A_0)$. Moreover $\bar{u}_d^\varepsilon > 0$ using the fact that $u_d^\varepsilon > 0$ and $\bar{\omega}_1 > 0$.

Finally, still using the fact that $\bar{u}_d^\varepsilon = \bar{\omega}_1$ on $[0, \sigma/2]$, we observe that

$$\left(\frac{\bar{u}_{d,xx}^\varepsilon}{\bar{u}_d^\varepsilon} + \frac{\mu}{x^2}\right)_{|[0,\sigma/2]} = \left(\frac{\bar{\omega}_{1,xx}}{\bar{\omega}_1} + \frac{\mu}{x^2}\right)_{|[0,\sigma/2]} = -\bar{\lambda}_1 \in L^\infty(0, \sigma/2).$$

And we can deduce that

$$\frac{\bar{u}_{d,xx}^\varepsilon}{\bar{u}_d^\varepsilon} + \frac{\mu}{x^2} \in L^\infty(0, 1).$$

So $\bar{u}_d^\varepsilon$ belongs to the set described in (3.1).

Now let us show that it is sufficient to steer the solution near $\bar{u}_d^\varepsilon$ instead of u_d: we first estimate

$$\|u_d^\varepsilon - \bar{u}_d^\varepsilon\|^2 = \int_0^1 \xi_\sigma(x)^2\Big(\bar{\omega}_1(x) - u_d^\varepsilon(x)\Big)^2 dx \le \int_0^\sigma \Big(\bar{\omega}_1(x) - u_d^\varepsilon(x)\Big)^2 dx.$$

Therefore it is possible to choose $\sigma > 0$ small enough so that

$$\|u_d^\varepsilon - \bar{u}_d^\varepsilon\|^2 \le \frac{\varepsilon^2}{4}.$$

Finally we obtain

$$\|u_d - \bar{u}_d^\varepsilon\| \le \|u_d - u_d^\varepsilon\| + \|u_d^\varepsilon - \bar{u}_d^\varepsilon\| \le \frac{\varepsilon}{2} + \frac{\varepsilon}{2} = \varepsilon. \qquad \square$$

3.2 Step 2

In this second step, for any u_d such that (3.1) holds, we select some $\alpha_\star = \alpha_\star(x)$ such that u_d becomes co-linear the first positive eigenfunction of the perturbed operator $A_0 + \alpha_\star I$.

Indeed, let us now consider u_0 non-zero and non-negative in $L^2(0, 1)$ and u_d as in (3.1). And define

$$\alpha_\star(x) := -\frac{u_{d,xx}}{u_d} - \frac{\mu}{x^2}, \qquad x \in (0, 1). \tag{3.5}$$

Since $\alpha_\star \in L^\infty(0, 1)$, we can define the perturbed operator $A := A_0 + \alpha_\star I$ with domain $D(A) := D(A_0)$. As in Lemma 2.5, we denote by $(-\lambda_k)_{k\ge1}$ and $\{\omega_k\}_{k\ge1}$

the eigenvalues and the corresponding orthonormal eigenfunctions of A. Here, for ω_1, we choose the *positive* eigenfunction associated to the first eigenvalue. Then, applying Lemma 2.7, we have

$$\lambda_1 = 0 \quad \text{and } \omega_1(x) = \frac{u_d(x)}{\|u_d\|} > 0, \qquad x \in (0, 1), \tag{3.6}$$

so u_d and ω_1 are co-linear.

3.3 Step 3

In this last step, we are now ready to choose the (static) bilinear control that allows us to achieve our goal. It is constructed in the following way as a small perturbation of $\alpha_\star$: we set

$$\alpha(x) := \alpha_\star(x) + \delta \tag{3.7}$$

where $\delta \in \mathbb{R}$ will be chosen later.

Observe that, adding δ to $\alpha_\star$ generates a shift on the eigenvalues corresponding to $\alpha_\star$ (they change from $(-\lambda_k)_{k\geq 1}$ to $(-\lambda_k + \delta)_{k\geq 1}$) whereas the eigenfunctions $\{\omega_k\}_{k\geq 1}$ remain the same.

The solution of (1.4) corresponding to the choice of α given in (3.7) can be written in Fourier series representation as

$$u(x,t) = \sum_{k=1}^{+\infty} e^{(-\lambda_k+\delta)t} \langle u_0, \omega_k\rangle \omega_k(x) = e^{\delta t}\langle u_0, \omega_1\rangle \omega_1(x) + r(x,t)$$

where

$$r(x,t) := \sum_{k=2}^{+\infty} e^{(-\lambda_k+\delta)t} \langle u_0, \omega_k\rangle \omega_k(x).$$

Since $u_d = \|u_d\|\omega_1$, we obtain

$$\begin{aligned} \|u(\cdot,t) - u_d\| &\leq \left\| e^{\delta t}\langle u_0, \omega_1\rangle \omega_1(x) - \|u_d\|\omega_1 \right\| + \|r(x,t)\| \\ &= \left| e^{\delta t}\langle u_0, \omega_1\rangle - \|u_d\| \right| + \|r(x,t)\|. \end{aligned}$$

Next we recall that $-\lambda_k \leq -\lambda_2$ for all $k \geq 2$. So

$$\|r(x,t)\|^2 = \sum_{k=2}^{+\infty} e^{2(-\lambda_k+\delta)t} |\langle u_0, \omega_k\rangle|^2 \leq e^{2(-\lambda_2+\delta)t} \sum_{k=2}^{+\infty} |\langle u_0, \omega_k\rangle|^2 \leq e^{2(-\lambda_2+\delta)t}\|u_0\|^2.$$

For $\varepsilon > 0$ fixed, let us choose $T_\varepsilon > 0$ such that

$$e^{-\lambda_2 T_\varepsilon} = \varepsilon \frac{\langle u_0, u_d\rangle}{\|u_0\|\,\|u_d\|^2} \quad \text{i.e.} \quad T_\varepsilon = \frac{-1}{\lambda_2} \ln \left(\varepsilon \frac{\langle u_0, u_d\rangle}{\|u_0\|\,\|u_d\|^2} \right)$$

which is possible since $\lambda_1 = 0$ is simple so $\lambda_2 \neq 0$. Since $u_0 \in L^2(0,1)$, $u_0 \geq 0$, $u_0 \neq 0$ and $\omega_1 > 0$ (see (3.6)), we get

$$\langle u_0, \omega_1\rangle = \int_0^1 u_0(x)\omega_1(x)dx > 0. \tag{3.8}$$

It is then possible to choose δ_ε such that

$$e^{\delta_\varepsilon T_\varepsilon} = \frac{\|u_d\|}{\langle u_0, \omega_1\rangle} = \frac{\|u_d\|^2}{\langle u_0, u_d\rangle},$$

that is

$$\delta_\varepsilon = \frac{1}{T_\varepsilon} \ln \left(\frac{\|u_d\|^2}{\langle u_0, u_d\rangle} \right).$$

We conclude that, for $\alpha(x) = \alpha_\star(x) + \delta_\varepsilon$,

$$\|u(\cdot, T_\varepsilon) - u_d\| \leq e^{(-\lambda_2+\delta_\varepsilon)T_\varepsilon}\|u_0\| = e^{-\lambda_2 T_\varepsilon} \frac{\|u_d\|^2}{\langle u_0, u_d\rangle}\|u_0\| = \varepsilon.$$

So we proved that system (1.4) is non-negatively globally approximately controllable in $L^2(0,1)$ by means of static controls $\alpha = \alpha(x)$ in $L^\infty(0,1)$. Moreover, by the maximum principle stated in Lemma 2.6, the corresponding solution to (1.4) remains non-negative at all times. □

4 Proof of Theorems 1.2 and 1.3

4.1 Proof of Theorem 1.2

The result directly follows from the proof of Theorem 1.1. It is sufficient to observe that inequality (3.8) of step 3 still holds true under the assumptions of Theorem 1.2. Indeed

$$\langle u_0, \omega_1\rangle = \int_0^1 u_0(x)\omega_1(x)dx = \int_0^1 u_0(x)\frac{u_d(x)}{\|u_d\|}dx = \frac{1}{\|u_d\|}\langle u_0, u_d\rangle > 0.$$

□

4.2 Proof of Theorem 1.3

The proof of Theorem 1.3 follows the lines of the proof of Theorem 1.1. So let us first establish the corresponding preliminar results.

Consider $V(x)$ satisfying (1.5). We introduce the associated functional space:

$$H_0^{1,V}(0,1) := \{z \in L^2(0,1) \cap H^1_{loc}((0,1]) \mid z(0) = 0 = z(1) \\ \text{and} \int_0^1 \left(z_x^2 - V(x)z^2\right) dx < +\infty\}.$$

As for the inverse-square potential, thanks to Hardy inequality (1.2), it is easy to see that $H_0^{1,V}(0,1) = H_0^1(0,1)$ when $\mu < 1/4$. $H_0^{1,V}(0,1)$ defines a new functional space only in the critical case $\mu = \mu_\star = 1/4$.

As for Lemma 2.1, using the improved Hardy-Poincaré inequalities in [30, Theorem 2.2], we can prove that

Lemma 4.1 *Let $V(x)$ be given such that* (1.5) *holds. Then $H_0^{1,V}(0,1) \hookrightarrow L^2(0,1)$ with compact embedding.*

With the same argument of proof, one can also show that Lemma 2.2 is still true for any $z \in H_0^{1,V}(0,1)$.

Next we define the *unperturbed operator*:

$$\begin{cases} D(A_0) := \{z \in H^2_{loc}((0,1]) \cap H_0^{1,V}(0,1) \mid z_{xx} + V(x)z \in L^2(0,1)\}, \\ A_0 z := z_{xx} + V(x)z. \end{cases} \tag{4.1}$$

From [30], we know that A_0 is a closed, self-adjoint, dissipative operator with dense domain in $L^2(0,1)$ and that Lemma 2.3 still holds true.

Concerning, the eigenfunction associated to the first eigenvalue, we prove:

Lemma 4.2 *Consider $V(x)$ satisfying* (1.5). *The eigenfunction $\bar{\omega}_1$ corresponding to the first eigenvalue $-\bar{\lambda}_1$ of the above operator A_0 satisfies*

$$\bar{\omega}_1(x) > 0 \textit{ for all } x \in (0,1) \quad \textit{or} \quad \bar{\omega}_1(x) < 0 \textit{ for all } x \in (0,1).$$

Proof The proof of Lemma 2.4 was based on an explicit expression of the eigenfunctions of A_0 in terms of Bessel functions obtained in [24]. Here we quote a result from Davila-Dupaigne [11]:

Proposition 4.1 *Let $\Omega \subset R^n$ be a bounded smooth domain and consider $a \in L^1_{loc}(\Omega)$, $a \geq 0$. Assume that there exists $r > 2$ such that*

$$\gamma(a) := \inf_{\varphi \in C_c^1(\Omega)} \frac{\int_\Omega \left(|\nabla\varphi|^2 - a(x)\varphi^2\right)}{\left(\int_\Omega |\varphi|^r\right)^{2/r}} > 0. \tag{4.2}$$

Define H as the completion of $C_c^\infty(\Omega)$ with respect to the norm

$$\|\varphi\|_H^2 := \int_\Omega \Big(|\nabla \varphi|^2 - a(x)\varphi^2 \Big).$$

Then the operator $L := -\Delta - a(x)$ with domain $D(L) := \{u \in H \mid \Delta u - a(x)u \in H\}$ has a positive first eigenvalue

$$\bar{\lambda}_1 = \inf_{\varphi \in H\setminus\{0\}} \frac{\int_\Omega \Big(|\nabla \varphi|^2 - a(x)\varphi^2 \Big)}{\int_\Omega \varphi^2},$$

which is simple. The above quotient is attained at a positive $\bar{\varphi}_1 \in H$ that satisfies

$$\begin{cases} -\Delta \bar{\varphi}_1 - a(x)\bar{\varphi}_1 = \bar{\lambda}_1 \bar{\varphi}_1, & \text{in } \Omega, \\ \bar{\varphi}_1 = 0, & \text{on } \partial\Omega. \end{cases}$$

We see that the result simply follows from Proposition 4.1 applied with $n = 1$, $\Omega = (0, 1)$ and $a(x) = V(x)$. So it suffices to prove that $a(x) = V(x)$ satisfies assumption (4.2).

We recall the following improved Hardy-Poincaré inequality from [30]: for all $1 \le q < 2$, there exists $C_q > 0$ such that, for all $\varphi \in H_0^1(0, 1)$,

$$\int_0^1 \Big(|\nabla \varphi|^2 - \frac{\mu_\star}{x^2}\varphi^2 \Big) \ge C_q \|\varphi\|^2_{W_0^{1,q}(0,1)}.$$

Since $a = V$ satisfies assumption (1.5), we deduce that, for all $1 \le q < 2$ and for all $\varphi \in C_c^1(\Omega)$,

$$\int_0^1 \Big(|\nabla \varphi|^2 - a(x)\varphi^2 \Big) \ge C_q \|\varphi\|^2_{W_0^{1,q}(0,1)}.$$

Next we use classical Sobolev embeddings. For Ω bounded domain of $\mathbb{R}^n$ with Lipschitz boundary, we have: for all q such that $n < q < \infty$,

$$W^{1,q}(\Omega) \hookrightarrow C^{0,1-n/q}(\Omega).$$

Let us now choose (for example) $q = 3/2$ so that $1 \le q < 2$ and $1 = n < q < \infty$ and apply this to $\Omega = (0, 1)$. It follows in particular that

$$W^{1,q}(0, 1) \hookrightarrow C^0(0, 1).$$

So there exists $c > 0$ such that

$$\sup_{x \in [0,1]} |\varphi(x)| \le c \|\varphi\|_{W_0^{1,q}(0,1)}.$$

Finally, fix $r > 2$. Then

$$\left(\int_\Omega |\varphi|^r\right)^{2/r} \leq \sup_{x\in[0,1]} |\varphi(x)| \leq c\|\varphi\|_{W_0^{1,q}(0,1)} \leq \frac{c}{\sqrt{C_q}}\int_0^1 \left(|\nabla\varphi|^2 - a(x)\varphi^2\right)$$

and (4.2) follows. □

Next, for any $\alpha \in L^\infty(Q_T)$ given, we consider the *perturbed* operator

$$A := A_0 + \alpha I \text{ with domain } D(A) := D(A_0).$$

One can easily see that, under assumption (1.5), Proposition 2.1 together with the well-posedness Theorem 2.1 (replacing the space $H_0^{1,\mu}(0,1)$ by $H_0^{1,V}(0,1)$) and the spectral Lemma 2.5 are still true.

With similar proofs, one can see that the maximum principle for perturbed operators stated in Lemma 2.6 holds unchanged for the solutions of (1.6) whereas Lemma 2.7 is simply replaced by

Lemma 4.3 *Assume $V(x)$ is given such that* (1.5) *holds. Let $u \in D(A_0)$ be given such that $u > 0$ on* $(0, 1)$ *and such that*

$$\frac{u_{xx}}{u} + V(x) \in L^\infty(0,1).$$

Next consider the operator

$$A := A_0 + \alpha_\star I \textit{ with domain } D(A) := D(A_0),$$

and where $\alpha_\star$ is defined by

$$\alpha_\star(x) := -\frac{u_{xx}}{u} - V(x) \textit{ for } x \in (0,1).$$

Let $(-\lambda_k)_{k\geq 1}$ and $\{\omega_k\}_{k\geq 1}$ be the eigenvalues and eigenfunctions of A given in Lemma 2.5. Then $\lambda_1 = 0$ is simple and

$$\lambda_1 = 0 \quad \textit{and} \quad |\omega_1| = \frac{u}{\|u\|}.$$

Moreover, $u/\|u\|$ and $-u/\|u\|$ are the only eigenfunctions of A with norm 1 that do not change sign on $(0, 1)$.

This concludes the generalization of all preliminaries results. Finally, it is easy to see that the proof of Theorem 1.1 can now be rewritten replacing μ/x^2 by $V(x)$ and leads to Theorem 1.3. □

5 Appendix

This section is devoted to the statements of various technical results from literature that we use throughout this paper.

Let us first of all recall the following result from [23, Appendix A] that concerns the regularity of the negative and positive parts of a function:

Theorem 5.1 *Let* $\Omega \subset \mathbb{R}^n$ *and* $1 \leq s \leq \infty$. *Consider* $v : \Omega \to \mathbb{R}$ *such that* $v \in W^{1,s}(\Omega)$. *Then* $v^+, v^- \in W^{1,s}(\Omega)$ *and, for all* $1 \leq i \leq n$,

$$(v^+)_{x_i} = \begin{cases} v_{x_i} & \text{in } \{x \in \Omega \mid v(x) > 0\}, \\ 0 & \text{in } \{x \in \Omega \mid v(x) \leq 0\}, \end{cases}$$

and

$$(v^-)_{x_i} = \begin{cases} 0 & \text{in } \{x \in \Omega \mid v(x) \geq 0\}, \\ -v_{x_i} & \text{in } \{x \in \Omega \mid v(x) < 0\}. \end{cases}$$

Next we recall the following expression of the eigenfunctions of A_0 (defined in (2.5)) that have been computed in [24]:

Proposition 5.1 *Assume that* μ *is given such that* $\mu \leq 1/4$ *and define*

$$\nu := \sqrt{\frac{1}{4} - \mu}.$$

We denote by J_ν *the Bessel function of first kind of order* ν *and we denote* $0 < j_{\nu,1} < j_{\nu,2} < \cdots < j_{\nu,n} < \cdots \to +\infty$ *as* $n \to +\infty$ *the sequence of positive zeros of* J_ν. *Then the admissible eigenvalues* $(-\bar{\lambda}_n)_{n \geq 1}$ *of* A_0 *are determined by*

$$\forall n \geq 1, \qquad \bar{\lambda}_n = (j_{\nu,n})^2$$

and corresponding (normalized) eigenfunctions are given by

$$\forall n \geq 1, \qquad \Phi_n(x) = \frac{1}{|J_\nu'(j_{\nu,n})|} \sqrt{x} J_\nu(j_{\nu,n} x), \qquad x \in (0, 1).$$

References

1. Ball, J.M.: Strongly continuous semigroups, weak solutions, and the variation of constants formula. Proc. Am. Math. Soc. **63**, 370–373 (1977)
2. Baras, P., Goldstein, J.: Remarks on the inverse square potential in quantum mechanics. Differential Equations (Birmingham, Ala., 1983). North-Holland Mathematics Studies, vol. 92, pp. 31–35. North-Holland, Amsterdam (1984)

3. Baras, P., Goldstein, J.: The heat equation with a singular potential. Trans. Am. Math. Soc. **284**(1), 121–139 (1984)
4. Biccari, U., Zuazua, E.: Null controllability for a heat equation with a singular inverse-square potential involving the distance to the boundary function. J. Differ. Equ. **261**(5), 2809–2853 (2016)
5. Cannarsa, P., Floridia, G.: Approximate controllability for linear degenerate parabolic problems with bilinear control. In: Proceedings of Evolution equations and Materials with Memory 2010, Casa Editrice Università La Sapienza Roma, pp. 19–39 (2011)
6. Cannarsa, P., Floridia, G.: Approximate multiplicative controllability for degenerate parabolic problems with Robin boundary conditions. Commun. Appl. Ind. Math. **2**(2) (2011)
7. Cannarsa, P., Floridia, A.Y., Khapalov, G.: Multiplicative controllability for semilinear reaction-diffusion equations with finitely many changes of sign. J. Math. Pures Appl. (9) **108**(4), 425–458 (2017)
8. Cannarsa, P., Khapalov, A.Y.: Multiplicative controllability for reaction-diffusion equations with target states admitting finitely many changes of sign. Discret. Contin. Dyn. Syst. Ser. B **14**(4), 1293–1311 (2010)
9. Cannarsa, P., Martinez, P., Vancostenoble, J.: Carleman estimates for a class of degenerate parabolic operators. SIAM J. Control Optim. **47**(1), 1–19 (2008)
10. Cazacu, C.: Controllability of the heat equation with an inverse-square potential localized on the boundary. SIAM J. Control Optim. **52**(4), 2055–2089
11. Davila, J., Dupaigne, L.: Comparison results for PDEs with a singular potential. Proc. R. Soc. Edinb. Sect. A Math. **133**(01), 61–83 (2003)
12. Ervedoza, S.: Control and stabilization properties for a singular heat equation with an inverse-square potential. Commun. Partial Differ. Equ. **33**(10–12), 1996–2019 (2008)
13. Fernandez-Cara, E., Zuazua, E.: The cost of approximate controllability for heat equations: the linear case. Adv. Differ. Equ. **5**(4–6), 465–514 (2000)
14. Floridia, G.: Approximate controllability for nonlinear degenerate parabolic problems with bilinear control. J. Differ. Equ. **257**(9), 3382–3422 (2014)
15. Floridia, G.: Nonnegative controllability for a class of nonlinear degenerate parabolic equations, preprint
16. Fotouhi, M., Salimi, L.: Null controllability of degenerate/singular parabolic equations. J. Dyn. Control Syst. **18**(4), 573–602 (2012)
17. Hajjaj, A., Maniar, L., Salhi, J.: Carleman estimates and null controllability of degenerate/singular parabolic systems. Electron. J. Differ. Equ. **2016**(292), 1–25 (2016)
18. Hardy, G.H., Littlewood, J.E., Pólya, G.: Inequalities, 2nd edn. Cambridge University Press, Cambridge (1952)
19. Khapalov, A.: Bilinear control for global controllability of the semilinear parabolic equations with superlinear terms. Control of Nonlinear Dstributed Parameter Systems (College Station, TX, 1999). Lecture Notes in Pure and Applied Mathematics, vol. 218, pp. 139–155. Dekker, New York (2001)
20. Khapalov, A.: Global non-negative controllability of the semilinear parabolic equation governed by bilinear control. ESAIM Control Optim. Calc. Var. **7**, 269–283 (2002)
21. Khapalov, A.Y.: Controllability of the semilinear parabolic equation governed by a multiplicative control in the reaction term: a qualitative approach. SIAM J. Control Optim. **41**(6), 1886–1900 (2003)
22. Khapalov, A.Y.: Controllability of Partial Differential Equations Governed by Multiplicative Controls. Lecture Notes in Mathematics, vol. 1995. Springer, Berlin (2010)
23. Kinderlehrer, D., Stampacchia, G.: An introduction to variational inequalities and their applications. SIAM Class. Appl. Math. (2000)
24. Martinez, P., Vancostenoble, J.: The Cost of boundary controllability for a parabolic equation with inverse square potential (submitted)
25. Opic, B., Kufner, A.: Hardy-type Inequalities. Pitman Research Notes in Mathematics, vol. 219. Longman, Harlow (1990)

26. Ouzahra, M.: Approximate and exact controllability of a reaction-diffusion equation governed by bilinear control. Eur. J. Control **32**, 32–38 (2016)
27. Vancostenoble, J.: Improved Hardy-Poincaré inequalities and Sharp Carleman Estimates for Degenerate/Singular Parabolic Problems. Discrete and Continuous Dynamical Systems - Series S, vol. 4(3), pp. 761–790. American Institute of Mathematical Sciences, Providence (2011)
28. Vancostenoble, J.: Lipschitz stability in inverse source problems for singular parabolic equations. Commun. Partial Differ. Equ. **36**(8), 1287–1317 (2011)
29. Vancostenoble, J., Zuazua, E.: Null controllability for the heat equation with singular inverse-square potentials. J. Funct. Anal. **254**(7), 1864–1902 (2008)
30. Vázquez, J.L., Zuazua, E.: The Hardy inequality and the asymptotic behaviour of the heat equation with an inverse-square potential. J. Funct. Anal. **173**(1), 103–153 (2000)

MIX
Papier aus verantwortungsvollen Quellen
Paper from responsible sources
FSC® C105338

If you have any concerns about our products,
you can contact us on
ProductSafety@springernature.com

In case Publisher is established outside the EU,
the EU authorized representative is:
Springer Nature Customer Service Center GmbH
Europaplatz 3, 69115 Heidelberg, Germany

Printed by Libri Plureos GmbH
in Hamburg, Germany